Frontiers in Differential Geometry, Partial Differential Equations and Mathematical Physics

In Memory of Gu Chaohao

Frontiers in Differential Geometry, Partial Differential Equations and Mathematical Physics

In Memory of Gu Chaohao

Edited by

Molin Ge
Chern Institute of Mathematics, China

Jiaxing Hong
Fudan University, China

Tatsien Li
Fudan University, China

Weiping Zhang
Chern Institute of Mathematics, China

World Scientific

NEW JERSEY • LONDON • SINGAPORE • BEIJING • SHANGHAI • HONG KONG • TAIPEI • CHENNAI

Published by

World Scientific Publishing Co. Pte. Ltd.
5 Toh Tuck Link, Singapore 596224
USA office: 27 Warren Street, Suite 401-402, Hackensack, NJ 07601
UK office: 57 Shelton Street, Covent Garden, London WC2H 9HE

Library of Congress Cataloging-in-Publication Data
Frontiers in differential geometry, partial differential equations, and mathematical physics : in memory of Gu Chaohao / edited by Molin Ge (Chern Institute of Mathematics, China), Jiaxing Hong (Fudan University, China), Tatsien Li (Fudan University, China), & Weiping Zhang (Chern Institute of Mathematics, China).
pages cm
Includes bibliographical references.
ISBN 978-9814578073 (hardcover : alk. paper)
1. Gu, Chaohao. 2. Geometry, Differential. 3. Differential equations, Partial. 4. Mathematical physics. I. Ge, M. L. (Mo-Lin) II. Hong, Jiaxing, 1942– III. Li, Daqian. IV. Zhang, Weiping, 1964–
QA641.F77 2014
516.3'6--dc23

2013048729

British Library Cataloguing-in-Publication Data
A catalogue record for this book is available from the British Library.

Printed in Singapore by World Scientific Printers.

Preface

The late Professor Gu Chaohao (1926–2012) was a highly distinguished mathematician with worldwide prestige and reputation and a member of the Chinese Academy of Sciences. About one year has passed away since his death, this collection of papers is compiled in memory of Gu Chaohao on the subjects of Differential Geometry, Partial Differential Equations and Mathematical Physics that Gu Chaohao devoted and made great contributions to with all his intelligence during his lifetime.

We would like to express our sincere gratitude to those who kindly contribute their papers to this collection. All these contributors are close friends, colleagues and students of Gu Chaohao. We also thank Ms. Zhou Chunlian of Fudan University for her enthusiastic and effective work in editing the collection.

Special thanks should be given to Dr. K. K. Phua for his kind suggestion to edit and publish this collection with World Scientific Publishing Co. Pte. Ltd., and to the School of Mathematical Sciences, Fudan University for its encouragement and support.

July 2013
The Editors

Gu Chaohao (1926-2012)

Contents

A Profile of the Late Professor Gu Chaohao

Tatsien Li
School of Mathematical Sciences, Fudan University,
Shanghai 200433, China;
Shanghai Key Laboratory for Contemporary Applied Mathematics;
Nonlinear Mathematical Modeling and Methods Laboratory
dqli@fudan.edu.cn

Gu Chaohao, born in the city of Wenzhou, Zhejiang Province on May 15, 1926, and died in Shanghai on June 24, 2012, was a great Chinese mathematician, educationist, and academician of the Chinese Academy of Sciences. His research area covers partial differential equations, differential geometry and mathematical physics. He was also a winner of the State Top Science and Technology Award.

In the year 2009, the State Top Science and Technology Award, which represents China's highest honor in the circle of science and technology, was conferred to two outstanding scientists, and one of them was the academician of the Chinese Academy of Sciences, Prof. Gu Chaohao of Fudan University. He was awarded for his eminent achievements in the fields of differential geometry, partial differential equations and mathematical physics, and for his remarkable contributions to the mathematical development of China.

Gu Chaohao was born in Wenzhou of Zhejiang Province. His father's name is Gu Boren, and his mother's Lu Zhongzhen. Brought up by a widowed aunt, he was quiet and brilliant as a young boy, who exhibited great interest in learning and extraordinary intelligence above the peers. In 1943, he passed the entrance exam and entered the Longquan Branch School of Zhejiang University which was founded for the purpose of facilitating the study of young people in the southeast of China during the War of Resistance Against Japan when Zhejiang University moved inland to Guizhou Province. Despite the poor material conditions, Gu Chaohao enthusiastically attended the mathematical courses there and eagerly acquired the knowledge of college mathematics, which laid a solid mathematical foundation for his later research work. During the college years, he also developed a keen interest in physics and earnestly completed the study of the four main courses in Physics and Mechanics, which enhanced his understanding of physics and mechanics.

After the victory of the war against Japanese invasion, Zhejiang University moved back to the city of Hangzhou. In 1946, Gu Chaohao became a student of the famous mathematician Su Buqing, and started his research experiences in mathematics. His exceptional talent allowed him to learn the methods of scholarly research from two great mathematicians by participating in both the seminar on differential geometry conducted by Prof. Su Buqing and the seminar on the theory of functions and Fourier analysis conducted by Prof. Chen Jiangong, which ensured his solid academic background in geometry and mathematical analysis.

Ever since he was young, Gu Chaohao has witnessed the humiliating history of the mother country being invaded by foreigners and the miserable life of the vast majority of laboring people, and he touched upon a lot of enlightening books and journals, which determined his lofty aspiration of dedicating to the revolutionary cause. As early as in 1940, when he was studying at Wenzhou Middle School of Zhejiang Province as a junior student of Grade 3, he joined the Chinese Communist Party. During his study at Zhejiang University, he organized the 'Truth-Seeking Society' with some other students, acted as the leader of the Society and actively took part in all kinds of patriotic movements launched by progressive students. His extraordinary academic performance, truth-seeking actions as well as a warm and serving heart won trust from teachers and respect from fellow students. He was once elected a chief of the Students' Union of Zhejiang University with the highest number of votes, and on the eve of liberation he helped to save and maintain the scientific and technological institutions in Hangzhou.

In the summer of 1948, Gu Chaohao graduated from Zhejiang University and stayed at his Alma Mater as a teaching assistant, and under the guidance of Prof. Su Buqing there, he carried out research on differential geometry of generalized spaces. In particular, he proposed a novel and systematic approach to handle geometry in a space of K-spreads, while the relevant paper was published in English in Science Record (1950)[1], which showed his creative aptitude in mathematics. In 1952, Gu Chaohao was promoted to a lecturer, and he transferred to Fudan University in 1953 as a result of the restructuring of colleges and departments in China. In 1956, he was advanced to an associate professorship, and in the same year selected a National Advanced Worker. During the period, he did a series of in-depth study into the space of K-spreads, the affinely connected space and the Finsler space, etc., and consequently published many papers, which rapidly established him as a backbone figure in the Chinese school of differential geometry led by Su Buqing.

In 1957, Gu Chaohao went to Moscow State University in Soviet Union for advanced studies, and joined in the meantime two research groups led by the renowned Soviet Union differential geometricians С. П. Фиников and П. К. Рашевский, respectively. He often reported his results in the seminars organized by the two teachers, and studied infinite continuous transformation pseudogroups with their encouragement and help. As a result, he greatly expanded the research field developed by E. Cartan, a well-known geometrician of the 20th century, and accomplished his doctoral dissertation 'On Some General Properties of Transformation Pseudogroups and Their Applications in Differential Geometry' [3]. Having successfully passed the oral defense of his dissertation in June, 1959, Gu Chaohao got his doctorate degree in Physics and Mathematical Science from Moscow State University and returned to China in July, 1959. In the following year, he was promoted to a full professorship.

When Gu Chaohao's achievements in differential geometry had attracted great attention, he keenly observed the new requirements of advanced science and technology on mathematics, and in accordance with the demand of developing the scientific cause of the country, he resolutely devoted his energy to the study of partial differential equations, then a brand-new research field in China. Actually, when studying in Soviet Union, Gu Chaohao had already tried to find time to participate in the seminars conducted by academician, И. Г. Петровский — a famous scholar of partial differential equations, and by Prof. О. А. Олейник, etc. After returning to the homeland, he pointed out that research should be done against the practical background of high speed flight and should use the problem of supersonic flow past a curved wedge as a model. Then on the basis of the previous theories on partial differential equations, the research scope would gradually move from linear to nonlinear, local to global, lower dimension to higher dimension, given type to variable type and known boundary to free boundary. This prediction about the development trend of partial differential equations was not only verified by the mainstream progress of partial differential equations worldwide later on, but also guided and led a group of his students on a unique research path of partial differential equations.

He successively obtained systematic and thorough results of the world's advanced level on quasilinear hyperbolic systems, positive symmetric systems of partial differential equations, higher-dimensional partial differential equations of mixed type and so on.

In June, 1974, Chinese American physicist C. N. Yang visited Fudan University, and started several research programs on the gauge field in cooperation with Gu Chaohao. They jointly published some papers including 'Some Problems in Gauge Field Theories' [8], etc., and from then on, Gu Chaohao extended his research to the field of mathematical physics. He managed to quickly probe into this new domain, thanks to his profound knowledge of differential geometry and partial differential equations as well as his thorough understanding of the physical and mechanical subjects. In 1981, his English monograph 'On Classical Yang-Mills Fields' was published on Physics Reports [13] occupying the pages of an entire issue, and afterwards, he also did a lot of original and noteworthy work on theories of harmonic mappings with indefinite metric, extremal surfaces of mixed type, and solitons.

Always insisting on combination of teaching and research, Gu Chaohao has made a great contribution to the training of talents and the educational undertaking. Being a teacher at Fudan University for quite many years, he has set up many courses, educated a large number of doctoral and master students, including some mathematicians of global influence, and among the graduate students he directly guided, three were elected academician of the Chinese Academy of Sciences. He has served as Dean of Department of Mathematics, Fudan University from 1978 to 1981, vice President of Fudan University as well as Dean of the Graduate School, Fudan University from 1982 to 1986, Director of the Institute of Mathematics, Fudan University since 1986, and also President of University of Science and Technology of China appointed by the State Council from 1988 to 1993.

Gu Chaohao was elected member of the Chinese Academy of Sciences in 1980, and became a doctoral supervisor on both Fundamental Mathematics and Hydromechanics. He has been Deputy Director of Chinese Mathematical Society (CMS) from 1984 to 1987, and President of Shanghai Mathematical Society (SMS) from 1985 to 1988. He was appointed as the chief scientist of the key project 'Nonlinear Science' of national basic research from 1991 to 1996. In addition, he has also been deputy to the Third, Sixth and Seventh National People's Congresses (NPCs), member of the Fifth, Eighth and Ninth Chinese People's Political Consultative

Conferences (CPPCCs), and member of the Standing Committee of the Eighth and Ninth CPPCCs.

Gu Chaohao has paid visits to the U.S., France, Germany, the Soviet Union, Japan, etc. many times upon invitation, and during the academic exchanges, he left a deep impression on the international mathematical circle and was highly rated by the fellow researchers. Besides, he was once a member of the editorial board of the international mathematical journal Mathematical Physics, and editor or managing editor of many Chinese mathematical journals, such as Science China Series A, Chinese Annals of Mathematics, Chinese Science Bulletin, Journal of Partial Differential Equations, and so on. In 1994, he got in the International Higher Education Academy of Sciences (IHEAS) as an academician, and in 1995 when the Franklin Institute of the U.S. conferred special awards to three distinguished physicists including C. N. Yang and t'Hooft, he was invited to be a member of the International Scientific Advisory Committee (ISAC).

Gu Chaohao was a prolific author, as he has 130 or so mathematical papers in print among which about 100 were his independent individual work, compiled 5 textbooks and wrote 4 monographs, two of which were published abroad by Springer-Verlag Publishing Cooperation. His research findings won the Science Conference Award in 1978, a second-class and a third-class National Prize for Natural Sciences in 1982, and two first-class Scientific and Technological Progress Prizes of the State Education Commission in both 1985 and 1986. Moreover, he got the HUA Loo-Keng Mathematics Award in Mathematics in 1995, the Scientific and Technological Progress Prize of the Ho Leung Ho Lee Foundation in 1995, the first Shanghai Science and Technology Meritorious Award in 2002, and the Outstanding Science and Technology Achievement Prize of the Ho Leung Ho Lee Foundation in 2005. The textbook *Equations of Mathematical Physics* issued under his editorship won the National Prize of Excellent textbooks for Higher Education in 1987. He was granted the title of 'Advanced Scientific and Technological Worker in Institutions of Higher Learning' in 1990, and obtained the Shanghai Education Hero Award in 2007.

Gu Chaohao's illustrious achievements and contributions to scientific and technological development were further acknowledged and honored many times in 2009, the 60th anniversary of the establishment of the People's Republic of China (PRC). According to the assessment done by the system of Shanghai Association for Science and Technology (SAST), Gu Chaohao ranked first in the list of Ten Figures with Remarkable Contributions to Shanghai's Scientific Research since the Founding of New China sixty years before. Approved by the Minor Planet Center and International Committee on Nomenclature of Asteroids on August 6th, 2009, a minor planet encoded with the international serial number 171448 and discovered by the Xuyi Observation Station of Zijin Mountain Observatory affiliated to the Chinese Academy of Sciences was named 'Gu Chaohao Star', and the grand naming ceremony was held at Fudan University on October 21st. It's particularly worth mentioning that Gu Chaohao attained the 2009 State Top Scientific and Technological Award which is the supreme honor in the scientific and technological circle of the country, and on the morning of January 11th, 2010, the award certificate was bestowed on him at the National Science and Technology Award Conference solemnly held in Beijing, to recognize and commend his eminent research achievements and great contributions to the country.

Gu Chaohao got married in 1957, and his wife Hu Hesheng is an accomplished woman mathematician specializing in differential geometry and mathematical physics. Also a professor of Fudan University, she was elected member of Chinese Academy of Sciences in 1991.

Gu Chaohao died in Shanghai on June 24, 2012.

Differential geometry, partial differential equations and mathematical physics are the three hot, important and interrelated research fields in the present study of mathematical theories. Equipped with the strong knowledge and consummate attainments in mathematics and physics, Gu Chaohao has led a group of scholars around him to actively penetrate into this 'Golden Triangle' region of differential geometry, partial differential equations and mathematical physics with his scientific foresight and sagacity. He is skillful at locating the key problem in time, and creatively giving an original solution. He has broken a new path of his own in research with distinctive features, and his main research achievements can be summarized as follows:

1. Differential Geometry

(1) Differential Geometry of Generalized Spaces

Riemannian geometry appearing in the 19th century was established on the basis of defining the quadratic differential form of the square of the distance between two adjacent points in space. Since the 20th century, spurred by the successful application of Riemannian geometry to General Relativity, differential geometry of generalized spaces like the Finsler space, the Cartan space, the path space, the space of K-spreads, etc. has appeared. Chinese differential geometricians represented by Su Buqing have made prominent contributions to the study of differential geometry of generalized spaces [19], which includes a lot of work done by Gu Chaohao.

To the space of K-spreads defined by a set of entirely integrable partial differential equations and first proposed by the American mathematician J. Douglas, Gu Chaohao blazed a new trail by using the implicit function equations for study, so as to deduce the integrable conditions of differential equations of K-spreads in a quite new form, get the projective connection of spaces, and quite conveniently prove the theorems of plane axioms, which was a great contribution to the development of geometry on spaces of K-spreads. When The Journal of Mathematical Reviews Реферативный Журнал Математика was first issued in the Soviet Union in 1956, it published a long article introducing Gu Chaohao's work in this regard.

It is one of the central problems in classical differential geometry to investigate whether abstract geometric objects can be brought about in concrete spaces. The famous mathematician J. Nash proved in 1956 a very fundamental theorem in Riemannian geometry about the isometric embedding of any Riemannian manifold into the Euclidean space, while in the same year, Gu Chaohao proved the embedding of the compact Finsler manifold into the Minkowski space [2]. Forty years later, the demands of research in materials science set off a new upsurge of study on Finsler geometry, to which the work of Gu Chaohao naturally became a solid basis [22].

(2) Infinite Continuous Transformation Pseudogroups and Homogeneous Riemannian Space

The concept of continuous transformation pseudogroups in infinite dimension was first put forward by S. Lie, and then greatly developed by E. Cartan. In 1957, Su Buqing said to Gu Chaohao that Cartan's work had progressed significantly with the effort of later generations except the theory of transformation pseudogroups. So when Gu Chaohao was in the Soviet Union, he made use of the favorable research conditions there and focused on this subject to analyze the geometric meaning of structure constants, carefully examine the situation of isotropy groups being divided into direct products with invariant vectors, and apply it to the space with symplectic measure and the homogeneous Riemannian space, which generated quite meaningful results. As the comments on his oral defense of the doctoral dissertation mentioned, 'Gu Chaohao is the first one after E. Cartan that has greatly pushed forward the theory of transformation pseudogroups.'

Besides, determining linear groups that can be regarded as isotropy groups play a key role in the theory of infinite continuous transformation pseudogroups. E. Cartan has used the representation theory of semi-simple Lie algebra in complex field to determine all the irreducible (complex) linear groups that can be isotropy groups of infinite continuous groups. It was Gu Chaohao who finally settled the case of real irreducible linear groups.

The above mentioned two aspects of work have greatly advanced E. Cartan's methods and results.

2. Partial Differential Equations

(1) Quasilinear Hyperbolic Systems

The study on discontinuous solutions of gas dynamic equations corresponding to shock waves could date back to as early as G.F.B. Riemann (1860). In the 1940s and 1950s, there appeared a classical work *Supersonic Flow and Shock Waves* by R. Courant and K. O. Friedrichs, and the research on discontinuous solutions to quasilinear hyperbolic systems excited a lot of interest. The case of one single equation has been consummately resolved in theory in the 1950s, while to the case of a system of equations, results were obtained only with the data of piecewise constants and a straight boundary. Concerning some important practical problems, like the supersonic flow past a sharp weage in a plane, with a curved boundary and non-constant data, it became an urgent problem that called for an immediate solution to determine the flow field with shock waves. In the early 1960s, Gu Chaohao has successfully solved the above flow problem with regard to isentropic and non- isentropic situations respectively [5], and about 16 years later, the American mathematician D. G. Schaeffer obtained the same result again. Moreover, Gu Chaohao and his cooperators have systematically settled the construction problem of local discontinuous solutions to the initial value problems of reducible hyperbolic systems with piecewise smooth initial data (the generalized Riemann problem) [4]. These are the earliest results about the theory of local discontinuous solutions to quasilinear hyperbolic systems, and in these researches, the thinking pattern of suitably transforming unknown shock waves into known boundaries and trying to reduce the multi-valued singularity of central waves always played a part in the study of discontinuous solutions. Based on the above work, his students established the complete theory of local discontinuous solutions to general quasilinear hyperbolic systems with two independent variables [20].

Gu Chaohao also did a pioneering work on the theory of global classical solutions to quasilinear hyperbolic systems which has been greatly developed till now. In a paper published in 1960, he had given a sufficient condition for the existence of global classical solution on $t \geq 0$ for the Cauchy problem of quasilinear reducible hyperbolic systems, and later, in 1964 the renowned American mathematician P. D. Lax offered a result showing the necessity of this condition.

(2) Positive Symmetric Systems

The theory of positive symmetric systems was proposed by the well-known American mathematician K. O. Friedrichs in 1958. It goes beyond the limit of classical classification of partial differential equations, uses the method of energy integration, and handles the problem of determining solutions to a major category of partial differential equations with a unified view, which has a wide range of applications. However, the previous theory is limited to the linear situation and stays at the level of generalized solutions. During the years 1963 and 1964, Gu Chaohao first established the theory of higher-order differentiable solutions to positive symmetric systems, which paved the way for getting the classical solution to partial differential equations with this theory, and subsequently developed the theory of quasilinear positive symmetric systems. In addition, he clarified the possibility and the specific method of reducing many important partial differential equations to positive symmetric systems, which considerably substantiated the theoretical framework and extended the application fields of positive symmetric systems, thus elevating the theory to a new height [6].

(3) Mixed-Type Equations

Partial differential equations of mixed type have developed under the drive of transonic flow, and since 1923 when the Italian mathematician F. G. Tricomi published the first article on equations of mixed type, people have paid great attention to this sort of equations. Yet till the end of the 1950s, there had been no significant progress in study of the equations due to the complexity of problems and lack of effective tools. Most of the research concentrated only on individual problems for some special second-order linear equations with two independent variables (e.g. the Tricomi equation, etc.) and the conclusions were rather fragmented. In the middle of the 1960s, Gu Chaohao successfully applied the theory of positive symmetric systems that he had developed to the study of mixed-type equations, and reduced a wide category of second-order higher-dimensional linear equations and quasilinear equations of mixed type (especially the Busemann equation closely related to conical supersonic flow) to positive symmetric systems, which resulted in a large number of solvable problems, demonstrated the well-posedness of classical solutions, and incisively revealed the essential implications of the lower order term in mixed-type equations to the establishment of boundary conditions and the regularity of solutions [7] [12]. This breakthrough to the theory of higher-dimensional equations of mixed type changed the research situation of mixed-type equations, and won high evalution in the world's mathematical circle. The American journal Math Reviews [82d:35067] commented that Gu Chaohao's above work was a 'nice' contribution to the theory of higher-dimensional equations of mixed type. Friedrichs maintained that Gu Chaohao had fulfilled his long-cherished wish of applying positive symmetric systems to mixed-type equations. In 1976, when the delegation of American mathematicians visited China, they praised Gu Chaohao's work as being 'quite novel and rather important' in their report. In the 1980s, Gu Chaohao further improved the theory of higher-dimensional mixed-type equations by extending the study from the second order to higher order, and therefore pioneered the research on higher-order higher-dimensional equations of mixed type [14].

Gu Chaohao and his cooperators also did the calculation work together on supersonic flow past a blunt body, which provided the data needed in engineering.

(4) Nonlinear Partial Differential Equations in Differential Geometry and Mathematical Physics

The Harmonic mapping as the extreme value of a certain action is a generalization to the concepts of geodesic, the minimum surface, etc. Previous studies were limited to the problem of harmonic mappings in Riemannian space with positive definite metric, corresponding to the solutions of a certain type of elliptic systems, while in 1980 Gu Chaohao first investigated the harmonic mapping from the two-dimensional Minkowski space R^{1+1} to any complete Riemann maniford, then reduced the problem to solving the initial value problem of a second-order semilinear hyperbolic system, the right-hand side of which is composed of the quadratic terms of the first-order partial derivatives of the solution, and with the aid of geometric speculations, proved the global existence of classical solutions, which consequently showed that if a one-dimensional nonlinear σ-model is non-singular at a certain moment, then it is non-singular either in the past or in future [10]. This surprisingly profound result illuminated a new research direction in the field of harmonic mappings, that is, the wave mapping, and as a classical citation in the research field of wave mapping, it provoked strong reactions around the world and induced the follow-up researches of some world's top mathematicians [24]–[28]. J. Elles, a famous British mathematician and also the founder of harmonic mappings, affirmed Gu Chaohao's status as a forerunner in this field [23]. Gu Chaohao and Madame Y. Choquet-Bruhat, an academician of French Academy of Sciences, worked cooperatively over the problem of hyperbolic harmonic mapping from the four-dimensional Minkowski space R^{3+1} to the symmetric space, and got the global existence of solutions with small initial data.

Since the middle of 1980s, Gu Chaohao has systematically studied time-like extremal surfaces in the Minkowski space R^{3+1} in relation to string movement, and he is the first one to explore extremal surfaces of mixed type including time-like, light-like and space-like parts at the same time [15] [17]. The research was also concerned with quasilinear equations of mixed type, and Gu Chaohao started from the zero length analytic curve to propose a unified construction method which not only led to the global existence of exact solution, but also proved the analyticity of solution in the hyperbolic domain (the C^2 solution should be analytic). Besides, he gave some boundary problems of extremal surfaces of mixed type, especially the problem of producing extremal surfaces through given boundary curve, and proved the existence of solutions when the boundary is an unclosed zero-length curve.

3. Mathematical Physics

(1) Mathematical Theory on Gauge Field

The gauge field theory originated from the electromagnetic field was introduced by physicists C. N. Yang and R. L. Mills in 1954, and was thus also known as the Yang-Mills field. In the late 1960s, on the basis of the gauge field theory, physicists S. L. Glashow, S. Weinberg and A. Salam established a unified theory about weak interaction and electromagnetic interaction that has been proved by experiments and they won the Nobel Prize for physics. Later on, two more Nobel Prizes for physics were conferred to related study on gauge field. As the concept of gauge potential corresponds to the connection on a fiber bundle in mathematics, the research on gauge field drew further attention from the scientific circle.

When C. N. Yang visited Fudan University in 1974, he found that Gu Chaohao had his own opinions about the gauge field and could not only understand the language of physics he used, but also express profound mathematical ideas with the language acceptable by physicists. Therefore, he began a 'very fruitful and effective cooperation' as he thought with Gu Chaohao.

The Yang-Mills equations are a set of highly complicated second-order nonlinear hyperbolic equations about the gauge potential, concerning whose initial value problem, Gu Chaohao and C. N. Yang provided the earliest proof of the existence and uniqueness of its local solution in 1975 [8].

C. N. Yang had used his own idea of the path phase factor to study the gauge field, and Gu Chaohao invented the approach of the (closed) loop phase factor in 1976, which successfully applied the holonomy group theory in a fiber bundle to the research of gauge field, and proved that using the phase factor of some standard loop and the gauge field intensity could uniquely determine the gauge potential. This method has been widely applied in later research.

Working with Hu Hesheng, Gu Chaohao also used Lie group theory to fully determine the general structure and classification of the spherically symmetric gauge field and give the specific expression of the gauge potential, which contributed to specifically determining the gauge field [11]. Gu Chaohao gave the decomposition of the gauge field of general compact Li group with respect to the Higgs field, so that he got the magnetic monopole and topological charge, and offered the value and geometric explanation of the topological charge. C. N. Yang has compared Gu Chaohao's work on gauge field to 'standing on a high mountain and looking downward, which provides a panoramic view' [21].

Due to his outstanding contribution to the research of gauge field, Gu Chaohao was specially invited to write a comprehensive report for the prestigious journal Physics Reports in 1981. He is actually the first Chinese scientist from mainland China that has enjoyed such an honor, and a page of Chinese abstract was added to his article as an exception.

(2) Soliton

The soliton theory originated from the research on solitary waves that maintain the wave shapes during the process of propagation and after interaction. From the late 1980s, Gu Chaohao started to enter the popular and important research field of modern nonlinear science. Many nonlinear partial differential equations with soliton solutions can be regarded as the integrable condition of a system of linear equations (called a Lax pair), so in principle one can use the method of Darboux transformation to construct a series of new solutions by integrating a known solution once and repeating the algebraic calculation each time. However, in the past one could only do the checking calculation for each specific equation, which was rather tedious, or even practically unfeasible. Gu Chaohao and Hu Hesheng worked together, and with their ingenious conception gave a universal method, which could apply to a series of important equations of mathematical physics with soliton solutions, such as the KdV hierarchy, the MKdV-SG hierarchy and the more widely used ADNS system, and thus greatly developed the method of Darboux transformation [16].

Based on the research on the Darboux matrix of the second order AKNS system, Gu Chaohao further provided the explicit construction of the unitary solution (uniton) of harmonic mappings in a plane, so as to arrive at the systematic results of related differential geometry problems. Gu Chaohao and the American mathematician D. Sattinger worked on their own and respectively gave the explicit expression of the Darboux matrix of the general nth order AKNS system, which offered an effective way of getting the explicit solutions of nonlinear equations in the nth order AKNS system, and could be widely applied. He further established the higher dimensional AKNS system and the more prevalent higher dimensional Yang-Mills system which contain a large number of nonlinear partial differential equations in higher dimension arising from practical problems, and the Darboux matrix given by Gu Chaohao is essential to obtaining explicit solutions of these systems [18].

Gu Chaohao and Hu Hesheng also worked on properties of soliton solutions and uncovered some interesting new phenomena, such as the soliton waves with periodic oscillation, inelastic scattering from interaction of solitons, unlimited collision and adherence to each other, the spiral soliton solutions of nonlinear σ-models, etc., which added new subjects to the study of soliton waves and attracted wide attention.

References

1. Gu Chao-hao, New treatment of geometries in a space of K-spreads, Science Record, 3(1950), 41–51.
2. Gu Chao-hao, Embedding of Finsler spaces into Minkowski spaces (in Chinese), Acta Math. Sinica 6(1956), 215–232.
3. Гу Чао-хао, Некоторые общие свойства псевдогрупп преобразований и их применения к дифференциальной геометрии, Докт. Дисс., МГУ, 1959.
4. Gu Chao-hao, Li Ta-tsien, Hou Zong-yi, Cauchy problem for quasilinear hyperbolic systems with discontinuous initial data (I) (II) (III) (in Chinese), Acta Math. Sinica 11(1961), 314–323; 324–327; 12(1962), 132–143.

5. Gu Chao-hao, A boundary value problem for hyperbolic system and applications (in Chinese), Acta Math. Sinica 13(1963), 32–48.
6. Gu Chao-hao, Some developments of the theory of positive symmetric systems and applications (in Chinese), Collection of Math. Papers (Inst. of Math., Fudan Univ.) (1964) 42–58.
7. Gu Chao-hao, A class of mixed type equations in multi-dimensional spaces (in Chinese), Advances in Chinese Math. (1965) 276–282; Scientia Sinica, 8(1965) 1574–1581.
8. Gu Chao-hao, C. N. Yang, Some problems on the gauge field theories (I), (II), (III), Scientia Sinica, 18(1975), 483–501; 20(1977), 47–55; 20(1977), 177–185.
9. Gu Chao-hao, Loop phase factor approach to gauge fields (in Chinese), Physica Energiae Fortis et Physica Nuclearis, 2(1978), 97–107.
10. Gu Chao-hao, On the Cauchy problem for harmonic maps defined on tow-dimensional Minkowski space. Comm. Pure Appl. Math., 33(1980) 727–737.
11. Gu Chao-hao, Hu He-sheng, On the spherically symmetric gauge fields, Comm. Math. Phys., 79(1981), 75–91.
12. Gu Chao-hao, On partial differential equations of mixed type in n independent variables, Comm. Pure Appl. Math., 34(1981), 333–345.
13. Gu Chao-hao, On classical Yang-Mills fields, Phys. Reports, 80(1981), 253–377.
14. Gu Chao-hao, On a class of mixed partial differential equations of higher order. Chin. Ann. of Math., 3(1985), 503–514.
15. Gu Chao-hao, The external surfaces in the 3-dimensional Minkowski space, Acta. Math. Sinica, New Series, 1(1985), 173–180.
16. Gu Chao-hao, Hu He-sheng, The unified explicit form of Backlund transformations for generalized hierarchies of KdV equations. Letter in Mathematical Physics, 11(1986), 325–335.
17. Gu Chao-hao, Extremal surfaces of mixed type on Minkowski spaces $\mathbb{R}^{n+1}$, Proc. of 1988 Conference on Variarional Problems, Paris.
18. Gu Chao-hao, Hu He-sheng, Zhou Zixiang, Darboux Transformation in Integrable System, Theory and Their Applications to Geometry, Springer, 2004.
19. Su Buqing, Selected Articles of Su Buqing (in Chinese), Zhejiang Science and Technology Press, 1991, 16–32.
20. Li Ta-tsien, Yu Wen-ci, Boundary Value Problems for Quasilinear Hyperbolic Systems,Duke University Mathematics Series V, 1985.
21. C. N. Yang, Selected Lectures of C. N. Yang (in Chinese), Nankai University Press, 1989, 38–43.
22. Shen Zhongmin, On Finsler geometry of submanifolds, Math. Ann., 311 (1998), 529–576,.
23. Eells J., Lemaire L., Selected topics in harmonic maps, Bulletin of the London Mathematical Society, 20(1988), 385–524.
24. Tao T., Global regularity of wave maps I,II, International Math. Research Motes, (2001), 299–328; Comm. Math. Phys., 224(2001), 443–544.
25. Christodoulou D., Tahvildar-Zadeh A. S., On the regularity of spherically symmetric wave maps, Comm. Pure Appl. Math., 46(1993), 1041–1091.

26. Terng C. L., Uhlenbeck K. K., 1+1 wave maps into symmetric spaces, Comm. Anal. Geom., 12(2004), 348–386.
27. Choquet-Bruhat Y., Global wave maps on black holes, Math. Physics Studies, 24(2003), 469–482.
28. Shatah J. Strauss W., Breathers as Harmonic geometric wave maps, Physica D, 99(1996), 113–133.

List of Publications of Gu Chaohao

Papers

1. New treatment of geometries in a space of K-spreads, Science Record 3 (1950), 41–51.
2. On the descriptive geometry of a space of K-spreads, Science Record 3 (1950), 53–59.
3. A note on bounded Schlicht functions, Science Record 3 (1950), 157–159.
4. On the theory of subspaces in a space of K-spreads, Science Record 4 (1951), 31–36.
5. Theory of spaces of K-spreads represented by implicit functions (in Chinese), Scientia Sinica 2 (1951), 1–19.
6. On the theory of subspaces in spaces of K-spreads (in Chinese), Scientia Sinica 2 (1951), 165–178.
7. On the first and second variation of the volume integral in a space with multi-dimensional measure (in Chinese), (with Su Buchin), Acta Math. Sinica (1952), 231–245.
8. The abscissa of uniform convergence of a Laplace integral, (with M. I. Yuh and K. K. Chen), Jour. of London Math. Society (1952), 356–359.
9. A solution of the problem of stability for nonlinear equations of second order (in Chinese), Acta Math. Sinica (1954), 347–357.
10. On the embedding of the spaces of paths (in Chinese), Acta Math. Sinica (1955), 369–381.
11. On the parallel displacement of area elements and nonholonomic manifolds (in Chinese), Acta Math. Sinica (1955), 383–392.
12. On the system of paths on surfaces in projective space (in Chinese), Jour. of Fudan Univ. (1955), 42–45.
13. The curves and surfaces of discontinuity of the solutions to the wave equation (in Chinese), Jour. of Fudan Univ. (1955), 55–57.
14. Embedding of Finsler spaces into Minkowski spaces (in Chinese), Acta Math. Sinica (1956), 215-232.

15. Lie derivatives of higher order and invariant groups of nonlinear geometric objectives (in Chinese), Jour. of Fudan Univ. (1956), 29–34.

16. On embedding of affinely connected spaces (in Chinese), Acta Math. Sinica (1956), 464–471.

17. On the isomorphism of geometrical objects (in Chinese), Jour. of Fudan Univ. (1956), 89–92.

18. Pairs of connections and integral manifolds of partial differential equations of second order (I) (in Chinese), Acta Math. Sinica (1956), 153-162.

19. Pairs of connections and integral manifolds of partial differential equations of second order (II) (in Chinese), Acta Math. Sinica (1956), 163–169.

20. Pairs of connections and integral manifolds of partial differential equations of second order (III) (in Chinese), Acta Math. Sinica (1956), 426–432.

21. A generalization of totally geodesic spaces (in Chinese), Jour. of Fudan Univ. (1957), 111–116.

22. Embedding of some special affinely connected spaces (in Chinese), Jour. of Fudan Univ. (1957), 357–361.

23. On Finsler spaces admitting a group of motions of the greatest order, Science Record (New series) 1 (1957), 21–24.

24. On the global embedding of Finsler spaces (in Chinese), Acta Math. Sinica (1958), 272–275.

25. On the global embedding of affinely connected space (in Chinese), Science Record (New series) 2 (1958), 7–9.

26. On the reducibility of the infinite continuous group of E. Cartan (in Russian), Izv. Vyssh. Uchebn. Zaved. Mat. 4 (1958), 60–66.

27. Semi-reducible homogeneous Riemannian spaces (in Russian), (with G. E. Kruchkovich)Dokl. Akad. Nauk. USSR (1958), 1183–1186.

28. Some classes of homogeneous Riemannian spaces (in Russian), Dokl. Akad. Nauk. USSR (1958), 171–174.

29. Homogeneous Riemannian spaces with invariant vector fields (in Chinese), Jour. of Fudan Univ. (1959), No. 4, 2–25, (1960), No. 1, 19–37.

30. On the group of motions for spaces with sympletic measure (in Russian), Nauk. Dokl. Vyssh. Shkoli, Phys. Mat. Nauk. 2 (1959).

31. Some general properties of the pseudogroups of transformations and applications to differential geometry (in Russian), Dissertation for the degree of doctor of science at the Moskow State University (1959).

32. The developments of differential geometry in China for the past ten years, Scientia Sinica 8 (1959), 1238–1242.

33. Cauchy problem for a class of quasilinear partial differential equations in several independent variables (in Chinese), Collection of Mathematical Papers of Fudan Univ. (1960), 40–45.

34. Cauchy problem for hyperbolic system with discontinuous initial data (in Chinese), (with Li Tatsien and Hou Zongyi), Collection of Science and Technology Papers, Shanghai (1960), 55–65.

35. Cauchy problem for typical hyperbolic systems with discontinuous initial data (in Chinese), (with the group of nonlinear PDEs), Collection of Mathematical Papers of Fudan Univ. (1960), 1–16.

36. Geometrical interpretation for integral manifolds of partial differential equations of second order (in Chinese), Jour. of Fudan Univ. (1960), 14–18.

37. On global solutions without shocks for quasilinear hyperbolic systems (in Chinese), Collection of Mathematical Papers of Fudan Univ. (1960), 36–39.

38. On irreducible isotropic groups of first order for infinite continuous pseudogroups (in Chinese), Collection of Science and Technology Papers, Shanghai (1960), 1–27.

39. On the generation of shock waves for quasilinear equations with discontinuous coefficients (in Chinese), Collection of Mathematical Papers of Fudan Univ. (1960), 33–35.

40. Quasi-linear equations with discontinuous coefficients (in Chinese), Collection of Mathematical Papers of Fudan Univ. (1960), 29–32.

41. Supersonic flows around plane wings with sharp leading edge (in Chinese), (with the group of nonlinear PDEs), Collection of Mathematical Papers of Fudan Univ. (1960), 17–28.

42. Cauchy problem for quasilinear hyperbolic systems with discontinuous initial data (I) (in Chinese), (with Li Tatsien and Hou Zongyi), Acta Math. Sinica (1961), 314–323.

43. Cauchy problem for quasilinear hyperbolic systems with discontinuous initial data (II) (in Chinese), (with Li Tatsien and Hou Zongyi), Acta Math. Sinica (1961), 324–327.

44. Cauchy problem for quasilinear hyperbolic systems with discontinuous initial data (III) (in Chinese), (with Li Tatsien and Hou Zongyi), Acta Math. Sinica (1962), 132–143.

45. Supersonic flows around some body (in Chinese), Jour. of Fudan Univ. (1962), 11–14.

46. A boundary value problem for hyperbolic system and applications (in Chinese), Acta Math. Sinica (1963), 32–48.

47. A generalization of positive symmetric system of equations (in Chinese), Jour. of Fudan Univ. (1963), 31–38.

48. Differentiable solutions to positive symmetric systems of partial differential equations (in Chinese), Acta Math. Sinica (1964), 503–516.

49. Group invariant positive symmetric systems of high order (in Chinese), Acta. Nat. Sci. in Univ. (1964), 39–44.

50. Some developments of the theory of positive symmetric systems and applications (in Chinese), Collection of Mathematical Papers (Inst. of Math., Fudan Univ.) (1964), 42–58.

51. A class of mixed type equations in multi-dimensional spaces (in Chinese), Advances in Chinese Math. (1965), 276–282, Scientia Sinica (1965), 1574–1581.

52. Some boundary value problems for hyperbolic symmetric systems (in Chinese), Advances in Chinese Math. (1965), 272–276.

53. Gravitational collapse of inhomogeneous spherically symmetrical bodies (in Chinese), Jour. of Fudan Univ. (1973), No. 3, 72-78.

54. Some problems on the gauge field theories (I), (with C. N. Yang), Jour. of Fudan Univ. (1975), 27–41, Scientia Sinica (1975), 483–501.

55. U1 gauge fields in an SUn gauge field and their dual charges (in Chinese), Jour. of Fudan Univ. (1976), No. 3,4, 161–168, Physica Energiae Fortis et Physica Nuclearis (1978), 295–304.

56. Global gauge fields of electromagnetic U(1) group (in Chinese), Scientia Sinica (1976), 320–328.

57. Some problems on the gauge field theories (II), (with C. N. Yang), Jour. of Fudan Univ. (1976) , 146–160; Scientia Sinica (1977), 47–55.

58. Loop phase factor approach to gauge fields (in Chinese), Jour. of Fudan Univ. (1976), No. 2, 51–60, Physica Energiae Fortis et Physica Nuclearis (1978), 97–107.

59. Determination of spherically symmetric gauge fields (in Chinese), Jour. of Fudan Univ. (1977), No. 2, 30–36.

60. Some problems on the gauge field theories (III), (with C. N. Yang), Scientia Sinica (1977), 177–185.

61. Geometric interpretation of instantons in Euclidean space, (with C. N. Yang et al.), Jour. of Fudan Univ. (1977), No. 4, 8-12; Scientia Sinica (1978), 767–772.

62. On spherically symmetric SU2 gauge fields (in Chinese), (with Hou Boyu), Jour. of Fudan Univ. (1977), No. 1, 92–99.

63. Riemannian spaces with local duality and gravitational instantons, (with C. N. Yang et al.), Jour. of Fudan Univ. (1977), 1-7; Scientia Sinica (1978), 475–482.

64. Spherically symmetric SU2 gauge fields and the description of monopoles (in Chinese), (with Hu Hesheng), Acta of Phys. Sinica (1977), 155–168.

65. The structure of gauge fields with Higgs fields of compact groups (in Chinese), Jour. of Fudan Univ. (1977), No. 3, 28–34.

66. Boundary value problems for mixed type equations in n variables (in Chinese), Chinese Science Bulletin (1978), 335–339.

67. Boundary value problems for quasilinear positive symmetric systems and applications to mixed type equations (in Chinese), Acta Math. Sinica (1978), 119–129.

68. On the boundary conditions for partial differential equations, Jour. of Fudan Univ. (1978), 1–5; Scientia Sinica 23 (1980), 940–944.

69. On the mathematical structure of gauge fields (in Chinese), Proc. of Symposium of Chinese Math. Society (1978).

70. Solving free boundary problem for quasilinear equations by difference method (in Chinese), (with Chen Shuxing), Acta Math. Appl. Sinica (1978), 250–265.

71. Some global solutions and asymptotic properties of quasilinear symmetric hyperbolic system (in Chinese), Acta Math. Sinica (1978), 130–134.

72. The symmetry broken and dual charges of gauge fields with compact gauge groups (in Chinese), Jour. of Fudan Univ. (1978), No. 2, 1–7; Chinese Science Bulletin (1978), No.10, 598–600.

73. On the variation problem for gauge conditions (in Chinese), Chinese Science Bulletin (1979), 492–495.

74. Conformally flat spaces and solutions to Yang-Mills equations, Phys. Rev. D21 (1980), 970–971.

75. Loop-space formulation of gauge theories, (with L. L. Chau Wang), Phys. Rev. Lett. 45 (1980), 2004–2007.

76. On the solutions of a class of linear partial differential equations, Comm. in PDEs 5 (1980), 985–997.

77. On the Cauchy problem for harmonic maps defined on two-dimensional Minkowski space, Comm. Pure Appl. Math. 33 (1980), 727–737.

78. On the initial boundary value problems for harmonic maps from 2-dimensional Minkowski space, Manus. Math. 33 (1980), 51–58.

79. On the mixed partial differential equations in n independent variables, Journees "Equations aux Derivees Partielles", Saint-Jean-de-Monts (1980).

80. On partial differential equations of mixed type in n independent variables, Comm. Pure Appl. Math. 34 (1981), 333–345.

81. On the boundary value problems for mixed partial differential equations in n independent variables, Nonlinear Partial Differential Equations (II), edited by H. Brezis and J. L. Lions, College de France Seminar (1981), 200–203.

82. On the spherically symmetric gauge fields, (with Hu Hesheng), Comm. Math. Phys. 79 (1981), 75–91.

83. On the structure of Yang-Mills fields, Hadronic Jour. 4 (1981).

84. Analycity of solutions to a class of mixed PDEs, Preprint No. 549, Institute of Mathematics, Bonn Univ. (1982).

85. Harmonic maps from Minkowski plane and applications in physics, Proc. of Symposium in Gauge Fields and Related Problems in Physics (1982), 41–44.

86. On a class of mixed partial differential equations of higher order, Chin. Ann. of Math. 3 (1982), 503–514.

87. On parallel Yang-Mills fields, Seminar on Differential Geometry, edited by S. T. Yau (1982), 443–453.

88. On the motion of a string in a curved space-time, Proc. of 1982 Grossmann Symposium (1982), 139–142.

89. Some problems on nonlinear hyperbolic systems, Proc. of 1980 Beijing DD Symposium (1982), 467–479.

90. Harmonic maps of manifolds with indefinite metric, Proc. of DD3 Symposium (1983).

91. On the harmonic maps from $\mathbb{R}^{1,1}$ to $S^{1,1}$, J. Reine Angw. Math. 346 (1984), 101–107.

92. Hyperbolic and mixed equations in differential geometry, Report at Plenary Meeting of the 50th Anniversary Conference of Chinese Mathematical Society (1985).

93. Mathematical aspects of the classical Yang-Mills theory, Proc. of Symposium on Yang-Mills Gauge Theories, Comm. in Theor. Phys. 4 (1985), 675–680.

94. On some developments of harmonic maps, (with Hu Hesheng), Advances in Chinese Math. (1985).

95. The extremal surfaces in the 3-dimensional Minkowski space, Acta Math. Sinica (1985), No. 2, 173–180.

96. The unified explicit form of Backlund transformations for generalized hierarchies of the KdV equations, (with Hu Hesheng), Lett. Math. Phys. 11 (1986), 325–335.

97. On the Bäcklund transformations for generalized hierarchies of compound MKdV-SG equations, Lett. Math. Phys. 12 (1986), 31–41.

98. On the determination of nonlinear partial differential equations admitting integrable systems, (with Hu Hesheng), Scientia Sinica (1986), No. 8, 704–719.

99. Some developments of the theory of partial differential equations of mixed type, (with Hong Jiaxing), Teubne-Texte Band 90 (1986), 102–135.

100. A global study of extremal surfaces in 3-dimensional Minkowski space, DD6 Proc., Springer-Verlag, Lect. Notes in Math. 1225 (1987).

101. On the Darboux matrices of B?cklund transformations for AKNS system, (with Zhou Zixiang), Lett. Math. Phys. 13 (1987), 179–187.

102. A class of boundary problems for extremal surfaces in Minkowski 3-space, J. Reine Angw. Math. 385 (1988), 195–202.

103. The mixed PDEs for amplifying spiral waves, Lett. Math. Phys. 16 (1988), 69–76.

104. A one dimensional N fermion problem with factorized S matrix, (with C. N. Yang), Comm. Math. Phys. 122 (1989), 105–116.

105. Existence globale d'applications harmonique sur l'espace-temps de Minkowski M3, (with Y. Choquet Bruhat), C. R. Acad. Sci. Paris t.308 (1989), 167–170.

106. Explicit form of Bäcklund transformations for GL(N), U(N) and O(2N) principal chiral fields, (with Zhou Zixiang), Nonlinear Evolution Equations: Integrability and Spectral Methods, Proc. of Workshop on Evolution Equations, Como, 1988, Manchester Univ. Press (1990), 115–123.

107. Extremal surfaces of mixed type on Minkowski space Rn+1, Variational Methods, Proc. of Conf. on Variational Problems, Paris, 1988, Birkhauser (1990), 283–296.

108. The Darboux matrices of Bäcklund transformations for NN soliton systems, Proc. of Symposium of Integrable Systems, Nankai Lectures on Math. Phys., World Scientific, Singapore (1990), 162–168.

109. Darboux transformation for a class of integrable systems in n variables, Analyse, Varieties et Physique, Proc. of Colloque International en l'honneur d'Yvonne Choquet-Bruhat, Paris, 1992, Kluwer Academic Publishers (1992).

110. On the interaction of solitons for a class of integrable systems in the space-time $\mathbb{R}^{n+1}$, Lett. Math. Phys. 26 (1992), 192–207.

111. Explicit solutions to the intrinsic generalization for the wave and sine-Gordon equations, (with Hu Hesheng), Lett. Math. Phys. 29 (1993), 1–11.

112. The inversion, permutability and periodicity of Darboux transformations (in Chinese), Jour. of USTC 23 (1993), 9–14.

113. The minimal surface and stationary conical flow in R3, (with Lin Junmin), Differential Geometry, Proc. of International Symposium on Differential Geometry, Shanghai, 1991, World Scientific, Singapore (1993), 79–86.

114. The soliton behavior of the principal chiral fields, (with Hu Hesheng), Proc. of XXI DGM Conf., Tianjin, 1992, World Scientific, Singapore (1993), 501–510.

115. A class of integrable systems and solitons in higher dimensional space-time $\mathbb{R}^{n+1}$, Proc. of the Workshop on Qualitative Aspect and Applications of Nonlinear Equations, Trieste, 1993, edited by H. Beirao da Veiga and Ta-Tsien Li, World Scientific (1994), 11–25.

116. Complete extremal surfaces of mixed type in 3-dimensional Minkowski space, Chin. Ann. of Math. 5B (1994), 385–400.

117. On Darboux transformations for soliton equations in high-dimensional space-time, (with Zhou Zixiang), Lett. Math. Phys. 32 (1994), 1–10.

118. Some developments of the theory of mixed PDEs, (with Hong Jiaxing), Partial Differential Equations in China, edited by C.H.Gu, X.X.Ding & C.C.Yang, Kluwer Academic Publishers (1994), 50–66.

119. Explicit construction of harmonic maps from R2 to U(N), (with Hu Hesheng), Chin. Ann. of Math. 16B (1995), 139–152.

120. Generalized self-dual Yang-Mills flows, explicit solutions and reductions, Proc. of KdV'95, Acta Applicandae Mathematicae (1995).

121. Integrable evolution systems based on generalized self-dual Yang-Mills equations and their soliton-like solutions, Lett. Math. Phys. (1995), 61–74.

122. Mixed partial differential equations in China (in Chinese), Collection of Papers for the 60th Anniversary of Chinese Math. Society (1995).

123. Unitons of harmonic maps from $\mathbb{R}^2$ to U(p, q), Lett. Math. Phys. 46 (1998), 347–357.

124. Construction of unitons via purely algebraic algorithm, (with Hu Hesheng), Chin. Ann. Math. 20B(1999), 1–6.

125. The group property of Darboux transformations from $\mathbb{R}^2$ to U(N), Lett. Math. Phys. 48 (1999), 181–185.

126. Constructive factorizations of unitons via singular Darboux transformations, (with Dong Yuxin and Shen Yibing), Sci. China A43 (2000), 149–157.

127. Some remarks on singular Darboux transformations for unitons, (with Hu Hesheng), Chin. Ann. of Math. 22B (2001), 263–266.

128. On time-like surfaces of positive constant Gaussian curvature and imaginary principal curvatures, (with Hu Hesheng and Inoguchi) J. Geom. Phys. 41 (2002), 296–311.

129. Darboux transformation and solitons of Yang-Mills-Higgs equations in R2,1, Sci. in China A45 (2002), 707–715.

130. On the classification of initial data for nonlinear wave equations, Chin. Ann. of Math. 23B (2002), 205–208.

131. Gravitation collapse of spherical symmetry with non-uniform density, Translated from J. Fudan Univ. Nat. Sci. 1973, no. 1, 73–78. Front. Math. China 1 (2006), no. 2, 161-168.

132. On the Cauchy problem for harmonic maps defined on two-dimensional Minkowski space, A Collection of Essays in Commemoration of Ho Leung Ho Lee Foundation (2005-2006), Edited by the Selection Board of Ho Leung Ho Lee Foundation, China Science and Technology Press, 2009, 75–89.

Monographs

1. Differential Geometry in Homogeneous Spaces (in Chinese), Shanghai Scientific and Technical Publishers, 1964.

2. On Classical Yang-Mills fields, Phy. Rep. 80(1980), No. 4.

3. Soliton Theory and Its Applications (in Chinese, Edited by Gu Chaohao), Zhejiang Science and Technology Publishing House, 1990; English Edition (Revised), Springer and Zhejiang Science and Technology Publishing House, 1995.

4. Darboux Transformation in Soliton Theory and Its Geometric Applications (in Chinese), (with Hu Hesheng and Zhou Zixiang), Shanghai Scientific and Technical Publishers, 1999 (First Edition) and 2005 (Second Edition).

5. Darboux Tranformation in Integrable Systems, Throry and Their Applications to Geometry, (with Hu Hesheng and Zhou Zixiang), Springer, 2005.

Text Books

1. Equations of Mathematical Physics (in Chinese), (with Xu Zhengfan, Li Tatsien, Hou Zongyi and Li Likang), Shanghai Scientific and Technical Publishers, 1960.

2. Equations of Mathematical Physics (in Chinese), (with Li Tatsien, Chen Shuxing, Zheng Songmu and Tan Yongji), Higher Education Publishers, 1979.

3. Partial Differential Equations (in Chinese), Shanghai Scientific and Technological Literature Publishing House, 1991.

4. Equations of Mathematical Physics (in Chinese), (with Li Tatsien, Chen Shuxing, Shen Weixi, Qin Tiehu and Shi Jiahong), Shanghai Scientific and Technical Publishers, 1987.

5. Applied Partial Differential Equations (in Chinese), (with Li Tatsien and Shen Weixi), Higher Education Publishers, 1993.

In Memory of Professor Gu Chaohao

Xiaqi Ding
Institute of Applied Mathematics, AMSS, Academia Sinica
xqding@amt.ac.cn

Professor Chaohao Gu is one of the mathematicians I respect most. He contributed to many important branches of mathematics including differential geometry, partial differential equations and mathematical physics. In my memory and through my contact with him, I found that he shared something in common with the famous Chinese mathematician Professor Luogeng Hua. He could find an entirely new approach and construct new results for some problems that had already been studied thoroughly by others.

In the summer of 1954, Professor Xinmo Wu opened a course on partial differential equations in mathematical physics at Beijing University which was initiated and organized by the Ministry of Education of the People's Republic of China. The audience was excellent young teachers selected and sent by universities all over the country, as well as Professor Gu. Professor Wu employed many meaningful problems from famous classic books as exercises. Then the participants reported their answers on the blackboard. One of the exercises was the 4th problem of the D'Alembert hyperbolic equation, i.e. to construct the solution when the solution is known on the two non-characteristic lines, in a characteristic rectangle, intersecting with two characteristic sides at the same vertex. In J.Hadamard's related paper and audience's answers on this problem, the solution was constructed in the region between the two support lines. But Gu's solution was different, as he determined the solution in the whole characteristic rectangle, which gave a new insight to the problem.

Another time, when I discussed with him about the 2-dimensional Riemann problem of hyperbolic conservation laws, I said, "It's natural that people consider the Riemann problem as to construct the solution when the distinct initial constant states are given in the four quadrants of the initial plane." Professor Gu suddenly suggested that people could consider the Riemann problem with the initial constant states given in two parts of the initial plane separated by a broken line (with a turning point). This idea inspired me and afterwards I asked some of my students to consider these kinds of problems in their Ph.D. theses.

Academician Petrovskii, President of the Moscow University of the Soviet Union, visited China in 1954. He gave lectures in Beijing and Shanghai on the same topic, which was constructing 2-dimensinal discontinuous planes and one-dimensional discontinuous lines of the space-time four dimensional wave equation. After the lectures, the audience did not have any comments or pose any questions. But Professor Gu managed to give the necessary and sufficient conditions such that discontinuities occur. It was really astonishing.

Professor Gu became famous very early in his career. His personal success in the research field is undoubtedly great. He also trained many elite students including some academicians and influenced many scholars in modern China. Professor Gu had a pleasant, amiable manner and was very approachable, making it easy for people to get close to him. We will always remember the famous mathematician, Professor Chaohao Gu!

In Memory of Professor Gu Chaohao

Gongqing Zhang (Kung Ching Chang)
Peking University
kcchang@math.pku.edu.cn

Although I had heard a lot about Professor Gu Chaohao when I was still an undergraduate student, it was not until after the Cultural Revolution that I had the opportunity to know him personally. In 1979, Professor Gu went to work with Professor C. N. Yang at the State University of New York at Stony Brook, and there were several weeks when he visited the Courant Institute of Mathematical Sciences at NYU. At that time, I was a visiting scholar at the Courant Institute, so we had much time to communicate with each other. Professor Gu lived in a quiet building not far from the institute with an American Chinese physicist. In the evenings, concerned for the safety of Professor Gu, several Chinese visiting scholars and I would often accompany him on his walk home. China had recently adopted the policy of opening up to the outside world, and it was urgent to establish connections between the Chinese mathematical circle and the international mathematical community. As an influential figure from China, Professor Gu was laden with a heavy responsibility to take this opportunity to meet with a broad spectrum of mathematicians frequently and introduce China to them. On one hand, he actively carried out academic exchanges with fellow scholars; on the other hand, he also tried to build up platforms for young Chinese scholars' future foreign exchange programs. He impressed me as being very energetic, highly efficient, and fast paced. Over the years, I have met with Professor Gu frequently at professional functions such as academic conferences, various review activities, oral defense sessions, meetings, and mutual visits. On occasion, we also happened to be together abroad: New York, Paris, Berkley, and San Diego to name a few. Through direct contacts with Professor Gu and based on our formal and casual conversations, I knew him further: he was quite knowledgeable and his research fields ranged widely. He was a principled man and always acted on behalf of the country. He attached great importance to the unity of the national mathematical community, especially the unity among mathematicians in universities, to which he contributed greatly.

The mathematical research of Professor Gu covers many fields. He started as a geometer, and when he studied in the former Soviet Union, he speedily finished his dissertation on differential geometry. For his excellent academic performance, he was awarded a doctoral degree by the former Soviet Union. Before returning to China, he realized that the subjects discussed at the seminars conducted by I. M. Gelfand at the Moscow University on quasi-linear hyperbolic partial differential equations were closely related to advanced sciences and high technologies, so he devoted a lot of time and energy to the study of these types of equations. Back in China, he led a group of students, including Professor Li Daqian, to quickly catch up to the frontier of this research field and achieved remarkable results. During the later years of the Cultural Revolution, in collaboration with Professor C. N. Yang, he brought his strength in differential geometry into full play and guided a group of young scholars to study the field of gauge theory. After the 1980s, he shifted his focus to mathematical physics, in particular to the study of solitons. He became a leader in the national research of nonlinear sciences. When he was over fifty, he did some original work on the Cauchy problem of harmonic mappings on two-dimensional Minkowski space and the Darboux transformation in integrable systems, etc. Since China adopted the policy of reform and opened its door to the western world, the two research groups on differential geometry and partial differential equations at Fudan University have not only maintained their traditional strengths in their respective areas, but also advanced with time to develop new research directions under the guidance of Professor Gu. Over the years, they have obtained many important results on diverse problems, which owes to the great foresight and long-term planning of Professor Gu. In modern China, mathematicians like Professor Gu and his peers have undergone the roughest course in scientific research, while Professor Gu among the others was the one who shouldered the dual responsibilities of being a scholar and an administrator at the same time. For such a man, it definitely requires talent, diligence, perseverance, and an ardent love for mathematics to achieve so many eminent results in research as well as to train so many excellent scholars.

Professor Gu always put the overall situation as top priority, which I can speak from my personal experience. In the 1980s and 1990s, the qualification of establishing a doctoral program or selecting a doctoral supervisor was to be assessed and determined by the Discipline Appraisal Group subordinate to the Academic Degree Commission of the State Council. The number of doctoral programs and doctoral supervisors became two important indices that reflected the rank of a university. However, the quota given each time for adding new doctoral programs could not meet the demands, and the call for an increase in the quota was vehement. In fact, the number of doctoral programs to be added was allotted according to the number of subordinate disciplines in the catalogue of subjects, and in mathematical sciences then, "Mathematics" was the primary discipline which included five subordinate disciplines, namely, "Fundamental Mathematics", "Applied Mathematics", "Probability and Statistics", "Computational Mathematics", and "Operational Research and Control Theory". (The method of classification was not necessarily reasonable, but it had been established and could not be changed). In the early 1990s, the Ministry of Education held a meeting and decided to revise the catalogue of subjects by allowing some more subordinate disciplines to the primary one. As the number of subordinate disciplines increased, the number of doctoral programs could be proportionately added, so the dilemma could be resolved. The problem with which we were confronted was whether to further divide the subordinate disciplines of mathematics. To relieve the pressure of limited mathematical doctoral programs in our country, it could be an expedient measure to divide "Fundamental Mathematics" further into several minor subordinate disciplines. In fact, it was said that many other disciplines had prepared to do so. But if we really put it into practice, the results would be a more specific training objective for doctoral students of mathematics and a lower requirement for the breadth of knowledge. In discussions, we all expressed our opinions; both the obvious and the more subtle aspects of the problem were taken into consideration. However, no conclusion was reached. Since Professor Gu was our group leader in revising the catalogue of mathematical disciplines, we waited on him to make a final decision. After a period of deliberation, during which time he analyzed the status of national mathematical studies, the developmental tendency of mathematics internationally in modern times, as well as the demand of training advanced research personnel of mathematics, etc. He then resolutely claimed that the training objective should not be too narrow and an expedient measure was unfeasible, so the subordinate disciplines of "Fundamental Mathematics" should not be further divided. Every participant supported his unselfish view in the long-term interest of our country and respected his courage to undertake any pressure.

At the beginning of this century, the economic situation of China was excellent. Many students turned their interest to majors that would bring immediate financial rewards. Because of this, there once appeared a phenomenon that the quality of domestic students who studied Fundamental Mathematics declined. Professor Gu detected this problem early, and in those years, every time we met, he would mention this and asked us to take it seriously in the hope of working out a solution. At that time, Professor Gu was nearly eighty, and he was still concerned with cultivating qualified successors of the Chinese mathematical society.

When the reform and open door policy was first adopted in China, most Chinese institutions of higher education needed to catch up with the latest trends of international development in teaching and research; however, by learning more about the western world and looking at the training in China from the past, they endeavored to explore a way of nurturing mathematical talents that was suited for China. Since many Chinese institutions of higher learning shared a lot of things in common and faced many similar problems, they had to rely on sincere co-operations and mutual help in coping with the difficulties. Professor Gu Chaohao paid special attention to the unity of mathematical scholars in these institutions.

The 1980s was a period of recovery for Chinese mathematics. Professor S. S. Chern began by placing an emphasis on the study on "two differentials" (i.e. differential equations and differential geometry), and organized several national academic activities, such as the seven times "Two Differentials" Conferences, and the "Nankai Academic Year" activities went on successively for several years, etc. The main supporters of such activities were Peking University, Fudan University, Nankai University as well as many other colleges, universities, and institutions. As they had the same goal to reach, the participants got along well with each other. Since the "two differentials" were rightly the strengths of Fudan University, led by Professor Gu, teachers of differential geometry and partial differential equations at Fudan University devoted themselves to these activities and made great contributions.

In the early 1980s, it was a common practice for most academic institutions to invite foreign experts to visit and deliver lectures, while domestic scholars would just listen and learn. As I recall, academic exchanges among universities in China were perhaps initiated by Fudan University, and at the beginning of 1983, Professor Gu invited me to Fudan University to lecture on "Critical Point Theory and Differential Equations in Geometry" for over half a month. He would come and preside over each class in person. Honestly, I was then a newcomer to the area. Facing such a large audience of teachers and fellow researchers, I felt pressured to prepare carefully for the lectures and to deepen my understanding of the contents, which was both challenging and worthwhile. During that time, I also had the opportunity to meet many young teachers then in the Department of Mathematics, Fudan University and learned a lot from them. This laid a foundation for our later cooperation with each other in academic activities. The initiative propelled Peking University and other Chinese universities in frequent domestic academic exchanges.

Professor Gu held the harmonious relationship in the mathematical community in high regards. In 1985, the annual conference of the Chinese Mathematical Society (CMS) was held in Shanghai. Fudan University assumed the role of the organizing unit for the conference with the detailed works arranged under the leadership of Professor Gu. Before the conference, as I was on a business trip to Shanghai, the CMS, for convenience, assigned me to report the names of the invited foreign guests to the preparatory group of the conference at Fudan University. For lack of communication in advance, it did not go very smoothly. In order to avoid unnecessary misunderstandings, Professor Gu talked with me for quite a while and asked me to convey his ideas to the CMS. Therefore, no more problems arose and the preparatory work proceeded well. During the conference, Professor Gu was personally in charge of the work of hosting foreign guests. At that time, there was no elevated highway in Shanghai, and the traffic was extremely crowded. It might take more than two hours to drive from the Hongqiao Airport to Fudan University during rush hours. As the mobile phone was not a common device of communication at that time, after they got on a taxi at the airport, the staff and foreign guests lost contact with the organizing group of the conference on the way, which made Professor Gu very anxious. Fortunately, the reception was consummately accomplished due to the meticulous management of Professor Gu, who allowed no slip. The organization part of the whole conference was impeccably done, which made the activity a successful demonstration of unity.

Previously, people said there was a division between the Southern School and the Northern School in the mathematical circle of China, but since the reform and open door policy, I have never felt such estrangement at all in my contact with Professor Gu. He was impartial and always considered matters in a comprehensive and detailed way. He paid close attention to issues of the national mathematical community, and even for a name list, he would take all aspects into account and make a final decision after careful consideration. He highly respected mathematicians of the older generation. For example, if Professor S. S. Chern wanted to discuss with him for something, Professor Gu would always go to Tianjin in person. As he came to attend meetings in Beijing, he often paid visits to Professor Jiang Zehan, Professor Duan Xuefu and Professor Cheng Minde, if time permitted. He took many of my generation as his students and always showed concern and support for us. So we have also regarded him as a senior scholar of noble character and high prestige, whom we respected and trusted like a teacher. Later, when I took on the jobs of serving the national mathematical community, I would often listen to his opinions when encountering imperative matters, and every time, he would offer much pertinent advice, help, and support without fail.

At the end of October 2005, Professor Gu invited me to give a series of lectures on the "W. Z. Yang Lectureship" at Fudan University. When I arrived in Shanghai, I learned that Professor Gu was hospitalized, and the doctor would not allow him to go out, so Professor Hu Hesheng came to preside over the lecture instead. One day, I mentioned the wish of visiting Professor Gu at the hospital, and to my surprise, the next afternoon without advance notice he came directly to my hotel accompanied by a secretary. Looking at his sickly complexion, I felt very apologetic and regretted proposing to visit him. After I briefly inquired about his health condition, Professor Gu quickly changed the topic to the issue of unity within the national mathematical community. He thought that we could set aside our differences by focusing on the education reform, which is of common interest to all our universities. I totally agreed and hoped to discuss further with him after his recovery. Unfortunately, from then on, Professor Gu's health condition deteriorated, and my wish consequently fell through.

Now Professor Gu has left us, this is a great loss to our mathematical community, and we'll deeply cherish the memory of him forever.

Stability of E-H Mach Configuration in Pseudo-Steady Compressible Flow

Shuxing Chen

School of Mathematical Sciences, Fudan University & Key Laboratory of Mathematics for Nonlinear Sciences (Fudan University), Ministry of Education, Shanghai 200433, China

sxchen@fudan.edu.cn

The paper is devoted to the study of Mach reflection in pseudo-steady compressible flow. The stability of E-H type Mach configuration under perturbation is proved. The problem is reduced to a generalized Tricomi problem of a nonlinear mixed type equation, then the existence of the generalized Tricomi problem implies the required stability.

1. Introduction

The stability of Mach configuration is a key problem in the theory of Mach reflection. The Mach configuration is constructed by three shocks (incident shock, reflected shock and Mach stem) issuing from a triple point and a contact discontinuity. In the problem of Mach reflection the supersonic coming flow loses its energy when it passes the incident shock and the reflected shock. If the energy loss is not strong enough the flow may still be supersonic across these two shocks. Meanwhile, the flow behind the Mach stem is always subsonic. Therefore, the whole flow behind the reflected shock and the Mach stem can be supersonic and subsonic flow separated by a contact discontinuity, or different subsonic flow separated a contact discontinuity. To simplify notations we call the first case by E-H Mach configuration, and call the latter by E-E type Mach configuration. The study of the E-H type Mach configuration is always related to the study of nonlinear mixed type equations. In Ref. 4 we proved the stability of E-H Mach configuration for steady flow. In this paper we will continue the study and prove the stability of E-H Mach configuration for pseudo-steady flow.

The E-H Mach configuration may also appear in the problems of shock reflection by a ramp. For such problems one can use pseudo-steady flows to describe the motion of the fluid, if the moment, when the coming shock meet the ramp, is taken as the time t=0. In such problems the reflection mode can be regular reflection, simple Mach reflection, complex Mach reflection or double Mach reflection etc., determined by the parameters of the coming flow and the inclined angle of the ramp (see Ref. 1). The study in Ref. 1 indicates that in the shock reflection for pseudo-steady flow the E-H Mach configuration appears very often, so that its stability must be clearly understood.

The stability of the E-E Mach configuration for pseudo-steady flow has been proved in Ref. 3. The setting of the problem, its mathematical reduction and some notations will be often employed in this paper. On the other hand, the method in Ref. 4 to treat the difficulties caused by the change of the type of the equations also plays key role in the discussion of this paper.

This paper is organized as follows. In Section 2 we reduce the stability of the E-H Mach configuration for the pseudo-steady flow to the existence of a free boundary value problem of a mixed type equation. In Section 3 we linearize the problem and give some related estimates. In Section 4 we give the existence of the free boundary value problem derived in Section 2, so that obtain the required stability. Since much discussion in this paper is similar to the related part in Ref. 3 or Ref. 4, we will only show the main steps and omit many technical details, which can be found in these two papers.

2. Nonlinear Free Boundary Value Problem

Let (ξ,η) be self-similar coordinates, let $(u,v),p,\rho,s$ denote the velocity, pressure, density and entropy respectively, and set $U=u-\xi, V=v-\eta$, then the system of the pseudo-steady compressible flow can be described by

$$\begin{pmatrix} \rho U & & 1 & \\ & \rho U & & \\ 1 & & a^{-2}\rho^{-1}U & \\ & & & U \end{pmatrix} \frac{\partial}{\partial\xi}\begin{pmatrix} U\\ V\\ p\\ s\end{pmatrix} + \begin{pmatrix} \rho V & & & \\ & \rho V & 1 & \\ & 1 & a^{-2}\rho^{-1}V & \\ & & & V \end{pmatrix} \frac{\partial}{\partial\eta}\begin{pmatrix} U\\ V\\ p\\ s\end{pmatrix} + \begin{pmatrix} \rho U\\ \rho V\\ 2\\ 0\end{pmatrix} = 0. \tag{2.1}$$

Denote by $\eta = \psi(\xi)$ the equation of the possible shock, by $[\ \cdot\ ]$ the jump of the quantity inside the bracket, and by e the inner energy, let $\tilde{E} = e + \frac{1}{2}(U^2 + V^2)$, then the Rankine–Hugoniot conditions on the shock are

$$\begin{bmatrix} \rho U \\ p + \rho U^2 \\ \rho UV \\ \rho \tilde{E} U + pU \end{bmatrix} \psi_\xi - \begin{bmatrix} \rho V \\ \rho UV \\ p + \rho V^2 \\ \rho \tilde{E} V + pV \end{bmatrix} = 0. \tag{2.2}$$

Assume that there is a flat Mach configuration constructed by the incident shock $S_1^0 : \eta = \psi_1^0 \xi$, the reflected shock $S_2^0 : \eta = \psi_2^0 \xi$, the Mach stem $S_3^0 : \eta = \psi_3^0 \xi$ and the contact discontinuity $D^0 : \eta = 0$. In the domain Ω_0 (ahead the incident shock S_1^0) the state of the supersonic flow is $\mathbf{U} = \mathbf{U_0^0}$. Meanwhile, the state in the domain Ω_1 (between S_1^0 and S_2^0) is $\mathbf{U} = \mathbf{U_1^0}$, the state in the domain $\Omega_{2,3}$ (between $S_{2,3}^0$ and D^0) is $\mathbf{U} = \mathbf{U_{2,3}^0}$. Now if $\mathbf{U_0^0}, \mathbf{U_1^0}, \mathbf{S_1^0}$ are perturbed to $\mathbf{U_0}, \mathbf{U_1}, \mathbf{S_1}$, then the location of other shocks and contact discontinuity, as well as the states in corresponding domains will also be perturbed. As in steady case for the downward stream of the subsonic flow an condition on pressure control is required. In this paper we also take a cutoff boundary $\xi = L$, and give the additional control condition as $\dfrac{\partial p}{\partial s} = 0$ on the downward part of the supersonic flow, where $\dfrac{\partial}{\partial s}$ is the derivative along characteristics. Denote by Ω_0^e be a neighborhood of Ω_0, by N_ε be the neighborhood of $\mathbf{U}_0^0$

$$N_\varepsilon = \{\mathbf{U_0}(\xi, \eta); \mathbf{U}_0(0,0) = \mathbf{U}_0^0, \|\mathbf{U}_0 - \mathbf{U}_0^0\|_{C^{2,\alpha_0}(\Omega_0^e)} < \varepsilon\}, \tag{2.3}$$

and let $b > 0$, L_ε be the neighborhood of $\mathbf{S}_1^0$

$$L_\varepsilon = \{\psi_1(\xi); \psi_1(0) = 0, \psi_1'(0) = \psi_1^0, \|\psi_1(\xi) - \psi_1^0 \xi\|_{C^{2,\alpha_0}(-b,0)} < \varepsilon\}, \tag{2.4}$$

where α_0 is a number in $(0, 1)$ determined later. Then the stability of the Mach configuration can be described by the following proposition.

Proposition 1. *Assume that* $(\mathbf{U}_0^0, S_1^0, \mathbf{U}_1^0, S_2^0, \mathbf{U}_2^0, D^0, \mathbf{U}_3^0, S_3^0)$ *forms an E-H type flat Mach configuration. Assume that the flat Mach configuration is weak, and for given* Ω_0^e *with* $e > 0$ *and small number* $\varepsilon > 0$*, the upstream flow* $\mathbf{U_0} \in N_\varepsilon$ *and* $\psi_1(\xi) \in L_\varepsilon$*, then there is a perturbed Mach configuration* $\{\mathbf{U}_0, S_1, \mathbf{U}_1, S_2, \mathbf{U}_2, D, \mathbf{U}_3, S_3\}$ *in a neighborhood of the triple intersection, which satisfies the Euler system* (2.1) *and Rankine–Hugoniot conditions* (2.2) *, and the pressure control on the cut-off boundary.*

The proposition 2.1 on the stability of Mach configuration can be reduced to the existence of a free boundary value problem. Like the treatment in Ref. 2, we define a function μ satisfying

$$(U\frac{\partial}{\partial\xi}+V\frac{\partial}{\partial\eta})\mu-2\mu=0. \tag{2.5}$$

Then by introducing a generalized Lagrange transformation the proof of the Proposition 2.1 is transformed to discuss the following problem (FB):

$$\text{The system}\begin{cases}\dfrac{\partial}{\partial x}(\dfrac{1}{\mu\rho U})=\dfrac{\partial}{\partial y}(\dfrac{V}{U}),\\ \dfrac{\partial}{\partial x}(\tilde{E}+\dfrac{p}{\rho})+\dfrac{\rho}{U}(U^2+V^2)=0,\\ \dfrac{\partial}{\partial x}(\dfrac{U}{\mu}+\dfrac{p}{\mu\rho U})-\dfrac{\partial}{\partial y}(\dfrac{pV}{U})+\dfrac{3}{\mu}=0,\\ \dfrac{\partial}{\partial x}(\dfrac{V}{\mu})+\dfrac{\partial p}{\partial y}+\dfrac{3}{\mu}\dfrac{V}{U}=0,\\ U\dfrac{\partial}{\partial x}\mu-2\mu=0.\ \text{ in } \omega_2\cup\omega_3,\end{cases} \tag{2.6}$$

$$\text{R–H conditions}\begin{cases}[\dfrac{1}{\rho U}]\dfrac{\chi'}{\mu}=-[\dfrac{V}{U}],\\ [\dfrac{1}{\rho U}(p+\rho U^2)]\dfrac{\chi'}{\mu}=-[\dfrac{pV}{U}],\\ [V]\dfrac{\chi'}{\mu}=[p],\\ [\tilde{E}+\dfrac{p}{\rho}]=0,\\ \mu=1\end{cases}\quad \text{on } \Gamma_2 \text{ and } \Gamma_3, \tag{2.7}$$

$$p \text{ and } V/U \text{ are continuous} \quad \text{on } \Gamma_D, \tag{2.8}$$

$$\frac{\partial p}{\partial x}=0 \quad \text{on } \Gamma_L. \tag{2.9}$$

Here x, y are the new coordinates, $\omega_2, \omega_3, \Gamma_2, \Gamma_3, \Gamma_D$ and Γ_L are the image of $\Omega_2, \Omega_3, S_2, S_3, D$ and $\{\xi = L\} \cap \{\eta > 0\}$ respectively. Particularly, $\Gamma_2, \Gamma_3, \Gamma_D$ is $\{y = \gamma_2(x), 0 < x < L\}, \{y = \gamma_3(x), 0 < x < L\}\{y = 0, 0 < x < L\}$, Γ_L is $\{x = L, y < 0\}$

The problem (FB) can be solved by a bootstrap way: to solve a corresponding boundary value problem in a domain with approximate fixed boundary and to determine the approximate boundary. When this iterative process is convergent, the problem is then solved. Here the boundary conditions on the fixed boundary come from the Rankine–Hugoniot conditions. The boundary value problem with fixed boundary takes the form (NL):

$$
\begin{cases}
\text{The system (same as that in (2.7))} \\
G^a \triangleq [p][\frac{1}{\rho U}] + [WU][W] = 0 \ \text{ on } \Gamma_2 \text{ and } \Gamma_3, \\
G^b \triangleq [p][U + \frac{p}{\rho U}] + [WU][pW] = 0 \ \text{ on } \Gamma_2 \text{ and } \Gamma_3, \\
G^c \triangleq [\tilde{E} + \frac{p}{\rho}] = 0 \ \text{ on } \Gamma_2 \text{ and } \Gamma_3, \\
\mu = 1 \qquad \text{on } \Gamma_2 \text{ and } \Gamma_3, \\
p \text{ and } V/U \text{ are continuous } \ \text{on } \Gamma_D, \\
\frac{\partial p}{\partial x} = 0 \quad \text{on } \Gamma_L.
\end{cases}
\tag{2.10}
$$

Meanwhile, the approximate boundary is determined by

$$
\begin{cases}
\frac{d\xi_i}{dx} = \frac{[\frac{V}{U}]_i}{[\frac{1}{\rho U}]_i}, \\
\xi_i(0) = 0,
\end{cases}
\qquad i = 2, 3.
\tag{2.11}
$$

In the following discussion the appearance of the supersonic flow behind the reflected shock gives great influence to the wave configuration. In the domain $y > 0$ the system can be reduced to

$$
\begin{cases}
D_R W - eD_I p + \frac{\lambda_R}{\rho U^2} = 0, \\
D_I W + eD_R p + \frac{\lambda_I}{\rho U^2} = 0, \\
\frac{\partial U}{\partial x} + d_W \frac{\partial W}{\partial x} + d_p \frac{\partial p}{\partial x} + 1 = 0, \\
\frac{\partial p}{\partial x} - \frac{\gamma p}{\rho} \frac{\partial \rho}{\partial x} = 0, \\
U \frac{\partial \mu}{\partial x} - 2\mu = 0,
\end{cases}
\tag{2.12}
$$

where $W = \frac{U}{V}$, $\lambda_R = \frac{\mu a^2 \rho V}{U^2 - a^2}$, $\quad \lambda_I = \frac{\mu a \rho U \sqrt{a^2 - U^2 - V^2}}{U^2 - a^2}$,
$D_R = \frac{\partial}{\partial x} + \lambda_R \frac{\partial}{\partial y}$, $D_I = \lambda_I \frac{\partial}{\partial y}$, $e = \frac{\mu}{a \rho U^2} \sqrt{a^2 - U^2 - V^2}$,
$d_W = \frac{U^2 W}{U + UW^2}$, $d_p = \frac{1}{\rho(U + UW^2)}$.

Similarly, in the supersonic region let

$$\lambda_\pm = \frac{\mu a^2\rho V \pm \mu a\rho U\sqrt{U^2+V^2-a^2}}{U^2-a^2},$$

$$\ell_\pm = (\frac{1}{U}(-\frac{\lambda_\pm}{\rho} - \mu V), \mu, \lambda_\pm, 0, 0).$$

Then multiplying the system by $\ell_\pm$ we obtain the system

$$\begin{cases} D_\pm W \pm e_1 D_\pm p + Q_\pm = 0, \\ \dfrac{\partial U}{\partial x} + d_W \dfrac{\partial W}{\partial x} + d_p \dfrac{\partial p}{\partial x} + 1 = 0, \\ \dfrac{\partial p}{\partial x} - \dfrac{\gamma p}{\rho}\dfrac{\partial \rho}{\partial x} = 0, \\ U\dfrac{\partial \mu}{\partial x} - 2\mu = 0 \end{cases} \tag{2.13}$$

in $y < 0$, where $D_\pm = \partial_\xi \pm \lambda_\pm \partial_\eta$, $e_1 = \dfrac{\sqrt{U^2+V^2-a^2}}{a\rho U^2}$, $Q_\pm = \dfrac{\lambda_\pm}{\mu\rho U^2}$. The last three equations in (2.11) or (2.12) can be integrated, and the integration leads to

$$\begin{cases} \dfrac{p}{\rho^\gamma} = M_1(y), \\ \dfrac{U^2}{2}(1+W^2) + \dfrac{\gamma p}{(\gamma-1)\rho} = M_2(y) + \int U(1+W^2)dx, \\ \mu = \mu_0(y)e^{\int 2U dx} \end{cases} \tag{2.14}$$

where $M_1(y), M_2(y), \mu_0(y)$ are determined by the boundary conditions on Γ_2 and Γ_3.

Eliminating the unknown function W in (2.12) and (2.13) we obtain an equation of second order for p:

$$D_I(eD_Ip) + D_R(eD_Rp) - \mu e D_R p + D_R Q_I - D_I Q_R = 0 \qquad \text{in } y > 0, \tag{2.15}$$

$$2e_1 D_- D_+ p + (2\mu e_1 + D_- e_1)D_+ p + D_+ e_1 D_- p + D_- Q_+ - D_+ Q_- = 0 \quad \text{in } y < 0, \tag{2.16}$$

where we denote by Q_R, Q_I the quantities $\dfrac{\lambda_R}{\mu\rho U^2}, \dfrac{\lambda_I}{\mu\rho U^2}$ for notational simplicity.

Let us give the detailed expressions of $D_R Q_I, D_I Q_R, D_\pm Q_\mp$. For instance,

$$\begin{aligned} D_R Q_I &= Q_{Ip} D_R p + Q_{IW} D_R W + Q_{IU} D_R U + Q_{I\rho} D_R \rho \\ &= Q_{Ip} D_R p + Q_{IW} e D_I p - Q_{IW} \mu Q_R + Q_{IU} D_R U + Q_{I\rho} D_R \rho. \end{aligned}$$

By using (2.14) we have

$$D_R\rho = \frac{\rho}{\gamma+1}D_R p - \frac{\rho^{\gamma+1}}{\gamma+1}\lambda_R M_1'(y).$$

On the other hand, from

$$\frac{\partial^2 U}{\partial x\partial y} = -\frac{\partial d_W}{\partial y}\frac{\partial W}{\partial x} - \frac{\partial}{\partial x}(d_W\frac{\partial W}{\partial y}) + \frac{\partial d_W}{\partial x}\frac{\partial W}{\partial y} - \frac{\partial d_p}{\partial y}\frac{\partial p}{\partial x} - \frac{\partial}{\partial x}(d_p\frac{\partial p}{\partial y}) + \frac{\partial d_p}{\partial x}\frac{\partial p}{\partial y}$$

we can obtain

$$\begin{aligned}\frac{\partial U}{\partial y} &= -d_W\frac{\partial W}{\partial y} + d_W\frac{\partial W}{\partial y}|_{(x,\gamma_2(x))} - d_p\frac{\partial p}{\partial y} + d_p\frac{\partial p}{\partial y}|_{(x,\gamma_2(x))}\\ &\quad + \int(-\frac{\partial d_W}{\partial y}(eD_Ip - Q_R + \frac{\lambda_R}{\lambda_I}eD_Rp + \frac{\lambda_R}{\lambda_I}Q_I)\\ &\quad + \frac{\partial d_W}{\partial x}(\frac{1}{\lambda_I}(-eD_Rp - Q_I)) - \frac{\partial d_p}{\partial y}\frac{dp}{dx} + \frac{\partial d_p}{\partial x}\frac{\partial p}{\partial y})dx.\\ &= \frac{d_W}{\lambda_I}(eD_Rp + Q_I) - d_p\frac{\partial p}{\partial y} + d_W\frac{\partial W}{\partial y}|_{(x,\gamma_2(x))} + d_p\frac{\partial p}{\partial y}|_{(x,\gamma_2(x))}\\ &\quad + \int(-\frac{\partial d_W}{\partial y}(eD_Ip - Q_R + \frac{\lambda_R}{\lambda_I}eD_Rp + \frac{\lambda_R}{\lambda_I}Q_I)\\ &\quad + \frac{\partial d_W}{\partial x}(\frac{1}{\lambda_I}(-eD_Rp - Q_I)) - \frac{\partial d_p}{\partial y}\frac{dp}{dx} + \frac{\partial d_p}{\partial x}\frac{\partial p}{\partial y})dx.\end{aligned}$$

Substituting the expression of $D_R\rho$, D_RU into D_RQ_I we obtain

$$D_RQ_I = h_xD_xp + h_yD_yp + h_a + h_b + h_c, \tag{2.17}$$

where h_x, h_y, h_a are nonlinear functions of $\mathbf{U}$, h_b is an integral of a quadratic form of $D\mathbf{U}$, and h_c is a differential form of $\mathbf{U}$ on the boundary Γ_2 (or Γ_3). Similar expression is available for D_IQ_R and $D_\pm Q_\mp$. Therefore, we can combine the differential forms in (2.14), (2.15) to obtain

$$\begin{cases} D_I(eD_Ip) + D_R(eD_Rp) + A_{21}\partial_x p + A_{22}\partial_y p = F_2 & \text{in } y>0,\\ D_-D_+p + A_{31}\partial_x p + A_{32}\partial_y p = F_3 \quad \text{in } y<0, \end{cases} \tag{2.18}$$

where

$$F_i = F_{ia} + F_{ib} + F_{ic},$$

and for $i = 2, 3$, F_{ia} is a nonlinear function of $\mathbf{U}$, F_{ib} is an integral of a quadratic form of $D\mathbf{U}$, and F_{ic} is a differential form of $\mathbf{U}$ on the boundary Γ_i. The connection condition on the line $y = 0$ is

$$\begin{cases} p_+ = p_-,\\ \frac{e_1}{\lambda_+ - \lambda_-}(\lambda_+D_-p + \lambda_-D_+p)|_{y=-0} = \frac{e}{\lambda_I}(\lambda_ID_Ip + \lambda_RD_Rp)|_{y=+0}. \end{cases} \tag{2.19}$$

The second order equation (2.15), (2.16) for the unknown functions is a nonlinear mixed type equation with discontinuous coefficients. In the classification of mixed type equations it belongs to the Lavrentiev–Bitsadze equation.

Let D be the tangential differential operator along the boundary Γ_2 or Γ_3, then from the boundary conditions $G^a = 0, G^b = 0, G^c = 0, \mu = 0$ we have

$$\begin{aligned} DG^a &= \alpha^a DU + \beta^a DW + \gamma^a Dp + \theta^a D\rho, \\ DG^b &= \alpha^b DU + \beta^b DW + \gamma^b Dp + \theta^b D\rho, \\ DG^c &= \alpha^c DU + \beta^c DW + \gamma^c Dp + \theta^c \delta\rho \end{aligned} \tag{2.20}$$

where the expressions of $\alpha^a, \cdots, \gamma^c$ can be obtained by direct computations. In view of that the matrix $\mathbf{M} = \begin{pmatrix} \alpha^a & \beta^a & \theta^a \\ \alpha^b & \beta^b & \theta^b \\ \alpha^c & \beta^c & \theta^c \end{pmatrix}$ is nonsingular, then $DW, DU, D\rho$ can be expressed by Dp. Then by using the equations given in (2.12) or (2.13), we obtain oblique derivative conditions for p on the boundary Γ_2 and Γ_3. The conditions can be simply written as

$$E_1 \partial_x p + E_2 \partial_y p = G_2, \qquad \text{on } \Gamma_2 \tag{2.21}$$

$$H_1 \partial_x p + H_2 \partial_y p = G_3, \qquad \text{on } \Gamma_3. \tag{2.22}$$

Hence the nonlinear problem (NL) is reduced to a generalized Tricomi problem (GT) for a mixed type equation of second order with unknown function p:

$$\begin{cases} \text{equation (2.15), (2.16)} \quad \text{in } \omega_2 \cup \omega_3, \\ E_1 \partial_x p + E_2 \partial_y p = G_2 \qquad \text{on } \Gamma_2, \\ H_1 \partial_x p + H_2 \partial_y p = G_3 \qquad \text{on } \Gamma_3, \\ \text{connection conditions (2.18)}, \\ p(0,0) = p_0^0, \\ \dfrac{\partial p}{\partial x} = 0 \quad \{x = L\} \cap \{y > 0\}. \end{cases} \tag{2.23}$$

Our next task is to solve the problem (GT).

3. Linearization and Corresponding Estimates

The crucial point to solve the problem (GT) is to study the corresponding linear problem and derive related estimates. Linearizing the problem (2.19) we obtain:

$$\begin{cases} D_I(eD_I\delta p) + D_R(eD_R\delta p) + A_{21}\partial_x\delta p + A_{22}\partial_y\delta p = f_2 \qquad \text{in } \; y > 0, \\ D_-D_+\delta p + A_{31}\partial_x\delta p + A_{32}\partial_y\delta p = f_3 \qquad \text{in } \; y < 0, \\ E_1\partial_x\delta p + E_2\partial_y\delta p = g_2, \qquad \text{on } \Gamma_2 \\ H_1\partial_x\delta p + H_2\partial_y\delta p = g_3, \qquad \text{on } \Gamma_3, \\ \delta p|_{y=+0} = \delta p|_{y=-0}, \\ \dfrac{e_1}{\lambda_+ - \lambda_-}(\lambda_+ D_-\delta p + \lambda_- D_+\delta p)|_{y=-0} = \dfrac{e}{\lambda_I}(\lambda_I D_I\delta p + \lambda_R D_R\delta p)|_{y=+0}, \\ \delta p(0,0) = 0, \\ \dfrac{\partial\delta p}{\partial x} = 0 \quad \text{on } \; x = L, \; y < 0. \end{cases} \tag{3.1}$$

We emphasize that when all coefficients are frozen at the origin, the principal part of (3.1) coincide with that for the standard Lavrentiev–Bitsadze equation. Indeed, from the expression of $D_I, D_R, D_\pm, e, e_1$ we know that

$$D_I^0 e_0 D_I^0 = e_0(\lambda_I^0)^2(\frac{\partial}{\partial y})^2,$$

$$D_R^0 e_0 D_R^0 = e_0(\frac{\partial}{\partial x})^2,$$

$$D_+^0 D_-^0 = (\frac{\partial}{\partial x})^2 - (\lambda_+^0)^2(\frac{\partial}{\partial y})^2.$$

Therefore, by using the transformation $y_1 = y/|\lambda_I^0|$ in $y > 0$ and the transformation $y_1 = y/\lambda_+^0$ in $y < 0$, we are led to $(\text{sgn } y_1)\partial_{xx}\delta p + \partial_{y_1 y_1}\delta p$, which is the principal part of the Lavrentiev–Bitsadze mixed type equation.

Applying the method to prove Theorem 5.3 of Ref. 4 we can reduce the problem (3.1) to a nonlocal boundary value problem for an elliptic equation in ω_2. The nonlocal problem takes the form

$$\begin{cases} D_I(eD_I\delta p) + D_R(eD_R\delta p) + A_{21}\partial_x\delta p + A_{22}\partial_y\delta p = f_2 \qquad \text{in } \; \omega_2, \\ E_1\partial_x\delta p + E_2\partial_y\delta p = g_2, \qquad \text{on } \Gamma_2 \\ \dfrac{\partial\delta p}{\partial x} = 0 \;\; \{x = L\} \cap \{y > 0\}, \\ \delta p(0,0) = 0, \\ \dfrac{\partial\delta p}{\partial x} = 0 \quad \text{on } \; x = L, \; y < 0 \\ A^\sharp\dfrac{\partial\delta p}{\partial y}(h_\ell, 0) + A^\flat\dfrac{\partial\delta p}{\partial y}(h_r, 0) + B^\sharp\dfrac{\partial\delta p}{\partial x}(h_\ell, 0) + B^\flat\dfrac{\partial\delta p}{\partial x}(h_r, 0) = D, \end{cases} \tag{3.2}$$

where $A^\sharp, A^\flat, B^\sharp, B^\flat$ and D are given functions depending the coefficients of the problem (3.1), D satisfies

$$\|D\|_{C^\alpha(0,L)} \leq C(\|g_3\|_{C^\alpha(0,L)} + \|f_3\|_{C^\alpha(\omega_2)}). \tag{3.3}$$

In (3.2), $h_r = h_r(x), h_\ell = h_\ell(x)$ are coordinates of the intersections of the characteristics issuing from a point $(x, \gamma_3(x))$ on Γ_3. When the background flat Mach configuration is weak, the reflected shock is near to the corresponding characteristics issuing from the origin. The above fact implies that $\dfrac{h_r(x)}{h_\ell(x)}$ is small. Hence the condition on $y = 0$ can be rewritten as

$$A^\sharp \frac{\partial \delta p}{\partial y}(x,0) + B^\sharp \frac{\partial \delta p}{\partial x}(x,0) = -A^\flat \frac{\partial \delta p}{\partial y}(\beta(x),0) - B^\flat \frac{\partial \delta p}{\partial x}(\beta(x),0) + D(h_\ell^{-1}(x)), \tag{3.4}$$

where $|\dfrac{\beta(x)}{x}|$ is sufficiently small. Therefore, we can establish an estimate for the problem (3.2)

$$\begin{aligned} \|\delta p\|_{C^{1,\alpha}(\omega_2)} \leq C(&\|f_2\|_{C^\alpha(\omega_2)} + \|g_2\|_{C^\alpha(0,L)} \\ &+\|D(h_\ell^{-1}(x)) - A_\flat \frac{\partial \delta p}{\partial y}(\beta(x),0) \\ &-B^\flat \frac{\partial \delta p}{\partial x}(\beta(x),0)\|_{C^\alpha(0,L)}), \end{aligned} \tag{3.5}$$

which yields

$$\begin{aligned} \|\delta p\|_{C^{1,\alpha}(\omega_2)} \leq C(&\|f_2\|_{C^\alpha(\omega_2)} + \|g_2\|_{C^\alpha(0,L)} + \|f_3\|_{C^\alpha(\omega_2)} \\ &+\|g_3\|_{C^\alpha(0,L)} + \delta\|\delta p\|_{C^{1,\alpha}(\omega_2)}), \end{aligned} \tag{3.6}$$

where δ is small. Then we are led to the estimate for the solution of the problem (3.1)

$$\|\delta p\|_{C^{1,\alpha}(\omega_{2,3})} \leq C(\|f_2\|_{C^\alpha(\omega_2)} + \|g_2\|_{C^\alpha(0,L)} + \|f_3\|_{C^\alpha(\omega_2)} + \|g_3\|_{C^\alpha(0,L)}). \tag{3.7}$$

4. Solution to the Nonlinear Problem

In this section we will use iteration and the estimate (3.7) to solve the free boundary value problem (FB). First we construct a sequence $\{\mathbf{U}^{(n)}\}$ of approximate solutions. The first term $\mathbf{U}^{(1)}$ is taken as $\mathbf{U}^0$, and $\delta\mathbf{U}^{(1)} = 0$, then $\delta p^{(n+1)}$ with $n \geq 1$ is defined as the solution of the linear problem:

$$\begin{cases} D_I^{(n)}(e^{(n)}D_I^{(n)}\delta p^{(n+1)}) + D_R^{(n)}(e^{(n)}D_R^{(n)}\delta p^{(n+1)}) \\ \qquad +A_{21}^{(n)}\partial_x\delta p^{(n+1)} + A_{22}^{(n)}\partial_y\delta p^{(n+1)} = \delta F_2^{(n)} \quad \text{in } y > 0, \\ D_-^{(n)}D_+^{(n)}\delta p^{(n+1)} + A_{31}^{(n)}\partial_x\delta p^{(n+1)} + A_{32}^{(n)}\partial_y\delta p^{(n+1)} = \delta F_3^{(n)} \quad \text{in } \ y < 0, \\ E_1^{(n)}\partial_x\delta p^{(n+1)} + E_2^{(n)}\partial_y\delta p^{(n+1)} = \delta G_2^{(n)}, \quad \text{on } \Gamma_2 \\ H_1^{(n)}\partial_x\delta p^{(n+1)} + H_2^{(n)}\partial_y\delta p^{(n+1)} = \delta G_3^{(n)}, \quad \text{on } \Gamma_3, \\ \delta p^{(n+1)}|_{y=+0} = \delta p^{(n+1)}|_{y=-0}, \\ \dfrac{e_1^{(n)}}{\lambda_+^{(n)} - \lambda_-^{(n)}}(\lambda_+^{(n)}D_-^{(n)}\delta p^{(n+1)} + \lambda_-^{(n)}D_+^{(n)}\delta p^{(n+1)})|_{y=-0} \\ \qquad = \dfrac{e^{(n)}}{\lambda_I^{(n)}}(\lambda_I^{(n)}D_I^{(n)}\delta p^{(n+1)} + \lambda_R^{(n)}D_R^{(n)}\delta p^{(n+1)})|_{y=+0}, \\ \delta p^{(n+1)}(0,0) = 0, \\ \dfrac{\partial\delta p^{(n+1)}}{\partial x} = 0 \quad \text{on } \ x = L, \ y < 0, \end{cases} \tag{4.1}$$

and $p^{(n+1)} = p^{(n)} + \delta p^{(n+1)}$. In (4.1) $\delta F_i^{(n)}, \delta G_i^{(n)}$ with $i = 2, 3$ are given according to the Newton iteration scheme. That is

$$\delta F_i^{(n)} = (\frac{\partial F_i}{\partial \mathbf{U}})^{(n)}\delta\mathbf{U}^{(n)} + (\frac{\partial F_i}{\partial \mathbf{P}})^{(n)}\delta D\mathbf{U}^{(n)} + F_i^{(n)}. \tag{4.2}$$

$$\delta G_i^{(n)} = (\frac{\partial G_i}{\partial \mathbf{U}})^{(n)}\delta\mathbf{U}^{(n)} + G_i^{(n)}. \tag{4.3}$$

In (4.3) $\dfrac{\partial F_i}{\partial \mathbf{P}}$ denotes the derivatives of F_i with respect to $D\mathbf{U}$, which are looking as variables.

Once $p^{(n+1)}$ is obtained, other components can be successively determined easily. That is, $W^{(n+1)}$ can be determined by

$$D_I^{(n)}W^{(n+1)} = -e^{(n)}D_R^{(n)}p^{(n+1)} - Q_I^{(n)} \qquad y > 0, \tag{4.4}$$

$$D_+^{(n)}W^{(n+1)} = -e_1^{(n)}D_+^{(n)}p^{(n+1)} - Q_+^{(n)} \qquad y < 0 \tag{4.5}$$

with the boundary conditions on $\Gamma_{2,3}$ as shown in (2.11). Afterwards, $U^{(n+1)}, \rho^{(n+1)}, \mu^{(n+1)}$ can be determined by using (2.13).

To prove the convergence of the sequence $\{p^{(n)}\}$ we notice that (4.2) implies

$$\|\delta F_i^{(n)}\|_{C^\alpha(\omega_i)} \le C(\|\delta \mathbf{U}^{(n)}\|^2_{C^\alpha(\Gamma_i)} + \|(\delta \mathbf{U}^{(n)}, \delta D\mathbf{U}^{(n)}\|^2_{C^\alpha(\omega_i)}),$$

$$\le C(\|\delta \mathbf{U}^{(n)}\|^2_{C^\alpha(\Gamma_i)} + \|\delta \mathbf{U}^{(n)}\|^2_{C^{1,\alpha}(\omega_i)}) \tag{4.6}$$

$$\|\delta G_i^{(n)}\|_{C^\alpha(\Gamma_i)} \le C\|\delta \mathbf{U}^{(n)}\|^2_{C^\alpha(\Gamma_i)}. \tag{4.7}$$

Then by using (3.7) we have the boundedness of the sequence $\{p^{(n)}\}$ in $C^{1,\alpha}(\omega_2 \cup \omega_3)$, provided the perturbation and L_0 are taken sufficiently small. Furthermore, for sufficiently small perturbation one can derive the convergence of the sequence $\{p^{(n)}\}$. In accordance, the convergence of $\{W^{(n)}\}$,$\{U^{(n)}\}$,$\{\rho^{(n)}\}$ and $\{\mu^{(n)}\}$ can also be obtained.

The limit p of $\{p^{(n)}\}$ is the solution of (3.1). Meanwhile, in view of (4.4),(4.5) we know the limit W of the sequence $\{W^{(n)}\}$ satisfies

$$\begin{cases} D_I W + eD_R p + Q_I = 0 & y > 0, \\ D_+ W + e_1 D_+ p + Q_+ = 0 & y < 0. \end{cases} \tag{4.8}$$

Combining with the second order equations in (3.1) satisfied by p we know W also satisfies

$$\begin{cases} D_R W - eD_I p + Q_R = 0 & y > 0, \\ D_- W - e_1 D_- p + Q_- = 0 & y < 0. \end{cases} \tag{4.9}$$

Moreover, U, ρ, μ also satisfy the other equations in (2.11). Besides, the boundary value conditions on Γ_2, Γ_3 and the connection conditions on $y = 0$ are satisfied by using the limiting process. Hence $\mathbf{U}$ is the solution of the problem (NL).

Finally, the bootstrap way mentioned in the end of Section 1 offers a convergent iteration process, and the limit solve the free boundary problem. Since the process of the proof is similar to Ref. 3 and Ref. 4, so the details are omitted here.

As mentioned in Section 2 the conclusion of Proposition 2.1 implies the stability of E-H Mach configuration of the pseudo-steady flow.

Acknowledgments

The paper is partially supported by National Natural Science Foundation of China (11031001) and the Doctorial Foundation of National Educational Ministry (20090071110002).

References

1. G. Ben-Dor, *Shock Waves Reflection Phenomena*. Springer-Verlag. New York, 1992.
2. Shuxing Chen, *Stability of Mach configuration*, Comm. Pure Appl. Math. **59** (2006), 1-35.
3. Shuxing Chen, *Mach configuration in pseudo-stationary compressible flow*, Jour. Amer. Math. Soc., **21** (2008), 63-100.
4. Shuxing Chen, *E-H Type Mach configuration and its stability*, Comm. Math. Phys. (2012).
5. R. Courant and K. O. Friedrichs, *Supersonic Flow and Shock Waves*. Interscience Publishers Inc., New York, 1948.
6. C. M. Dafermos, *Hyperbolic Conservation Laws in Continuum Physics*, Springer-Verlag, Berlin Heiderberg New York, 2022.
7. D. Gilbarg and N. S. Trudinger, *Elliptic Partial Differential Equations of Second Order*, Second Edition, Grundlehren der Mathematischen Wissenschaften, **224**, Springer, Berlin-New York, 1983.
8. J. K. Hunter and M. Brio, *Weak shock reflection*, J. Fluid Mech., **410** (2000), 235-261.
9. A. Majda, *One perspective on open problems in multi-dimensional conservation laws*, IMA vol. Math. Appl. **29**(1991), 217-237.
10. C. S. Morawetz, *Potential theory for regular and Mach reflection of a shock at a wedge*, Comm. Pure Appl. Math. **47**(1994), 593-624.
11. John von Neumann, *Oblique Reflection of Shocks*. U.S. Dept. Comm. Off. of Tech. Serv., Washigton, D.C. PB-37079 (1943).
12. J. Smoller, *Shock waves and reaction-diffusion equations*, Springer-Verlag, New York, 1983.

Incompressible Viscous Fluid Flows with Slip Boundary Conditions and Their Numerical Simulations

Guo Ben-yu*

Department of Mathematics, Shanghai Normal University, Shanghai, 200234, China

Division of Computational Science of E-institute of Shanghai Universities, Scientific Computing Key Laboratory of Shanghai Universities, Shanghai Normal University

byguo@shnu.edu.cn

In this paper, we review recent results on incompressible viscous fluid flows with slip boundary conditions and their numerical simulations. We first consider the existence, stability and regularity of solutions of the primitive form of the n-dimensional Navier-Stokes equations with slip boundary conditions, and present the spectral method for solving such problems. Then we present the existence, stability and regularity of solutions of the vorticity-stream function form of the Navier-Stokes equations, and the corresponding spectral method. Finally we discuss other problems with divergence-free solutions.

1. Introduction

The Navier-Stokes equations play an important role in studying incompressible viscous fluid flows. Usually, we considered the primitive form of the Navier-Stokes equations with non-slip boundary conditions, see [22, 23, 30]. It is also interesting and important to consider slip boundary conditions. Ma and Wang [24, 25] discussed the case, in which the normal component of velocity and the normal derivative of tangential component of velocity vanish on the boundary, and analyzed the structural stability of the corresponding steady problems. Guermond and Quartapelle [8], and Orszag,

*This work is supported in part by NSF of China N.11171227, Fund for Doctoral authority of China N.20123127110001, Fund for E-institute of Shanghai Universities N.E03004, and Leading Academic Discipline Project of Shanghai Municipal Education Commission N.J50101.

Israeli and Deville [28] proposed numerical algorithms for solving some special problems with similar boundary conditions. Mucha [27] studied the steady Navier-Stokes equations in an infinite pipe, with more general slip boundary conditions. Recently, Guo [13] considered the n-dimensional unsteady Navier-Stokes equations with slip boundary conditions, and obtained some results on the existence, stability and regularity of its solutions.

There have been a lot of literatures concerning numerical solutions of the primitive form of the Navier-Stokes equations with non-slip boundary conditions, see [1, 2, 6, 10–12, 18, 21, 22, 26, 29, 30] and the references therein. In actual computation, the main difficulty is how to ensure the incompressibility of numerical solutions. For finite difference method, we approximated the compressibility by certain difference equation as in Kuo [21] and Ladyzhenskaya [22]. In particular, Chorin [4] proposed the artificial compressibility method, while Chorin [3] and Teman [31] provided the projection method. However, the above treatments bring additional errors. Moreover, it is not possible to deal with the boundary value of the pressure exactly, see Gresho [7]. For finite element method, we usually approximated a weak form of the continuity equation as in Girault and Raviart [6] and the references therein. For spectral method, we may adopt the artificial compressibility method or the projection method. Some authors also approximated the weak form of the continuity equation as in finite element method, see Bernardi and Maday [1] and the references therein. Certainly, it seems better to construct divergence-free base functions and then solve the Navier-Stokes equations directly. Whereas, this is a non-trivial job. But, for the Navier-Stokes equations with certain slip boundary conditions, we might design divergence-free base functions satisfying the physical boundary conditions exactly. With such base functions, we could approximate the Navier-Stokes equations directly. In other words, we need neither the artificial compressibility method nor the projection method, see Guo and Jiao [16].

In the study of incompressible viscous fluid flows, we are also interested in the vorticity-stream function form of the Navier-Stokes equations. In this case, the incompressibility is fulfilled automatically. In the most existing literatures concerning non-slip boundary conditions, we imposed homogeneous boundary conditions on both of the vorticity and the stream function. Indeed, it is not correct, since non-slip boundary conditions do not imply the homogeneous boundary condition on the vorticity. However, fortunately, this treatment is really physical for certain slip boundary conditions. Guo [13] derived some results on the existence, stability and regularity of solutions of the vorticity-stream function form of the Navier-Stokes equations, while Jia and Guo [20] proposed the corresponding spectral method.

In this paper, we review the recent results on the incompressible viscous fluid flows with certain slip boundary conditions, and their numerical simulations. In the next section, we first consider the existence, stability and regularity of solutions of the primitive form of the n-dimensional Navier-Stokes equations with slip boundary conditions. Then we present the spectral method for solving such problem numerically. In Section 3, we present the existence, stability and regularity of solutions of the vorticity-stream function form of the Navier-Stokes equations, as well as the corresponding spectral method. In the final section, we discuss some of other problems with divergence-free solutions.

2. Primitive Form of Navier-Stokes Equations with Non-slip Boundary Conditions

In this section, we consider the primitive form of the n-dimensional Navier-Stokes equations with slip boundary conditions, and the related spectral method providing highly accurate numerical results.

2.1. Solutions of Navier-Stokes equations

Let $\mathbf{x} = (x_1, x_2, \cdots, x_n)^T$, $\Omega = \{\ \mathbf{x} \mid |x_i| < 1 \text{ for } 1 \le i \le n\}$ with the boundary $\partial\Omega$, and $\bar{\Omega} = \Omega \cup \partial\Omega$. We denote by $\mathbf{U}(\mathbf{x},t)$ the velocity with the components $U^{(j)}(\mathbf{x},t)$, $1 \le j \le n$. $P(\mathbf{x},t)$ stands for the pressure. The positive constant ν is the kinetic viscosity, and the given function $\mathbf{f}(\mathbf{x},t)$ is the body force. $\mathbf{U}_0(\mathbf{x})$ describes the initial velocity. We denote by $\boldsymbol{n}$ the unit vector in the outward normal direction on $\partial\Omega$, and denote by $\boldsymbol{\tau}$ the unit vector in the tangential direction on $\partial\Omega$. For simplicity, let $\partial_z \mathbf{U}(\mathbf{x},t) = \frac{\partial \mathbf{U}}{\partial z}(\mathbf{x},t)$, $z = x_1, x_2, \cdots, x_n, t$. $\partial_{\boldsymbol{n}} \mathbf{U}(\mathbf{x},t)$ means the outward normal derivative of $\mathbf{U}(\mathbf{x},t)$ on $\partial\Omega$.

Let $T > 0$. The primitive form of the Navier-Stokes equations with slip boundary conditions is as follows,

$$\begin{cases} \partial_t \mathbf{U}(\mathbf{x},t) + (\mathbf{U}(\mathbf{x},t)\cdot\nabla)\mathbf{U}(\mathbf{x},t) - \nu\Delta\mathbf{U}(\mathbf{x},t) + \nabla P(\mathbf{x},t) = \mathbf{f}(\mathbf{x},t), & \\ \qquad\qquad\qquad\qquad\qquad\qquad\qquad\qquad\qquad \mathbf{x}\in\Omega,\ 0 < t \le T, & \\ \nabla\cdot\mathbf{U}(\mathbf{x},t) = 0, & \mathbf{x}\in\bar{\Omega},\ 0\le t\le T, \\ \mathbf{U}(\mathbf{x},t)\cdot\boldsymbol{n} = \partial_{\boldsymbol{n}}(\mathbf{U}(\mathbf{x},t)\cdot\boldsymbol{\tau}) = 0, & \mathbf{x}\in\partial\Omega,\ 0\le t\le T, \\ \mathbf{U}(\mathbf{x},0) = \mathbf{U}_0(\mathbf{x}), & \mathbf{x}\in\bar{\Omega}. \end{cases} \tag{1}$$

Let $\mathbf{L}^s(\Omega) = (L^s(\Omega))^n$ and $\mathbf{H}^r(\Omega) = (H^r(\Omega))^n$. For any vector functions $\mathbf{u}, \mathbf{v} \in \mathbf{L}^2(\Omega)$, the inner product and the norm are given by

$$(\mathbf{u}, \mathbf{v})_{\mathbf{L}^2(\Omega)} = \sum_{j=1}^{n} (u^{(j)}, v^{(j)})_{L^2(\Omega)}, \qquad ||\mathbf{v}||_{\mathbf{L}^2(\Omega)} = (\mathbf{v}, \mathbf{v})^{\frac{1}{2}}_{\mathbf{L}^2(\Omega)}.$$

For integer $r \geq 0$, we define the inner product $(\mathbf{u}, \mathbf{v})_{\mathbf{H}^r(\Omega)}$, the semi-norm $|\mathbf{v}|_{\mathbf{H}^r(\Omega)}$ and the norm $||\mathbf{v}||_{\mathbf{H}^r(\Omega)}$ similarly.

We shall use the following notations,

$$a(\mathbf{u}, \mathbf{w}) = \sum_{j=1}^{n} \sum_{i=1}^{n} \int_{\Omega} \partial_{x_i} u^{(j)}(\mathbf{x}) \partial_{x_i} w^{(j)}(\mathbf{x}) d\mathbf{x}, \qquad \forall\ \mathbf{u}, \mathbf{w} \in \mathbf{H}^1(\Omega),$$

$$b(\mathbf{u}, \mathbf{z}, \mathbf{w}) = \sum_{j=1}^{n} \sum_{i=1}^{n} \int_{\Omega} z^{(i)}(\mathbf{x}) \partial_{x_i} u^{(j)}(\mathbf{x}) w^{(j)}(\mathbf{x}) d\mathbf{x}, \quad \forall\ \mathbf{u}, \mathbf{z}, \mathbf{w} \in \mathbf{H}^1(\Omega).$$

If $\mathbf{w}(\mathbf{x}) \cdot \boldsymbol{n} = \partial_{\boldsymbol{n}}(\mathbf{w}(\mathbf{x}) \cdot \boldsymbol{\tau}) = 0$ on $\partial\Omega$, then

$$(\Delta \mathbf{u}, \mathbf{w})_{\mathbf{L}^2(\Omega)} + a(\mathbf{u}, \mathbf{w}) = 0.$$

If $\mathbf{u}, \mathbf{w}, \mathbf{z} \in \mathbf{H}^1(\Omega)$, $\nabla \cdot \mathbf{z}(\mathbf{x}) = 0$ on $\bar{\Omega}$ and $\mathbf{z}(\mathbf{x}) \cdot \boldsymbol{n} = 0$ on $\partial\Omega$, then

$$b(\mathbf{u}, \mathbf{z}, \mathbf{w}) + b(\mathbf{w}, \mathbf{z}, \mathbf{u}) = 0.$$

In order to derive the weak formulation of problem (1), we introduce the space

$$V(\Omega) = \{\ \mathbf{w} \mid \mathbf{w} \in \mathbf{H}^1(\Omega),\ \nabla \cdot \mathbf{w}(\mathbf{x}) = 0 \text{ on } \bar{\Omega} \text{ and } \mathbf{w}(\mathbf{x}) \cdot \boldsymbol{n} = 0 \text{ on } \partial\Omega\}.$$

$$\mathscr{H}(\Omega) = \{\ \mathbf{w} \mid \mathbf{w} \in \mathbf{L}^2(\Omega), \nabla \cdot \mathbf{w}(\mathbf{x}) = 0 \text{ on } \bar{\Omega} \text{ and } \mathbf{w}(\mathbf{x}) \cdot \boldsymbol{n} = 0 \text{ on } \partial\Omega\}.$$

$V'(\Omega)$ is the dual space of $V(\Omega)$.

The weak formulation of problem (1) is to seek $\mathbf{U} \in L^2(0, T; V(\Omega)) \cap L^\infty(0, T; \mathscr{H}(\Omega))$ such that

$$\begin{cases} (\partial_t \mathbf{U}(t), \mathbf{w})_{\mathbf{L}^2(\Omega)} + b(\mathbf{U}(t), \mathbf{U}(t), \mathbf{w}) + \nu a(\mathbf{U}(t), \mathbf{w}) = (\mathbf{f}(t), \mathbf{w})_{\mathbf{L}^2(\Omega)}, \\ \qquad\qquad\qquad\qquad\qquad\qquad \forall \mathbf{w} \in V(\Omega) \cap \mathbf{L}^n(\Omega),\ 0 < t \leq T, \\ \mathbf{U}(\mathbf{x}, 0) = \mathbf{U}_0(\mathbf{x}), \qquad\qquad\qquad \mathbf{x} \in \bar{\Omega}. \end{cases} \tag{2}$$

We have the basic result on the existence of solutions of problem (2), stated below.

Theorem 1. If $\mathbf{U}_0 \in \mathscr{H}(\Omega)$ and $\mathbf{f} \in L^2(0, T; V'(\Omega))$, then problem (2) has at least one solution.

We next consider the stability of problem (2). In many literatures, one took the boundedness of solutions as the stability of solutions. Indeed, it is not enough, since the stability should describe the ability of considered systems against the perturbations. As we know, nonlinear systems generally do not possess the usual stability. However, it might be stable in a generalized sense. The generalized stability of dynamical systems means that if certain norms of errors of data involved in underlying systems do not exceed some critical bounds, then the norms of errors of solutions are proportional to the norms of data errors, see [19].

Now, let $\mathbf{U}$ be a solution of (2). Suppose that $\mathbf{U}_0$ and $\mathbf{f}$ are disturbed by the errors $\widetilde{\mathbf{U}}_0$ and $\tilde{\mathbf{f}}$, respectively. Accordingly, the solution $\mathbf{U}$ becomes $\mathbf{U}+\widetilde{\mathbf{U}}$. In this case, we obtain from (2) that

$$\begin{cases} (\partial_t\widetilde{\mathbf{U}}(t),\mathbf{w})_{\mathbf{L}^2(\Omega)}+b(\widetilde{\mathbf{U}}(t),\mathbf{U}(t)+\widetilde{\mathbf{U}}(t),\mathbf{w})+b(\mathbf{U}(t),\widetilde{\mathbf{U}}(t),\mathbf{w}) & \\ \quad +\nu a(\widetilde{\mathbf{U}}(t),\mathbf{w})=(\tilde{\mathbf{f}}(t),\mathbf{w})_{\mathbf{L}^2(\Omega)}, & \forall\mathbf{w}\in V(\Omega)\cap\mathbf{L}^n(\Omega),\ 0<t\le T,\\ \widetilde{\mathbf{U}}(\mathbf{x},0)=\widetilde{\mathbf{U}}_0(\mathbf{x}), & \mathbf{x}\in\bar{\Omega}. \end{cases}$$

Theorem 2. Let $n=2$. If $\mathbf{U}\in L^2(0,T;V(\Omega))\cap L^\infty(0,T;\mathscr{H}(\Omega))$, then for any $\widetilde{\mathbf{U}}_0\in\mathscr{H}(\Omega)$ and any $\tilde{\mathbf{f}}\in L^2(0,T;V'(\Omega))$,

$$\begin{aligned} ||\widetilde{\mathbf{U}}(t)||^2_{\mathbf{L}^2(\Omega)}+\nu\int_0^t|\widetilde{\mathbf{U}}(\xi)|^2_{\mathbf{H}^1(\Omega)}d\xi &\le \exp(\frac{c}{\nu}\int_0^t|\mathbf{U}(\xi)|^2_{\mathbf{H}^1(\Omega)}d\xi)\\ \cdot(||\widetilde{\mathbf{U}}_0||^2_{\mathbf{L}^2(\Omega)}+\frac{c}{\nu}\int_0^t\exp(-\frac{c}{\nu}&\int_0^\xi|\mathbf{U}(\eta)|^2_{\mathbf{H}^1(\Omega)}d\eta)||\tilde{\mathbf{f}}(\xi)||^2_{V'(\Omega)}d\xi). \end{aligned} \tag{3}$$

Since the error estimate (3) involves the factor $\exp(\frac{c}{\nu}\int_0^t|\mathbf{U}(\xi)|^2_{\mathbf{H}^1(\Omega)}d\xi)$, which in turn depends on the norms $||\mathbf{U}_0||^2_{\mathbf{L}^2(\Omega)}$ and $||\mathbf{f}||^2_{L^2(0,T;V'(\Omega))}$, we know from Theorem 2 that problem (2) with $n=2$ is of the generalized stability for any solution $\mathbf{U}\in L^2(0,T;V(\Omega))\cap L^\infty(0,T;\mathscr{H}(\Omega))$ and for any errors $\widetilde{\mathbf{U}}_0\in\mathscr{H}(\Omega)$ and $\tilde{\mathbf{f}}\in L^2(0,T;V'(\Omega))$. But there is no restriction on the values of the norms $||\mathbf{U}_0||^2_{\mathbf{L}^2(\Omega)}$, $||\mathbf{f}||^2_{L^2(0,T;V'(\Omega))}$, $||\widetilde{\mathbf{U}}_0||^2_{\mathbf{L}^2(\Omega)}$ and $||\tilde{\mathbf{f}}||^2_{L^2(0,T;V'(\Omega))}$.

We also see from Theorem 2 that if $\widetilde{\mathbf{U}}_0(\mathbf{x})\equiv\tilde{\mathbf{f}}(\mathbf{x},t)\equiv 0$, then $\widetilde{\mathbf{U}}(\mathbf{x},t)\equiv 0$. This fact, together with Theorem 1, indicates that problem (2) with $n=2$ has a unique solution, as long as that $\mathbf{U}_0\in\mathscr{H}(\Omega)$ and $\mathbf{f}\in L^2(0,T;V'(\Omega))$.

Theorem 3. Let $n = 3$. If $\mathbf{U} \in L^2(0,T;V(\Omega)) \cap L^8(0,T;\mathbf{L}^4(\Omega)) \cap L^\infty(0,T;\mathscr{H}(\Omega))$, then for any $\widetilde{\mathbf{U}}_0 \in \mathscr{H}(\Omega)$ and any $\tilde{\mathbf{f}} \in L^2(0,T;V'(\Omega))$,

$$\begin{aligned}||\widetilde{\mathbf{U}}(t)||^2_{\mathbf{L}^2(\Omega)} + \nu \int_0^t |\widetilde{\mathbf{U}}(\xi)|^2_{\mathbf{H}^1(\Omega)} d\xi \le \exp(\frac{c}{\nu}\int_0^t ||\mathbf{U}(\xi)||^8_{\mathbf{L}^4(\Omega)} d\xi) \\ \cdot(||\widetilde{\mathbf{U}}_0||^2_{\mathbf{L}^2(\Omega)} + \frac{c}{\nu}\int_0^t \exp(-\frac{c}{\nu}\int_0^\xi ||\mathbf{U}(\eta)||^8_{\mathbf{L}^4(\Omega)} d\eta)||\tilde{\mathbf{f}}(\xi)||^2_{V'(\Omega)} d\xi).\end{aligned} \tag{4}$$

The estimate (4) shows that problem (2) with $n = 3$ is of the generalized stability for any solution $\mathbf{U} \in L^2(0,T;V(\Omega)) \cap L^8(0,T;\mathbf{L}^4(\Omega)) \cap L^\infty(0,T;\mathscr{H}(\Omega))$, and for any errors $\widetilde{\mathbf{U}}_0 \in \mathscr{H}(\Omega)$ and $\tilde{\mathbf{f}} \in L^2(0,T;V'(\Omega))$. It also implies the uniqueness of such solution. However, the existence of such global solution is still an open problem.

Finally, we list some results on the regularity of solutions of problem (2).

Theorem 4. Let $n = 2$. If $\mathbf{U}_0 \in V(\Omega) \cap \mathbf{H}^2(\Omega)$, $\mathbf{f} \in H^1(0,T;V'(\Omega))$, $\mathbf{f}(0) \in \mathscr{H}(\Omega)$ and $\partial_{\boldsymbol{n}}(\mathbf{U}_0 \cdot \boldsymbol{\tau}) = 0$ on $\partial\Omega$, then $\partial_t\mathbf{U} \in L^2(0,T;V(\Omega)) \cap L^\infty(0,T;\mathscr{H}(\Omega))$.

Theorem 5. Let $n = 2$. If the conditions of Theorem 4 hold and $\mathbf{f} \in L^2(0,T;\mathscr{H}(\Omega))$, then $U \in L^2(0,T;V(\Omega) \cap \mathbf{H}^2(\Omega))$. If, in addition, $\mathbf{f} \in L^\infty(0,T;\mathscr{H}(\Omega))$, then $\mathbf{U} \in L^\infty(0,T;V(\Omega) \cap \mathbf{H}^2(\Omega))$.

Theorem 6. Let $n = 3$, $\mathbf{U}_0 \in V(\Omega) \cap \mathbf{H}^2(\Omega)$, $\mathbf{f} \in H^1(0,T;V'(\Omega))$, $f(0) \in \mathbf{L}^2(\Omega)$ and $\partial_{\boldsymbol{n}}(\mathbf{U}_0 \cdot \boldsymbol{\tau}) = 0$ on $\partial\Omega$. If, in addition, one of the following conditions holds:

(i) ν is suitably large,

(ii) $||\mathbf{U}_0||_{\mathbf{H}^2(\Omega)}$ and $||\mathbf{f}||_{H^1(0,T;V'(\Omega))}$ are suitably small,

(iii) T is suitably small,

Then problem (2) has a unique solution such that $\partial_t\mathbf{U} \in L^2(0,T;V(\Omega)) \cap L^\infty(0,T;\mathscr{H}(\Omega))$.

The proofs of Theorems 1-6 could be found in [13].

2.2. Spectral method for primitive form of Navier-Stokes equations

In this subsection, we present the spectral method for solving problem (2). The key point is to construct divergence-free base functions satisfying the boundary conditions exactly, and to explore proper expressions of divergence-free functions.

Let $I = \{\ y \mid |y| < 1\}$ and $\chi(y)$ be a certain weight function. The inner product and the norm of the weighted space $L^2_\chi(I)$ are denoted by

$(u, v)_{L^2_\chi(I)}$ and norm $||v||_{L^2_\chi(I)}$, respectively. For simplicity, we omit the subscript χ in notations whenever $\chi(y) \equiv 1$.

Let $L_l(y)$ be the Legendre polynomials of degree l, $l \geq 0$. The set of $L_l(y)$ is a complete $L^2(I)$-orthogonal system, namely,

$$\int_I L_l(y) L_{l'}(y) dy = (l + \frac{1}{2})^{-1} \delta_{l,l'}, \qquad l, l' \geq 0,$$

where $\delta_{l,l'}$ is the Kronecker symbol.

We also introduce the polynomials

$$G_l(y) = \frac{(-1)^l}{2^{l-2}(l-1)!} \partial_y^{l-2}((1-y^2)^{l-1}), \qquad l \geq 2.$$

In fact,

$$\partial_y G_{l+1}(y) = -2L_l(y), \quad G_{l+1}(y) = \frac{2}{l(l+1)}(1-y^2)\partial_y L_l(y), \quad l \geq 1. \quad (5)$$

Obviously, $G_l(\pm 1) = 0$. Moreover,

$$\int_I G_{l+1}(y) G_{l'+1}(y)(1-y^2)^{-1} dy = \frac{8}{l(l+1)(2l+1)} \delta_{l,l'}, \qquad l \geq 1.$$

Now, let l_i be non-negative integers and $l = (l_1, l_2, \cdots, l_n)^T$. For any l with $l_j \geq 1$ and $l_i \geq 0 (i \neq j)$, we introduce the following polynomials,

$$\psi_l^{(j)}(\mathbf{x}) = G_{l_j+1}(x_j) \prod_{\substack{1 \leq i \leq n \\ i \neq j}} L_{l_i}(x_i).$$

Let $\chi_i(\mathbf{x}) = (1 - x_i^2)$. The set of all $\psi_l^{(j)}(\mathbf{x})$ is a complete $L^2_{\chi_j^{-1}}(\Omega)-$orthogonal system, namely,

$$\int_\Omega (1-x_j^2)^{-1} \psi_l^{(j)}(\mathbf{x}) \psi_{l'}^{(j)}(\mathbf{x}) d\mathbf{x} = \eta_l^{(j)} \delta_{l,l'}, \qquad l_j \geq 1 \text{ and } l_i \geq 0 \text{ for } i \neq j, \quad (6)$$

where $\delta_{l,l'}$ is the n-dimensional Kronecker symbol, and

$$\eta_l^{(j)} = \frac{2^{n+2}}{l_j(l_j+1)} \Big(\prod_{1 \leq \nu \leq n} (2l_\nu + 1) \Big)^{-1}, \qquad l_j \geq 1 \text{ and } l_i \geq 0 \text{ for } i \neq j.$$

We introduce the space

$$W(\Omega) = \{ \mathbf{v}(\mathbf{x}) \mid v^{(j)}(\mathbf{x}) \in L^2_{\chi_j^{-1}}(\Omega) \text{ for } 1 \leq j \leq n \}.$$

For any $\mathbf{v} \in W(\Omega)$, its components can be expanded as

$$v^{(j)}(\mathbf{x}) = \sum_{l_j=1}^{\infty} \sum_{\substack{1\le i\le n\\ i\neq j}} \sum_{l_i=0}^{\infty} \hat{v}_l^{(j)} \psi_l^{(j)}(\mathbf{x}), \qquad 1 \le j \le n, \tag{7}$$

where

$$\hat{v}_l^{(j)} = \frac{1}{\eta_l^{(j)}} \int_\Omega (1-x_j^2)^{-1} v^{(j)}(\mathbf{x}) \psi_l^{(j)}(\mathbf{x}) d\mathbf{x}.$$

By using (7), (5) and (6) successively, we derive that if $l_j \ge 1$ and $l_i \ge 0$ for $i \neq j$, then

$$\begin{aligned}\hat{v}_l^{(j)} &= \frac{2}{l_j(l_j+1)\eta_l^{(j)}} \int_\Omega v^{(j)}(\mathbf{x}) \partial_{x_j} L_{l_j}(x_j) (\prod_{\substack{1\le \nu\le n\\ \nu\neq j}} L_{l_\nu}(x_\nu)) d\mathbf{x} \\ &= -\frac{1}{2^{n+1}} (\prod_{1\le\nu\le n} (2l_\nu+1)) \int_\Omega \partial_{x_j} v^{(j)}(\mathbf{x}) (\prod_{1\le\nu\le n} L_{l_\nu}(x_\nu)) d\mathbf{x}.\end{aligned}$$

If, in addition, $\nabla \cdot \mathbf{v}(\mathbf{x}) = 0$ on $\bar{\Omega}$ and all $l_i \ge 1$, then

$$\sum_{j=1}^{n} \hat{v}_l^{(j)} = -\frac{1}{2^{n+1}} (\prod_{1\le\nu\le n} (2l_\nu+1)) \int_\Omega \nabla \cdot \mathbf{v}(\mathbf{x}) (\prod_{1\le\nu\le n} L_{l_\nu}(x_\nu)) d\mathbf{x} = 0. \tag{8}$$

Next, let $\mathcal{B}_j$ be the set consisting of all $l = (l_1, l_2, \cdots, l_n)^T$ with $l_j \ge 1$, and at least one component $l_i (i \neq j)$ vanishes. Then, we use (7), (5) and (8) successively to deduce that

$$\nabla \cdot \mathbf{v}(\mathbf{x}) = \sum_{j=1}^{n} \sum_{l_j=1}^{\infty} \sum_{\substack{1\le i\le n\\ i\neq j}} \sum_{l_i=0}^{\infty} \hat{v}_l^{(j)} \partial_{x_j} \psi_l^{(j)}(\mathbf{x}) = -2 \sum_{j=1}^{n} \sum_{l\in\mathcal{B}_j} \hat{v}_l^{(j)} (\prod_{1\le\nu\le n} L_{l_\nu}(x_\nu)).$$

This implies $\hat{v}_l^{(j)} = 0$ for all $l \in \mathcal{B}_j$. According to the previous statements, we assert that if $\mathbf{v} \in W(\Omega)$ and $\nabla \cdot \mathbf{v}(\mathbf{x}) = 0$ on $\bar{\Omega}$, then we could expand the components of $\mathbf{v}(\mathbf{x})$ as

$$v^{(j)}(\mathbf{x}) = \sum_{i=1}^{n} \sum_{l_i=1}^{\infty} \hat{v}_l^{(j)} \psi_l^{(j)}(\mathbf{x}), \qquad 1 \le j \le n, \tag{9}$$

where the coefficients $\hat{v}_l^{(j)}$ satisfy the equality

$$\sum_{j=1}^{n} \hat{v}_l^{(j)} = 0. \tag{10}$$

We now design the spectral scheme for solving problem (2). We set

$$\chi^{-1}(x) = (\chi_1^{-1}(x_1), \chi_2^{-1}(x_2), \cdots, \chi_n^{-1}(x_n)),$$

and define the non-isotropic weighted space

$$\mathbf{L}^2_{\chi^{-1}}(\Omega) = L^2_{\chi_1^{-1}}(\Omega) \otimes L^2_{\chi_2^{-1}}(\Omega) \otimes \cdots \otimes L^2_{\chi_n^{-1}}(\Omega),$$

equipped with the following inner product and norm,

$$(\mathbf{u}, \mathbf{v})_{\mathbf{L}^2_{\chi^{-1}}(\Omega)} = \sum_{j=1}^{n} (u^{(j)}, v^{(j)})_{L^2_{\chi_j^{-1}}(\Omega)}, \qquad ||\mathbf{v}||_{\mathbf{L}^2_{\chi^{-1}}(\Omega)} = (\mathbf{v}, \mathbf{v})^{\frac{1}{2}}_{\mathbf{L}^2_{\chi^{-1}}(\Omega)}.$$

We introduce the following finite-dimensional sets,

$$\mathcal{Q}_N^{(j)}(\Omega) = span\{\ \psi_l^{(j)}(\mathbf{x}) \mid 1 \le l_i \le N, 1 \le i \le n\}, \qquad 1 \le j \le n,$$

$$\mathcal{Q}_N(\Omega) = \mathcal{Q}_N^{(1)}(\Omega) \otimes \mathcal{Q}_N^{(2)}(\Omega) \otimes \cdots \otimes \mathcal{Q}_N^{(n)}(\Omega), \qquad V_N(\Omega) = V(\Omega) \bigcap \mathcal{Q}_N(\Omega).$$

The orthogonal projection $\tilde{P}_{N,\Omega} : \mathbf{L}^2_{\chi^{-1}}(\Omega) \to V_N(\Omega)$ is defined by

$$(\tilde{P}_{N,\Omega}\mathbf{v} - \mathbf{v}, \phi)_{\mathbf{L}^2_{\chi^{-1}}(\Omega)} = 0, \qquad \forall\, \phi \in V_N(\Omega).$$

The spectral scheme for solving (2) is to find $\mathbf{U}_N(t) \in V_N(\Omega)$ for all $0 \le t \le T$, such that

$$\begin{cases} (\partial_t \mathbf{U}_N(t), \phi)_{\mathbf{L}^2(\Omega)} + b(\mathbf{U}_N(t), \mathbf{U}_N(t), \phi) + \nu a(\mathbf{U}_N(t), \phi) = (\mathbf{f}(t), \phi)_{\mathbf{L}^2(\Omega)}, \\ \qquad\qquad\qquad\qquad\qquad\qquad\qquad\qquad \forall\, \phi \in V_N(\Omega),\ 0 < t \le T, \\ \mathbf{U}_N(\mathbf{x}, 0) = \tilde{P}_{N,\Omega}\mathbf{U}_0(\mathbf{x}), \qquad\qquad\qquad \mathbf{x} \in \bar{\Omega}. \end{cases} \tag{11}$$

In order to describe the errors of numerical solutions, we let $I_i = \{\ x_i \mid |x_i| < 1\}$ for $1 \le i \le n$, and introduce the following quantities,

$$A_r(\mathbf{v}) = \sum_{j=1}^{n} \Big(\sum_{\substack{1 \le i \le n \\ i \ne j}} \|\partial_{x_i}^r v^{(j)}\|_{L^2_{\chi_i^r}(\Omega/I_j; L^2_{\chi_j^{-1}}(I_j))} + \|\partial_{x_j}^r v^{(j)}\|_{L^2(\Omega/I_j; L^2_{\chi_j^{r-1}}(I_j))} \Big),$$

$$\begin{aligned} B_r(\mathbf{v}) = \sum_{j=1}^{n} \Big(& \sum_{\substack{1 \le i \le n \\ i \ne j}} \|\partial_{x_i}^{r-1}\partial_{x_j} v^{(j)}\|_{L^2_{\chi_i^{r-1}}(\Omega)} \\ & + \sum_{\substack{1 \le k \le n \\ k \ne i,j}} \|\partial_{x_i}\partial_{x_k}^{r-1} v^{(j)}\|_{L^2_{\chi_k^{r-1}}(\Omega/I_j; L^2_{\chi_j^{-1}}(I_j))} \\ & + \|\partial_{x_j}^r v^{(j)}\|_{L^2(\Omega/I_j; L^2_{\chi_j^{r-1}}(I_j))} + \|\partial_{x_i}\partial_{x_j}^{r-1} v^{(j)}\|_{L^2(\Omega/I_j; L^2_{\chi_j^{r-2}}(I_j))} \\ & + \|\partial_{x_i}^r v^{(j)}\|_{L^2(\Omega/I_i; L^2_{\chi_i^{r-1}}(I_i))} \Big). \end{aligned}$$

Furthermore,

$$\begin{aligned} D(\mathbf{v},t) &= |\mathbf{v}(t)|^2_{\mathbf{H}^1(\Omega)} + N^{-\frac{1}{2}} B_2^2(\mathbf{v}(t)),\\ R_r(\mathbf{v},t) &= B_r^2(\partial_t \mathbf{v}(t)) + \|\mathbf{v}(t)\|_{\mathbf{L}^2(\Omega)} |\mathbf{v}(t)|_{\mathbf{H}^1(\Omega)} B_r^2(\mathbf{v}(t))\\ &\quad + (\|\mathbf{v}(t)\|_{\mathbf{L}^2(\Omega)} + N^{-\frac{1}{2}} B_2(\mathbf{v}(t)))(|\mathbf{v}(t)|_{\mathbf{H}^1(\Omega)} + N^{-\frac{1}{2}} B_2(\mathbf{v}(t))) B_r^2(\mathbf{v}(t)),\\ \rho_r(\mathbf{v}) &= N^{-\frac{1}{2}} A_{r-1}^2(\mathbf{v}) + B_r^2(\mathbf{v}). \end{aligned}$$

We have the following results.

Theorem 7. Let $\mathbf{U}(\mathbf{x},t)$ and $\mathbf{u}_N(\mathbf{x},t)$ be the solutions of (2) and (11) with $n=2$, respectively. Then for integers $N \geq 2$ and $1 \leq r \leq N+1$,

$$\begin{aligned} &||\mathbf{U}(t) - \mathbf{U}_N(t)||^2_{\mathbf{L}^2(\Omega)} + \nu \int_0^t |\mathbf{U}(\xi) - \mathbf{U}_N(\xi)|^2_{\mathbf{H}^1(\Omega)} d\xi\\ &\leq cN^{3-2r} \Big(e^{\frac{c}{\nu} \int_0^t D(\mathbf{U},\xi)d\xi} \big(\frac{1}{\nu} N^{-\frac{1}{2}} \int_0^t e^{-\frac{c}{\nu}\int_0^\xi D(\mathbf{U},\eta)d\eta} R_r(\mathbf{U},\xi)d\xi + \rho_r(\mathbf{U}_0)\big)\\ &\quad + B_r^2(\mathbf{U}(t)) + \nu \int_0^t B_r^2(\mathbf{U}(\xi))d\xi \Big), \end{aligned} \tag{12}$$

provided that the norms involved at the right side of the above inequality are finite.

Theorem 8. Let $\mathbf{U}(\mathbf{x},t)$ and $\mathbf{u}_N(\mathbf{x},t)$ be the solutions of (2) and (11) with $n=3$, respectively. Then for integers $N \geq 2$ and $1 \leq r \leq N+1$,

$$\begin{aligned} &||\mathbf{U}(t) - \mathbf{U}_N(t)||^2_{\mathbf{L}^2(\Omega)} + \nu \int_0^t |\mathbf{U}(\xi) - \mathbf{U}_N(\xi)|^2_{\mathbf{H}^1(\Omega)} d\xi\\ &\leq cN^{3-2r} \Big(e^{\frac{c}{\nu} \int_0^t D(\mathbf{U},\xi)d\xi} \big(\frac{1}{\nu} N^{-\frac{1}{4}} \int_0^t e^{-\frac{c}{\nu}\int_0^\xi D(\mathbf{U},\eta)d\eta} R_r(\mathbf{U},\xi)d\xi + \rho_r(\mathbf{U}_0)\big)\\ &\quad + B_r^2(\mathbf{U}(t)) + \nu \int_0^t B_r^2(\mathbf{U}(\xi))d\xi \Big), \end{aligned} \tag{13}$$

provided that the norms involved at the right side of the above inequality are finite.

The proofs of Theorems 7 and 8 were given in [16]. The error estimates (12) and (13) demonstrate the spectral accuracy in space of the spectral scheme (11). The numerical results in [16] coincide with the analysis very well.

Remark 1. By using the divergence-free base functions satisfying the physical boundary conditions exactly, we could approximate the Navier-Stokes equations directly, see (11). In other words, we avoid to use the artificial compressibility method or the projection method. This fact simplifies the actual computation and the numerical analysis, and provides much more accurate numerical results. Moreover, thanks to the relation (10), we only need to evaluate the unknown coefficients of the expansions of arbitrary $n-1$ exponents of the numerical velocity $\mathbf{U}_N(\mathbf{x},t)$, for the n-dimensional Navier-Stokes equations. This feature also saves the computational time essentially.

Remark 2. We may consider the slip boundary conditions as

$$\mathbf{U}(\mathbf{x},t)\cdot\boldsymbol{n}=0, \qquad \partial_{\boldsymbol{n}}(\mathbf{U}(\mathbf{x},t)\cdot\boldsymbol{\tau})=\mathbf{g}(\mathbf{x},t), \qquad \mathbf{x}\in\partial\Omega,\ 0\le t\le T.$$

More precisely, let $S_i=\{\ \mathbf{x}\mid |x_i|=1\}$ for $1\le i\le n$, and

$$\partial_{\boldsymbol{n}}U^{(j)}(\mathbf{x},t)=g_i^{(j)}(\mathbf{x},t), \qquad \mathbf{x}\in S_i, i\ne j, 1\le i,j\le n.$$

Besides,

$$Q(\mathbf{g}(t),\mathbf{w})=\sum_{j=1}^{n}\sum_{\substack{1\le i\le n\\ i\ne j}}\int_{S_i} g_i^{(j)}(\mathbf{x},t)w^{(j)}(\mathbf{x})dS.$$

Then, the weak formulation of the Navier-Stokes equations with the above boundary conditions is to seek $\mathbf{U}\in L^2(0,T;V(\Omega))\cap L^\infty(0,T;\mathscr{H}(\Omega))$, such that

$$\begin{cases}(\partial_t\mathbf{U}(t),w)_{\mathbf{L}^2(\Omega)}+b(\mathbf{U}(t),\mathbf{U}(t),\mathbf{w})+\nu a(\mathbf{U}(t),\mathbf{w})\\ \qquad\qquad =Q(\mathbf{g}(t),\mathbf{w})+(\mathbf{f}(t),\mathbf{w})_{\mathbf{L}^2(\Omega)}, & \forall\ \mathbf{w}\in V(\Omega),\ 0<t\le T,\\ \mathbf{U}(\mathbf{x},0)=\mathbf{U}_0(\mathbf{x}), & \mathbf{x}\in\bar{\Omega}.\end{cases}$$

We can deal with the existence, stability and regularity of solutions of the above problem by an argument similar to the proofs of Theorems 2.1 and 4.1 of [13]. The corresponding spectral scheme is to find $\mathbf{u}_N(t)\ \in V_N(\Omega)$ for all $0\le t\le T$, such that

$$\begin{cases}(\partial_t\mathbf{u}_N(t),\phi)_{\mathbf{L}^2(\Omega)}+b(\mathbf{u}_N(t),\mathbf{u}_N(t),\phi)+\nu a(\mathbf{u}_N(t),\phi)\\ \qquad =Q(\mathbf{g}(t),\phi)+(\mathbf{f}(t),\phi)_{\mathbf{L}^2(\Omega)}, & \forall\ \phi\in V_N(\Omega),\ 0<t\le T,\\ \mathbf{u}_N(\mathbf{x},0)=\tilde{P}_{N,\Omega}\mathbf{U}_0(\mathbf{x}), & \mathbf{x}\in\bar{\Omega}.\end{cases}$$

With a modification of the proof of result (13) given in [16], we could prove the spectral accuracy in space of the above spectral scheme.

3. Vorticity-Stream Function Form of Navier-Stokes Equations with Non-slip Boundary Conditions

In study of incompressible viscous fluid flows, we prefer to the vorticity-stream function form of the Navier-Stokes equations sometimes. In this case, the incompressibility is satisfied automatically. This feature simplifies the theoretical analysis and the numerical simulation.

3.1. Solutions of vorticity-stream function form

For simplicity of statements, we focus on the case with $n = 2$. We denote the vorticity by $\omega(\mathbf{x},t)$ and the stream function by $\psi(\mathbf{x},t)$, namely,

$$\partial_{x_1}\psi(\mathbf{x},t) = -U^{(2)}(\mathbf{x},t), \qquad \partial_{x_2}\psi(\mathbf{x},t) = U^{(1)}(\mathbf{x},t),$$

$$\omega(\mathbf{x},t) = \nabla \times \mathbf{U}(\mathbf{x},t) = \partial_{x_1}U^{(2)}(\mathbf{x},t) - \partial_{x_2}U^{(1)}(\mathbf{x},t).$$

We consider the slip boundary conditions as before. Clearly, the condition $\mathbf{U}(\mathbf{x},t)\cdot \boldsymbol{n} = 0$ on $\partial\Omega$ implies $\partial_{\boldsymbol{\tau}}\psi(\mathbf{x},t) = 0$ on $\partial\Omega$. Since $\psi(\mathbf{x},t)$ is determined uniquely by $\mathbf{U}(\mathbf{x},t)$, apart from a constant, we may take $\psi(\mathbf{x},t) = 0$ on $\partial\Omega$. Next, the boundary conditions $\mathbf{U}(\mathbf{x},t)\cdot\boldsymbol{n} = 0$ and $\partial_{\boldsymbol{n}}(\mathbf{U}(\mathbf{x},t)\cdot\tau) = 0$ on $\partial\Omega$, lead to $\omega(\mathbf{x},t) = 0$ on $\partial\Omega$. A similar fact was also observed by Mucha [27].

The vorticity-stream function form is as follows,

$$\begin{cases} \partial_t\omega(\mathbf{x},t) + \partial_{x_2}\psi(\mathbf{x},t)\partial_{x_1}\omega(\mathbf{x},t) - \partial_{x_1}\psi(\mathbf{x},t)\partial_{x_2}\omega(\mathbf{x},t) - \nu\Delta\omega(\mathbf{x},t) \\ \quad = g(\mathbf{x},t) = \nabla\times\mathbf{f}(\mathbf{x},t), & \mathbf{x}\in\Omega,\ 0<t\le T, \\ -\Delta\psi(\mathbf{x},t) = \omega(\mathbf{x},t), & \mathbf{x}\in\bar{\Omega},\ 0\le t\le T, \\ \omega(\mathbf{x},t) = \psi(\mathbf{x},t) = 0, & \mathbf{x}\in\partial\Omega,\ 0\le t\le T, \\ \omega(\mathbf{x},0) = \omega_0(\mathbf{x}) = \nabla\times U_0(\mathbf{x}), & \mathbf{x}\in\bar{\Omega}. \end{cases} \tag{14}$$

Remark 3. In some existing literatures, one regarded (14) as the vorticity-stream function form of the Navier-Stokes equations with non-slip boundary conditions. But it is not correct, since the boundary condition $\mathbf{U}(\mathbf{x},t) = 0$ on $\partial\Omega$, does not implies $\omega(\mathbf{x},t) = 0$ on $\partial\Omega$. In fact, it is difficult to deal with the boundary value of vorticity for non-slip boundary conditions. We now introduce the following notations,

$$a(u,w) = \int_\Omega (\partial_{x_1}u(\mathbf{x})\partial_{x_1}w(\mathbf{x}) + \partial_{x_2}u(\mathbf{x})\partial_{x_2}w(\mathbf{x}))d\mathbf{x}, \qquad \forall u,w\in H^1(\Omega),$$

$$J(u,z,w) = \int_\Omega (\partial_{x_2}z(\mathbf{x})\partial_{x_1}u(\mathbf{x}) - \partial_{x_1}z(\mathbf{x})\partial_{x_2}u(\mathbf{x}))w(\mathbf{x})d\mathbf{x},$$
$$\forall u,w\in H^1(\Omega),\ z\in H^2(\Omega).$$

If $w(\mathbf{x}) = 0$ on $\partial\Omega$, then

$$(\Delta u, w)_{L^2(\Omega)} + a(u,w) = 0.$$

If $u(\mathbf{x}) = 0$ on $\partial\Omega$, then

$$J(u, z, w) + J(w, z, u) = 0.$$

The weak formulation of (14) is to seek $\omega \in L^2(0, T; H_0^1(\Omega)) \cap L^\infty(0, T; L^2(\Omega))$ and $\psi \in L^\infty(0, T; H_0^1(\Omega) \cap H^2(\Omega))$, such that

$$\begin{cases} (\partial_t \omega(t), w)_{L^2(\Omega)} + J(\omega(t), \psi(t), w) + \nu a(\omega(t), w) = (g(t), w)_{L^2(\Omega)}, \\ \qquad\qquad\qquad\qquad\qquad\qquad\qquad \forall w \in H_0^1(\Omega),\ 0 < t \le T, \\ a(\psi(t), w) = (\omega(t), w)_{L^2(\Omega)}, \qquad\qquad \forall w \in H_0^1(\Omega),\ 0 \le t \le T, \\ \omega(\mathbf{x}, 0) = \omega_0(\mathbf{x}), \qquad\qquad\qquad\qquad \mathbf{x} \in \bar{\Omega}. \end{cases} \tag{15}$$

We have the following result on the existence of solutions of problem (15).

Theorem 9. If $\omega_0 \in L^2(\Omega)$ and $g \in L^2(0, T; H^{-1}(\Omega))$, then problem (15) has at least one solution.

We next deal with the stability of problem (15). Let $(\omega, \psi)^T$ be a solution of (15). Suppose that ω_0 and g have the errors $\widetilde{\omega}_0$ and $\widetilde{g}$, respectively. They induce the error of $(\omega, \psi)^T$, denoted by $(\widetilde{\omega}, \widetilde{\psi})^T$. Then we obtain from (15) that

$$\begin{cases} (\partial_t \tilde{\omega}(t), w)_{L^2(\Omega)} + J(\tilde{\omega}(t), \psi(t) + \tilde{\psi}(t), w) + J(\omega(t), \tilde{\psi}(t), w) \\ \qquad + \nu a(\tilde{\omega}(t), w) = (\tilde{g}(t), w)_{L^2(\Omega)}, \qquad \forall w \in H_0^1(\Omega),\ 0 < t \le T, \\ a(\tilde{\psi}(t), w) = (\tilde{\omega}(t), w)_{L^2(\Omega)}, \qquad\qquad \forall w \in H_0^1(\Omega),\ 0 \le t \le T, \\ \tilde{\omega}(\mathbf{x}, 0) = \tilde{\omega}_0(\mathbf{x}), \qquad\qquad\qquad\qquad \mathbf{x} \in \bar{\Omega}. \end{cases}$$

Theorem 10. If $\omega \in L^2(0, T; H_0^1(\Omega)) \cap L^\infty(0, T; L^2(\Omega))$, then for any $\widetilde{\omega}_0 \in L^2(\Omega)$ and any $\tilde{g} \in L^2(0, T; H^{-1}(\Omega))$,

$$\begin{aligned} ||\widetilde{\omega}(t)||^2_{L^2(\Omega)} + \nu \int_0^t |\widetilde{\omega}(\xi)|^2_{H^1(\Omega)} d\xi \le \exp\Big(\frac{c}{\nu} \int_0^t |\omega(\xi)|^2_{H^1(\Omega)} d\xi\Big) \\ \cdot \Big(||\widetilde{\omega}_0(t)||^2_{L^2(\Omega)} + \frac{c}{\nu} \int_0^t \exp\Big(-\frac{c}{\nu} \int_0^\xi |\omega(\eta)|^2_{H^1(\Omega)} d\eta\Big) ||\tilde{g}(\xi)||^2_{L^2(\Omega)} d\xi\Big). \end{aligned} \tag{16}$$

Since the estimate (16) involves the factor $\exp(\frac{c}{\nu} \int_0^t |\omega(\xi)|^2_{H^1(\Omega)} d\xi)$, which in turn depends on the norms $||\omega_0||^2_{L^2(\Omega)}$ and $||g||^2_{L^2(0,T;H^{-1}(\Omega))}$, Theorem 10 implies the generalized stability of problem (15) for any solution $\omega \in L^2(0, T; H_0^1(\Omega)) \cap L^\infty(0, T; L^2(\Omega))$, and for any errors $\widetilde{\omega}_0 \in L^2(\Omega)$ and $\tilde{g} \in L^2(0, T; H^{-1}(\Omega))$. But there is no restriction on the values of the norms $||\omega_0||^2_{L^2(\Omega)}$, $||g||^2_{L^2(0,T;H^{-1}(\Omega))}$, $||\widetilde{\omega}_0||^2_{L^2(\Omega)}$ and $||\widetilde{g}||^2_{L^2(0,T;H^{-1}(\Omega))}$.

Obviously, if $\tilde{\omega}_0(\mathbf{x}) \equiv \tilde{g}(\mathbf{x}, t) \equiv 0$, then $\tilde{\omega}(\mathbf{x}, t) \equiv 0$. This fact with Theorem 9, shows that problem (15) admits a unique solution, as long as that

$\omega_0 \in L^2(\Omega)$ and $g \in L^2(0,T;H^{-1}(\Omega))$. It also tells us that problem (2) with $n=2$ has a unique solution $\mathbf{U} \in L^2(0,T;V(\Omega)\cap\mathbf{H}^2(\Omega))\cap L^\infty(0,T;\mathscr{H}(\Omega)\cap\mathbf{H}^1(\Omega))$, provided that $\mathbf{U}_0 \in V(\Omega)$ and $\mathbf{f} \in L^2(0,T;\mathbf{L}^2(\Omega))$.

The proofs of Theorems 9 and 10 could be found in [13].

3.2. Spectral method for the vorticity-stream function form

We now turn to the spectral method for solving problem (15). For any integer $N \geq 0$, $V_N(\Omega)$ stands for the set of all algebraic polynomials of degree at most N for the variables x_1 and x_2. Further, $V_N^0(\Omega) = H_0^1(\Omega) \cap V_N(\Omega)$. The orthogonal projection $P_{N,\Omega} : L^2(\Omega) \to V_N(\Omega)$ is defined by

$$(P_{N,\Omega}v - v, \phi)_{L^2(\Omega)} = 0, \qquad \forall\, \phi \in V_N(\Omega).$$

The spectral scheme for solving (15) is to seek $\omega_N(t) \in V_N^0(\Omega)$ and $\psi_N(t) \in V_N^0(\Omega)$ for all $0 \leq t \leq T$, such that

$$\begin{cases} (\partial_t\omega_N(t),\phi)_{L^2(\Omega)} + J(\omega_N(t),\omega_N(t),\phi) + \nu a(\omega_N(t),\phi) = (g(t),\phi)_{L^2(\Omega)}, \\ \qquad\qquad\qquad\qquad\qquad\qquad \forall\, \phi \in V_N^0(\Omega),\ 0 < t \leq T, \\ a(\psi_N(t),\phi) = (\omega_N(t),\phi)_{L^2(\Omega)}, \qquad \forall \phi \in V_N^0(\Omega),\ 0 \leq t \leq T, \\ \omega_N(\mathbf{x},0) = P_{N,\Omega}\omega_0(\mathbf{x}), \qquad \mathbf{x} \in \bar{\Omega}. \end{cases} \tag{17}$$

In order to describe the errors of numerical solutions, we introduce the following quantities,

$$A_r(v) = ||\partial_{x_1}^r v||_{L^2_{\chi_1^r}(\Omega)} + ||\partial_{x_2}^r v||_{L^2_{\chi_2^r}(\Omega)},$$

$$\begin{aligned} B_r(v) = {} & ||\partial_{x_1}^r v||_{L^2_{\chi_1^{r-1}}(\Omega)} + ||\partial_{x_2}^r v||_{L^2_{\chi_2^{r-1}}(\Omega)} \\ & + ||\partial_{x_1}^{r-1}\partial_{x_2} v||_{L^2_{\chi_1^{r-2}}(\Omega)} + ||\partial_{x_1}\partial_{x_2}^{r-1} v||_{L^2_{\chi_2^{r-2}}(\Omega)}. \end{aligned}$$

Furthermore,

$$D(v) = A_2^2(v) + B_2^2(v) + \|v\|^2_{H^{\frac{3}{4}}(I_1,H^{\frac{3}{4}}(I_2))} + \|v\|^2_{L^\infty(\Omega)},$$

$$\begin{aligned} R_r(v,w,t) = {} & B_{r-1}^2(\partial_t v(t)) + D(v(t))B_{r-1}^2(v(t)) \\ & + \|v(t)\|^2_{L^2(\Omega)} B_r(v(t)) + D(v(t))B_r^2(w(t)), \end{aligned}$$

$$\rho_{r,s}(v) = A_{r-1}^2(v) + B_{r-1}^2(v).$$

We have the following result.

Theorem 11. Let $(\omega(\mathbf{x},t),\psi(\mathbf{x},t))^T$ and $(\omega_N(\mathbf{x},t),\psi_N(\mathbf{x},t))^T$ be the solutions of (15) and (17), respectively. Then for integers $2 \le r \le N+1$,

$$\begin{aligned}
&||\omega(t)-\omega_N(t)||_{L^2(\Omega)} + \nu \int_0^t |\omega(\xi)-\omega_N(\xi)|^2_{H^1(\Omega)} d\xi \\
&\le cN^{2-2r}\Big(e^{\frac{c}{\nu}\int_0^t D(\omega(\xi))d\xi} \big(\rho_r(\omega_0) + \frac{1}{\nu}\int_0^t R_r(\omega,\psi,\xi) e^{-\frac{c}{\nu}\int_0^\xi D(\omega(\eta))d\eta} d\xi\big) \\
&\quad + B_{r-1}(\omega(t)) + \nu \int_0^t B_r(\omega(\xi))d\xi \Big),
\end{aligned} \tag{18}$$

provided that the norms involved at the right side of the above inequality are finite.

The above result is a special case of Theorem 3.2 of [20]. The estimate (18) implies the spectral accuracy in space of the spectral scheme (17). The numerical results in [20] coincide the analysis very well.

Remark 4. We may consider the stream function form of the two-dimensional Navier-Stokes equation as follows,

$$\begin{aligned}
&\partial_t \Delta\psi(\mathbf{x},t) + \partial_{x_2}\psi(\mathbf{x},t)\partial_{x_1}\Delta\psi(\mathbf{x},t) - \partial_{x_1}\psi(\mathbf{x},t)\partial_{x_2}\Delta\psi(\mathbf{x},t) - \nu\Delta^2\psi(\mathbf{x},t) \\
&\quad = -\nabla \times \mathbf{f}(\mathbf{x},t), \qquad\qquad \mathbf{x}\in\Omega,\ 0<t\le T.
\end{aligned}$$

In this case, the incompressibility is also fulfilled automatically. For the fluid flows with non-slip boundary conditions, we can derive the exact boundary condition as $\partial_{\boldsymbol{n}}\psi(\mathbf{x},t) = \psi(\mathbf{x},t) = 0$ on $\partial\Omega$. Some authors studied the solutions of this problem, and designed the related spectral schemes, e.g., see [14, 15]. Whereas, the above boundary condition is not valid for the fluid flows with slip boundary conditions. Therefore, the stream function form seems more suitable for the fluid flows with non-slip boundary conditions, while the vorticity-stream function form is more appreciate for the fluid flows with slip boundary conditions.

4. Other Problems with Divergence-free Solutions

There are other practical problems with divergence-free solutions.

As an example, we consider the Darwin model of approximation to the Maxwell equations arising in electro-magnetic fields. Let Ω be a cube and $S_i = \{\ \mathbf{x} \mid |x_i| = 1\}$ for $1 \le i \le 3$. $\mathbf{B}(\mathbf{x},t)$ denotes the magnetic field with the components $B^{(j)}(\mathbf{x},t),\ 1\le j\le 3$. $\mathbf{B}_0(\mathbf{x})$ describes the initial magnetic field. $\mathbf{J}(\mathbf{x},t)$ stands for the density of electric current with the components $J^{(j)}(\mathbf{x},t),\ 1 \le j \le 3$. The constant μ is the relative permeability.

As pointed out by Degond and Raviat [5], we need to solve three related problems at any time t. For instance, one of them is of the form

$$\begin{cases} -\Delta \mathbf{B}(\mathbf{x},t) = \mu \mathrm{Rot}\mathbf{J}(\mathbf{x},t), & \mathbf{x}\in\Omega,\ 0\le t\le T, \\ \nabla\cdot\mathbf{B}(\mathbf{x},t) = 0, & \mathbf{x}\in\bar{\Omega},\ 0\le t\le T, \\ \mathbf{B}(\mathbf{x},t)\cdot\mathbf{n} = \mathbf{B}_0(\mathbf{x})\cdot\mathbf{n}, & \mathbf{x}\in\partial\Omega,\ 0\le t\le T, \\ \mathrm{Rot}\mathbf{B}(\mathbf{x},t)\times\mathbf{n} = \mu\mathbf{J}(\mathbf{x},t)\times\mathbf{n}, & \mathbf{x}\in\partial\Omega,\ 0\le t\le T. \end{cases} \tag{19}$$

According to the boundary conditions, we assert that $B^{(j)}(\mathbf{x},t) = B_0^{(j)}(\mathbf{x})$ on S_j, $1\le j\le 3$. Moreover, a careful calculation shows that

$$\begin{aligned} &\partial_{x_1}B^{(2)}(\mathbf{x},t) = \mu J^{(3)}(\mathbf{x},t) + \partial_{x_2}B_0^{(1)}(\mathbf{x}), && \text{on } S_1, \\ &\partial_{x_1}B^{(3)}(\mathbf{x},t) = -\mu J^{(2)}(\mathbf{x},t) + \partial_{x_3}B_0^{(1)}(\mathbf{x}), && \text{on } S_1, \\ &\partial_{x_2}B^{(3)}(\mathbf{x},t) = \mu J^{(1)}(\mathbf{x},t) + \partial_{x_3}B_0^{(2)}(\mathbf{x}), && \text{on } S_2, \\ &\partial_{x_2}B^{(1)}(\mathbf{x},t) = -\mu J^{(3)}(\mathbf{x},t) + \partial_{x_1}B_0^{(2)}(\mathbf{x}), && \text{on } S_2, \\ &\partial_{x_3}B^{(1)}(\mathbf{x},t) = \mu J^{(2)}(\mathbf{x},t) + \partial_{x_1}B_0^{(3)}(\mathbf{x}), && \text{on } S_3. \\ &\partial_{x_3}B^{(2)}(\mathbf{x},t) = -\mu J^{(1)}(\mathbf{x},t) + \partial_{x_2}B_0^{(3)}(\mathbf{x}), && \text{on } S_3. \end{aligned} \tag{20}$$

Clearly, the right sides of the above six equalities are known functions. Furthermore, all $B^{(j')}(\mathbf{x},t)$ are the tangential components of $\mathbf{B}(\mathbf{x},t)$ on S_j for $j\ne j'$. Therefore,

$$\partial_{\mathbf{n}}(\mathbf{B}(\mathbf{x},t)\cdot\tau) = \mathbf{G}(\mathbf{x},t), \qquad \mathbf{x}\in\partial\Omega,\ 0\le t\le T,$$

where $\mathbf{G}(\mathbf{x},t)$ is a known vector function with the components as the right sides of (20). Next, let $\mathbf{B}(\mathbf{x},t) = \mathbb{B}(\mathbf{x},t) + \mathbf{B}_0(\mathbf{x})$. Then the problem (19) is reformulated to

$$\begin{cases} -\Delta \mathbb{B}(\mathbf{x},t) = \Delta\mathbf{B}_0(\mathbf{x}) + \mu \mathrm{Rot}\mathbf{J}(\mathbf{x},t), & \mathbf{x}\in\Omega,\ 0\le t\le T, \\ \nabla\cdot\mathbb{B}(\mathbf{x},t) = 0, & \mathbf{x}\in\bar{\Omega},\ 0\le t\le T, \\ \mathbb{B}(\mathbf{x},t)\cdot\mathbf{n} = 0, & \mathbf{x}\in\partial\Omega,\ 0\le t\le T, \\ \partial_{\mathbf{n}}(\mathbb{B}(\mathbf{x},t)\cdot\tau) = \mathbf{G}(\mathbf{x},t) - \partial_{\mathbf{n}}(\mathbf{B}_0(\mathbf{x})\cdot\tau), & \mathbf{x}\in\partial\Omega,\ 0\le t\le T. \end{cases} \tag{21}$$

We could deal with the above problem theoretically and solving them numerically by using the idea and techniques developed in the work [13, 16, 20].

We may use other orthogonal approximations for numerical simulations of incompressible fluid flows with non-slip boundary conditions, governed by

$$\begin{cases} \partial_t \mathbf{U}(\mathbf{x},t) + (\mathbf{U}(\mathbf{x},t)\cdot\nabla)\mathbf{U}(\mathbf{x},t) - \nu\Delta\mathbf{U}(\mathbf{x},t) + \nabla P(\mathbf{x},t) = \mathbf{f}(\mathbf{x},t), \\ \qquad\qquad\qquad\qquad\qquad\qquad\qquad\qquad \mathbf{x}\in\Omega,\ 0<t\le T, \\ \nabla\cdot\mathbf{U}(\mathbf{x},t) = 0, \qquad\qquad\qquad\qquad\quad \mathbf{x}\in\bar{\Omega},\ 0\le t\le T, \\ \mathbf{U}(\mathbf{x},t) = 0, \qquad\qquad\qquad\qquad\qquad \mathbf{x}\in\partial\Omega,\ 0\le t\le T, \\ \mathbf{U}(\mathbf{x},0) = \mathbf{U}_0(\mathbf{x}), \qquad\qquad\qquad\quad\ \mathbf{x}\in\bar{\Omega}. \end{cases} \tag{22}$$

It is noted that the boundary conditions $\nabla\cdot\mathbf{U}(\mathbf{x},t) = \mathbf{U}(\mathbf{x},t) = 0$ on $\partial\Omega$, imply that $\partial_{x_j}U^{(j)}(\mathbf{x},t) = 0$ for $|x_j| = 1,\ 1\le j\le n$.

Let $G_l(y)$ be the same as before, and

$$F_l(y) = \frac{(-1)^l}{2^{l-4}(l-2)!}(1-y^2)^2\partial_y^{l-4}((1-y^2)^{l-2}), \qquad l\ge 4.$$

In fact,

$$\partial_y F_{l+1}(y) = -2G_l(y), \qquad\qquad l\ge 3.$$

Let l_i be non-negative integers and $l = (l_1, l_2, \cdots, l_n)^T$. We introduce the polynomials

$$\psi_l^{(j)}(\mathbf{x}) = F_{l_j+1}(x_j)\prod_{\substack{1\le\nu\le n\\ \nu\ne j}} G_{l_\nu}(x_\nu).$$

Let $\chi_j(\mathbf{x})$ be the same as before, and

$$\omega_j(\mathbf{x}) = \chi_j^{-2}(\mathbf{x})\prod_{\substack{1\le\nu\le n\\ \nu\ne j}} \chi_\nu^{-1}(\mathbf{x}).$$

The set of all $\psi_l^{(j)}(\mathbf{x})$ is a complete $L^2_{\omega_j}(\Omega)$–orthogonal system. If $\mathbf{v}\in \mathbf{H}^1(\Omega)$, $\nabla\cdot\mathbf{v}(\mathbf{x}) = 0$ on $\bar{\Omega}$, and $\mathbf{v}(\mathbf{x}) = 0$ on $\partial\Omega$, then we could expand the components of $\mathbf{v}(\mathbf{x})$ as

$$v^{(j)}(\mathbf{x}) = \sum_{i=1}^{n}\sum_{l_i=3}^{\infty}\hat{v}_l^{(j)}\psi_l^{(j)}(\mathbf{x}), \qquad 1\le j\le n,$$

where the coefficients $\hat{v}_l^{(j)}$ satisfy the equality

$$\sum_{j=1}^{n}\hat{v}_l^{(j)} = 0.$$

According to the above property, we could design the spectral method for solving problem (22), and prove its spectral accuracy in space, see [17].

An interesting and open problem is how to generalize the techniques described in this review paper to solve various problems with divergence-free solutions, defined on non-rectangle domains.

References

[1] C. Bernardi and Y. Maday, Spectral Methods. In eds. P. G. Ciarlet and J. L. Lions, *Handbook of Numerical Analysis, Vol.5, Techniques of Scientific Computing,* pp. 209-486, Elsevier, Amsterdam (1997).

[2] C. Canuto, M. Y. Hussaini, A. Quarteroni and T. A. Zang, *Spectral Methods in Fluid Dynamics,* Springer-Verlag, Berlin (1988).

[3] A. J. Chorin, The numerical solution of the Navier-Stokes equations for an incompressible fliud, *Bull. Amer. Math. Soc.,* **73**(6), 928-931 (1967). ISSN 0273-0979.

[4] A. J. Chorin, Numerical solution of the Navier-Stokes equations, *J. Comp. Phys.,* **22**(104), 745-762(1968). ISSN 0025-5718.

[5] P. Degond and P. A. Raviat, An analysis of the Darwin model of approximation to Maxwell's equations, *Forum Math.*, **4**(4), 13-44(1992). ISSN 0933-7741.

[6] V. Girault and P. A. Raviart, *Finite Element Approximation of the Navier-Stokes equations, Lecture Notes in Mathematics, 794*, Springer-Verlag, Berlin (1979).

[7] P. M. Gresho, On pressure boundary conditions for the incompressible Navier-Stokes equations, *Int. J. for Numer. Meth. in Fluids,* **7**(10), 1111-1145 (1987). ISSN 1097-0363.

[8] J. L. Guermond and L. Quartapelle, Uncoupled $\omega - \psi$ formulation for plate flows in multiply connected domains, *Math. Model. Meth. in Appl. Sci.,* **7**(6), 731-767(1997). ISSN 1793-6314.

[9] Guo Ben-yu, A class of difference schemes of two-dimensional viscous fluid flow, *Research Report of SUST* (1965). Also see *Acta Mathematica Sinica,* **17**(4), 242-258 (1974). ISSN 1439-8516.

[10] Guo Ben-yu, Spectral method for Navier-Stokes equations, *Scientia Sinica,* **28A**(11), 1139-1153 (1985). ISSN 1006-9283.

[11] Guo Ben-yu, *Difference Methods for Partial Differential Equations,* Science Press, Beijing (1988).

[12] Guo Ben-yu, *Spectral Methods and Their Applications,* World Scientific, Singapore (1998).

[13] Guo Ben-Yu, Navier-Stokes equations with slip boundary conditions, *Math. Meth. in Appl. Sci.*, **31**(5), 607-626(2008). ISSN 1099-1476.

[14] Guo Ben-Yu and He Li-ping, The fully discrete Legendre spectral approximation of two-dimensional unsteady incompressible fluid flow in stream function form, *SIAM J. Numer. Anal.,* **35**(1), 146-176 (1998). ISSN 0036-1429.

[15] Guo Ben-yu, He Li-ping and Mao De-kang, On the two-dimensional Navier-Stokes equations in stream function form, *JMAA*, **205**(1), 1-31 (1997). ISSN 0022-247X.

[16] Guo Ben-yu and Jiao Yu-jian, Spectral method for Navier-Stokes equations with slip boundary conditions, *J. Sci. Comp.*, DOI: 10.1007/S10915-013-97295, ISSN 0885-7474.

[17] Guo Ben-yu and Jiao Yu-jian, Spectral method for Navier-Stokes equations with non-slip boundary conditions by using divergence-free base functions, (unpublished).

[18] Guo Ben-yu and Ma He-ping, Combined finite element and pseudospectral method for the two-dimensional evolutionary Navier-Stokes equations, *SIAM J. Numer. Anal.,* **30**(4), 1066-1083(1993). ISSN 0036-1429.

[19] Guo Ben-yu, B. D. Sleeman and Chen Sui-yang, On the discrete logistic model of biology, *Applicable Analysis*, **33**(3-4), 215-231(1989). ISSN 0003-6811.

[20] Jiao Yu-jian and Guo Ben-yu, Spectral method for vorticity-stream function form of Navier-Stokes equations with slip boundary conditions, *Math. Meth. in Appl. Sci.*, **35**(3), 257-267(2012). ISSN 1099-1476.

[21] Kuo Pen-yu, Numerical methods for incompressible viscous flow, *Scientia Sinica,* **20**(3), 287-304 (1977). ISSN 1006-9283.

[22] O. A. Ladyzhenskaya, *The Mathematical Theory of Viscous Incompressible Flow,* English translation, Second edition, Gordon and Breach, New York (1969).

[23] J. L. Lions, *Quèlques Méthodes de Résolution des Problèms aux Limités Non Linéaires*, Dunod, Paris (1969).

[24] T. Ma and S. H. Wang, The geometry of the stream lines of steady states of the Navier-Stokes equations. In eds. G. Q. Chen and E. Dibenedetto, *Nonlinear Partial Differential Equations, Contemp. Math., Vol 238,* pp. 193-202, AMS, Providence (1999).

[25] T. Ma and S. H. Wang, *Geometry Theory of Incompressible Flows with Applications to Fluid Dynamics, Mathematical Surveys and Monographs, Vol. 119,* AMS, Providence (2005).

[26] Y. Maday and A. Quarteroni, Spectral and pseudospectral approximations of the Navier-Stokes equations, *SIAM J. Numer. Anal.,* **19**(4), 761-780(1982). ISSN 0036-1429.

[27] P. B. Mucha, On Navier-Stokes equations with slip boundary conditions in an infinite pipe, *Acta Applicandae Mathematics,* **76**(1), 1-15(2003). ISSN 0167-8019.

[28] S. Orszag, M. Israeli, M. O. Deville, Boundary conditions for incompressible flows, *J.of Sci. Comp.,* **1**(1), 75-111(1986). ISSN 0885-7474.

[29] P. J. Roache, *Computational Fluid Dynamics,* 2nd Edition, Hermosa Publishers, Albuquerque (1976).

[30] R. Téman, *Navier-Stokes Equations,* North-Holland, Amsterdam (1977).

[31] R. Téman, Sur lapproximation de la solution des equations de Navier-Stokes par la méthode des fractionnarires II, *Arch. Rational Mech. Anal.,* **33**(5), 377-385(1969). ISSN 0003-9527.

Global Existence and Uniqueness of the Solution for the Generalized Schrödinger-KdV System

Boling Guo
Institute of Applied Physics and Computational Mathematics,
P. O. Box 8009, Beijing, China, 100088
gbl@iapcm.ac.cn

Bolin Ma
College of Mathematics, Physics and Information Engineering,
Jiaxing University, Zhejiang, China, 314001
blma@mail.zjxu.edu.cn

Jingjun Zhang
College of Mathematics, Physics and Information Engineering,
Jiaxing University, Zhejiang, China, 314001
zjj_math@yahoo.com.cn

In this paper, we consider a generalized Schrödinger-KdV system. By using energy method as well as various type commutator estimates, global existence and uniqueness of smooth solution for this system is obtained.

1. Introduction

In the present paper, we study the global existence and uniqueness of the Cauchy problem for the Schrödinger-KdV system given by

$$\begin{cases} iE_t + aE_{xx} = bnE + c|E|^2E, \\ n_t - \beta\Lambda^{2\alpha}n_x \pm (n^2)_x + (|E|^2)_x = 0, \end{cases} \tag{1}$$

with initial data $(E(0,x), n(0,x)) = (E_0, n_0)$, where $\alpha \in (\frac{1}{2}, 1)$, a, b, c and β are real numbers satisfying $ab\beta \neq 0$, and $\Lambda = (-\Delta)^{1/2} = (-\partial_{xx})^{1/2}$. In (1), $E = E(t,x) : \mathbb{R} \times \mathbb{R} \to \mathbb{C}$ is a complex-valued function, and $n = n(t,x) : \mathbb{R} \times \mathbb{R} \to \mathbb{R}$ is a real-valued function.

This system describes the interactions between short wave and long wave, and arises in fluid mechanics as well as plasma physics. The case $c = 0$ appears in the study of resonant interaction between short and long capillary-gravity waves on water of uniform finite depth (see Refs. 4, 8, 12, 13).

When $\alpha = 1$, system (1) reduces to the usual Schrödinger-KdV system, which has been studied by many researchers in the past decades, see e.g. Refs. 1, 2, 5, 6, 7, 14 and the references cited therein. On the other hand, if $E \equiv 0$, then (1) reduces to the generalized KdV equation, which was studied by Kenig, Ponce and Vega.[10,11] However, as far as we can know, there are no results on the system (1) with $\alpha < 1$, and this is the main motivation of this paper.

The main result of the paper is stated in the following theorem.

Theorem 1. *Assume* $ab\beta \neq 0$, $c \in \mathbb{R}$, *and* $(E_0, n_0) \in H^{k+1} \times H^{k+\alpha}$ *with* $k \geq 0$ *be an integer and* $1/2 < \alpha < 1$. *Then system* (1) *has a solution* (E, n) *satisfying*

$$(E, n) \in L^\infty(-T, T; H^{k+1} \times H^{k+\alpha})$$

for any $T > 0$. *Moreover, such solution is unique when* $k \geq 2$.

Theorem 1 is proved by using energy method together with approximate argument. An important step in this proof is to obtain uniform bounds independent of the approximate parameter for the solution of the regularized system in the above spaces. The main difficulty in deriving these uniform bounds is to get the L^∞ norm estimate for n_x. When $\alpha = 1$, one can overcome this difficulty only through integrating by parts. However, in our case, we use various new type commutator estimates to circumvent this obstacle, see the lemmas given in Section 3.

For the regular solutions of (1), the following three conserved quantities hold:

$$\mathcal{M}(t) := \int_{\mathbb{R}} |E(t,x)|^2 dx, \tag{2}$$

$$\mathcal{H}_1(t) := \int_{\mathbb{R}} |n(t,x)|^2 dx - \frac{2}{b} \int_{\mathbb{R}} \mathrm{Im}(E\overline{E_x})(t,x) dx, \tag{3}$$

$$\begin{aligned} \mathcal{H}_2(t) :=& a \int_{\mathbb{R}} |E_x(t,x)|^2 dx - \frac{b\beta}{2} \int_{\mathbb{R}} |\Lambda^\alpha n(t,x)|^2 dx \pm \frac{b}{3} \int_{\mathbb{R}} n^3(t,x) dx, \\ &+ b \int_{\mathbb{R}} (n|E|^2)(t,x) dx + \frac{c}{2} \int_{\mathbb{R}} |E(t,x)|^4 dx. \end{aligned} \tag{4}$$

From (4), one sees that if we want to get global bounds for $\|E_x\|_{L^2}$ and $\|\Lambda^\alpha n\|_{L^2}$, then a and $b\beta$ should have the opposite sign, namely, $ab\beta < 0$. In the case $ab\beta > 0$, we let $\tilde{E}(t,x) = E(t,-x)$ and $\tilde{n}(t,x) = n(t,-x)$, then $(\tilde{E},\tilde{n})$ solves

$$\begin{cases} i\tilde{E}_t + a\tilde{E}_{xx} = b\tilde{n}\tilde{E} + c|E|^2\tilde{E}, \\ \tilde{n}_t - (-\beta)\Lambda^{2\alpha}\tilde{n}_x \mp (\tilde{n}^2)_x - (|\tilde{E}|^2)_x = 0. \end{cases} \tag{5}$$

And we see that $ab(-\beta) < 0$ in (5), which reduces to (1). So in order to obtain global existence, we only need to consider (1) in the case $ab\beta < 0$.

From now on, for the sake of simplicity, we set $a = b = 1, \beta = -1$ in (1) and consider the following system

$$\begin{cases} iE_t + E_{xx} = nE + c|E|^2E, \\ n_t + \Lambda^{2\alpha}n_x + (n^2)_x + (|E|^2)_x = 0. \end{cases} \tag{6}$$

As we will see later, the sign for the term $(n^2)_x$ will not affect the global existence result for (1), so we take the positive sign for this term in the above system. In the following argument, we mainly focus on the system (6).

For $s \in \mathbb{R}$, H^s and $\dot{H}^s$ denote the usual inhomogeneous and homogeneous Sobolev space, respectively. For any function u, $\Lambda^s u$ is defined as

$$\Lambda^s u = \mathcal{F}^{-1}(|\xi|^s \mathcal{F}u)$$

with $\mathcal{F}u$ the Fourier transform of u. Throughout the paper, the term n_x^2 means $(n_x)^2$. Similarly, $|E|_x^2 := (|E|^2)_x$.

This paper is organized as follows. In the next section, we give a regularized system for (6) and prove the global existence of the solution for this system. In Section 3, we give the uniform bounds for the approximate solution. Theorem 1 is proved in Section 4.

2. A Regularized System

This section deals with the existence and uniqueness for a regularized system of (6). Note that in the second equation of (6), the nonlinear terms contain one order derivative. Note also that in $\mathcal{H}_2(t)$, E and n have different regularity. Combining these two facts, we define $\mathcal{L} := (I + \epsilon\Lambda)^{-1}$ $(\epsilon > 0)$ and consider the following regularized system

$$\begin{cases} iE_t^\epsilon + E_{xx}^\epsilon = (\mathcal{L}n^\epsilon)E^\epsilon + c|E^\epsilon|^2E^\epsilon, \\ n_t^\epsilon + \Lambda^{2\alpha}n_x^\epsilon + \mathcal{L}(\mathcal{L}n^\epsilon)_x^2 + \mathcal{L}(|E^\epsilon|^2)_x = 0 \end{cases} \tag{7}$$

equipped with smooth initial data $(E^\epsilon(0), n^\epsilon(0)) = (E_0^\epsilon, n_0^\epsilon) \in H^r \times H^r$ such that

$$(E_0^\epsilon, n_0^\epsilon) \to (E^0, n^0) \text{ in } H^{k+1} \times H^{k+\alpha}, \ \epsilon \to 0. \tag{8}$$

Here r is selected large enough (e.g., $r > k + 100$) such that all the calculations given in the next section are meaningful. For the operator $\mathcal{L}$, it is easy to see that the following properties hold:

(i) $\|\mathcal{L}u\|_{H^s} \leq \|u\|_{H^s}$, $\|\mathcal{L}u\|_{\dot{H}^s} \leq \|u\|_{\dot{H}^s}$, $\forall\ s \in \mathbb{R}$;
(ii) $\int_{\mathbb{R}^d} (\mathcal{L}u)\bar{v}dx = \int_{\mathbb{R}^d} u\overline{\mathcal{L}v}dx$;
(iii) $\mathcal{L}$ *commutes with usual differential operators, such as* $\partial_x, \partial_{xx}, \cdots$.

For system (7), we have the following global existence result.

Theorem 2. *Fix* $r > k + 100$. *Then for any given* $\epsilon > 0$, *the regularized system* (7) *has a unique smooth solution* (E^ϵ, n^ϵ) *such that* $(E^\epsilon, n^\epsilon) \in C(\mathbb{R}; H^r \times H^r)$.

To prove Theorem 2, we need the following two lemmas.

Lemma 1. *Let* $\gamma > 0$, $2 \leq p \leq \infty$ ($p \neq \infty$ *if* $\gamma = \frac{1}{2}$), *and* $\gamma \geq \frac{1}{2} - \frac{1}{p}$, *then there holds*

$$\|u\|_{L^p(\mathbb{R})} \leq C\|u\|_{L^2(\mathbb{R})}^{1-(\frac{1}{2}-\frac{1}{p})\frac{1}{\gamma}} \|\Lambda^\gamma u\|_{L^2(\mathbb{R})}^{(\frac{1}{2}-\frac{1}{p})\frac{1}{\gamma}} \tag{9}$$

for any $u \in H^\gamma(\mathbb{R})$.

Lemma 2. *Let* $s \geq 0$ *and* $1 < p < \infty$. *If* $u,\ v \in \mathcal{S}(\mathbb{R}^d)$, *the Schwarz class, then*

$$\|\Lambda^s(uv)\|_{L^p} \leq C\|u\|_{L^{p_1}}\|\Lambda^s v\|_{L^{p_2}} + \|\Lambda^s u\|_{L^{p_3}}\|v\|_{L^{p_4}} \tag{10}$$

with $\frac{1}{p_1} + \frac{1}{p_2} = \frac{1}{p_3} + \frac{1}{p_4} = \frac{1}{p}$ *and* $p_2, p_3 \in (1, \infty)$.

As for the system (1), system (7) also enjoys three conserved quantities which are given in Lemma 3 below.

Lemma 3. *Assume* $(E^\epsilon, n^\epsilon) \in C(\mathbb{R}; H^r \times H^r)$ *be a smooth solution of* (7), *then the following three quantities conserved for all time:*

$$\mathcal{M}^\epsilon(t) := \int_{\mathbb{R}} |E^\epsilon(t,x)|^2 dx, \tag{11}$$

$$\mathcal{H}_1^\epsilon(t) := \int_{\mathbb{R}} |n^\epsilon(t,x)|^2 dx - 2\int_{\mathbb{R}} \mathrm{Im}(E^\epsilon \overline{E_x^\epsilon})(t,x)dx, \tag{12}$$

$$\begin{aligned}\mathcal{H}_2^\epsilon(t) := &\int_{\mathbb{R}} |E_x^\epsilon(t,x)|^2 dx + \frac{1}{2}\int_{\mathbb{R}} |\Lambda^\alpha n^\epsilon(t,x)|^2 dx + \frac{1}{3}\int_{\mathbb{R}} (\mathcal{L}n^\epsilon)^3(t,x)dx,\\ &+ \int_{\mathbb{R}} ((\mathcal{L}n^\epsilon)|E^\epsilon|^2)(t,x)dx + \frac{c}{2}\int_{\mathbb{R}} |E^\epsilon(t,x)|^4 dx.\end{aligned} \tag{13}$$

Moreover, there holds that

$$\|E^\epsilon(t)\|_{H^1} + \|n^\epsilon(t)\|_{H^\alpha} \le C(\|E_0^\epsilon\|_{H^1}, \|n_0^\epsilon\|_{H^\alpha}),\ \forall\, t \in \mathbb{R}. \tag{14}$$

Proof. Multiplying the first equation of (7) by $\overline{E^\epsilon}$ and integrating the imaginary part over $\mathbb{R}$, then (11) follows. To obtain (12), we take inner product of the second equation of (7) with n^ϵ and get

$$\frac{1}{2}\frac{d}{dt}\|n^\epsilon\|_{L^2}^2 + \int_{\mathbb{R}} (\mathcal{L}n^\epsilon)|E^\epsilon|_x^2 dx = 0.$$

Since $(\mathcal{L}n^\epsilon)E^\epsilon = iE_t^\epsilon + E_{xx}^\epsilon - c|E^\epsilon|^2 E^\epsilon$, we have

$$\begin{aligned}\int_{\mathbb{R}} (\mathcal{L}n^\epsilon)|E^\epsilon|_x^2 dx = 2\mathrm{Re}\int_{\mathbb{R}} (\mathcal{L}n^\epsilon)E^\epsilon \overline{E_x^\epsilon} dx &= -2\mathrm{Im}\int_{\mathbb{R}} E_t^\epsilon \overline{E_x^\epsilon} dx\\ &= -\frac{d}{dt}\int_{\mathbb{R}} \mathrm{Im}(E^\epsilon \overline{E_x^\epsilon})dx.\end{aligned}$$

Combining the above two equalities gives (12). For (13), we integrate the second equation of (7) over $(-\infty, x)$ and obtain

$$m_t^\epsilon + \Lambda^{2\alpha} n^\epsilon + \mathcal{L}(\mathcal{L}n^\epsilon)^2 + \mathcal{L}|E^\epsilon|^2 = 0, \tag{15}$$

where $m^\epsilon := \int_{-\infty}^x n^\epsilon(t,y)dy$. Multiplying this equation by n_t^ϵ and noting that

$$\int_{\mathbb{R}} m_t^\epsilon n_t^\epsilon dx = \frac{1}{2}\int_{\mathbb{R}} (m_t^\epsilon)_x^2 dx = 0,$$

then there holds

$$\frac{d}{dt}\left(\frac{1}{2}\|\Lambda^\alpha n^\epsilon\|_{L^2}^2 + \frac{1}{3}\int_{\mathbb{R}} (\mathcal{L}n^\epsilon)^3(t,x)dx\right) + \int_{\mathbb{R}} (\mathcal{L}n^\epsilon)_t |E^\epsilon|^2 dx = 0.$$

On the other hand, multiplying the first equation of (7) by $\overline{E^\epsilon}$ and taking the real part, we have

$$\frac{d}{dt}\left(\frac{1}{2}\|E_x^\epsilon\|_{L^2}^2 + \frac{c}{4}\int_{\mathbb{R}} |E^\epsilon(t,x)|^4 dx\right) + \frac{1}{2}\int_{\mathbb{R}} (\mathcal{L}n^\epsilon)|E^\epsilon|_t^2 dx = 0.$$

Adding these two equalities, we obtain $\frac{d}{dt}\mathcal{H}_2^\epsilon(t) = 0$, so the quantity $\mathcal{H}_2^\epsilon(t)$ is conserved for all time.

Now we prove (14). First, the conservation of $\mathcal{H}_1^\epsilon(t)$ gives

$$\|n^\epsilon\|_{L^2}^2 \le |\mathcal{H}_1^\epsilon(0)| + 2\left|\int_{\mathbb{R}} \mathrm{Im}(E^\epsilon \overline{E_x^\epsilon})dx\right| \le C(1 + \|E_x^\epsilon\|_{L^2}), \tag{16}$$

where C depends on $\|E_0^\epsilon\|_{H^1}$ and $\|n_0^\epsilon\|_{H^\alpha}$. Next, by using (9) and (16), we have

$$\begin{aligned}\left|\frac{1}{3}\int_{\mathbb{R}}(\mathcal{L}n^\epsilon)^3 dx\right| &\le C\|n^\epsilon\|_{L^2}^{3-\frac{1}{2\alpha}}\|\Lambda^\alpha n^\epsilon\|_{L^2}^{\frac{1}{2\alpha}} \le C(1+\|E_x^\epsilon\|_{L^2}^{\frac{3}{2}-\frac{1}{4\alpha}})\|\Lambda^\alpha n^\epsilon\|_{L^2}^{\frac{1}{2\alpha}}\\ &\le \frac{1}{4}\|E_x^\epsilon\|_{L^2}^2 + \frac{1}{4}\|\Lambda^\alpha n^\epsilon\|_{L^2}^2 + C.\end{aligned}$$

where we have used the fact $\alpha > 1/2$ in the last inequality. Similarly, we can obtain

$$\begin{aligned}\left|\int_{\mathbb{R}}\mathcal{L}n^\epsilon|E^\epsilon|^2dx\right| \le C\|n^\epsilon\|_{L^2}\|E^\epsilon\|_{L^4}^2 &\le C(1+\|E_x^\epsilon\|_{L^2}^{\frac{1}{2}})\|E_x^\epsilon\|_{L^2}^{\frac{1}{2}}\\ &\le \frac{1}{4}\|E_x^\epsilon\|_{L^2}^2 + C,\end{aligned}$$

and

$$\frac{|c|}{2}\int_{\mathbb{R}}|E^\epsilon|^4dx \le C\|E^\epsilon\|_{L^2}^3\|E_x^\epsilon\|_{L^2} \le \frac{1}{4}\|E_x^\epsilon\|_{L^2}^2 + C.$$

Inserting these three estimates into the following inequality

$$\begin{aligned}\|E_x^\epsilon\|_{L^2}^2 + \frac{1}{2}\|\Lambda^\alpha n^\epsilon\|_{L^2}^2 \le |\mathcal{H}_2^\epsilon(0)| + \left|\frac{1}{3}\int_{\mathbb{R}}(n^\epsilon)^3dx\right| + \left|\int_{\mathbb{R}} n^\epsilon|E^\epsilon|^2dx\right|\\ +\frac{|c|}{2}\int_{\mathbb{R}}|E^\epsilon|^4dx,\end{aligned}$$

we thus obtain $\|E_x^\epsilon\|_{L^2} + \|\Lambda^\alpha n^\epsilon\|_{L^2} \le C$, which in turn, by (16), gives $\|n^\epsilon\|_{L^2} \le C$. This ends the proof of the lemma. □

Proof of Theorem 2. Since system (7) can be written in the following form

$$\begin{aligned}E^\epsilon &:= U(t)E_0^\epsilon - i\int_0^t U(t-t')(E^\epsilon\mathcal{L}n^\epsilon + c|E^\epsilon|^2E^\epsilon)dt',\\ n^\epsilon &:= V(t)n_0^\epsilon - \int_0^t V(t-t')\mathcal{L}((\mathcal{L}n^\epsilon)_x^2 + |E^\epsilon|_x^2)dt',\end{aligned}$$

where $U(t) = \exp(it\partial_{xx})$ and $V(t) = \exp(-t\Lambda^{2\alpha}\partial_x)$, then by the standard contraction method, it is easy to see that (7) admits a unique solution

(E^ϵ, n^ϵ) locally in time satisfying $(E^\epsilon, n^\epsilon) \in C([-T,T]; H^r \times H^r)$. Moreover, if T^* is the maximal existence time for such solution, then either $T^* = \infty$ or

$$\|E^\epsilon(t)\|_{H^r} + \|n^\epsilon(t)\|_{H^r} \to \infty \text{ as } |t| \to T^*.$$

Hence, in order to finish the proof of Theorem 2, we must show that $T^* = \infty$. To this end, it suffices for us to prove that

$$\|E^\epsilon(t)\|_{H^r} + \|n^\epsilon(t)\|_{H^r} \le C(T^*), \ \forall\, t \in (-T^*, T^*). \tag{17}$$

First, note that the estimate (14) implies that

$$\|E^\epsilon(t)\|_{L^\infty} + \|n^\epsilon(t)\|_{L^\infty} \le C(\|E_0^\epsilon\|_{H^1}, \|n_0^\epsilon\|_{H^\alpha}), \ |t| < T^*. \tag{18}$$

Next, from (7), and with the help of (10) and (18), the H^r-norm energy for (E^ϵ, n^ϵ) can be estimated as follows

$$\frac{d}{dt}(\|E^\epsilon\|_{H^r}^2 + \|n^\epsilon\|_{H^r}^2) \le C(\|E^\epsilon\|_{H^r}^2 + \|n^\epsilon\|_{H^r}^2).$$

Thus, (17) follows from the above inequality together with Gronwall's inequality. The proof of Theorem 2 is complete. □

3. Uniform Estimates for (E^ϵ, n^ϵ)

In order to obtain uniform estimates for (E^ϵ, n^ϵ), we need the following three calculus inequalities.

Lemma 4. *Let $s > 0$ and $1 < p < \infty$. If $u,\ v \in \mathcal{S}(\mathbb{R}^d)$, the Schwarz class, then*

$$\|\Lambda^s(uv) - u\Lambda^s v\|_{L^p} \le C(\|\nabla u\|_{L^{p_1}}\|\Lambda^{s-1}v\|_{L^{p_2}} + \|\Lambda^s u\|_{L^{p_3}}\|v\|_{L^{p_4}}) \tag{19}$$

with $\frac{1}{p_1} + \frac{1}{p_2} = \frac{1}{p_3} + \frac{1}{p_4} = \frac{1}{p}$ and $p_2, p_3 \in (1,\infty)$.

Lemma 5. *Let $0 < s < 1$, $u,\ v \in \mathcal{S}(\mathbb{R})$, then there holds that*

$$\|\Lambda^s(uv) - (\Lambda^s u)v - u(\Lambda^s v)\|_{L^p} \le C\|\Lambda^{s/2}u\|_{L^{p_1}}\|\Lambda^{s/2}v\|_{L^{p_2}}$$

with $p, p_1, p_2 \in (1,\infty)$ satisfying $\frac{1}{p} = \frac{1}{p_1} + \frac{1}{p_2}$.

Lemma 6. *Let $0 < s < 1$, $u,\ v \in \mathcal{S}(\mathbb{R})$, then there holds that*

$$\begin{aligned}\|\Lambda^s(uv)_{xx} &- (\Lambda^s u_{xx})v - u(\Lambda^s v_{xx}) - (s+2)(\Lambda^s u_x)v_x \\ &- (s+2)u_x(\Lambda^s v_x)\|_{L^p} \le C\|\Lambda^{\frac{2+s}{2}}u\|_{L^{p_1}}\|\Lambda^{\frac{2+s}{2}}v\|_{L^{p_2}}\end{aligned}$$

with $p, p_1, p_2 \in (1,\infty)$ satisfying $\frac{1}{p} = \frac{1}{p_1} + \frac{1}{p_2}$.

The proof of Lemma 4 can be found in Refs. 3, 9, 10. Using Coifman–Meyer multiplier theorem, it is not hard to obtain Lemma 5 and Lemma 6.

Proposition 1. *Assume that* $(E_0, n_0) \in H^1 \times H^\alpha$ *and* $(E^\epsilon, n^\epsilon) \in C(\mathbb{R}; H^k \times H^k)$ *is the solution of the regularized system* (7). *Then for* $t \in \mathbb{R}$, *there holds that*

$$\|E^\epsilon(t)\|_{H^1} + \|n^\epsilon(t)\|_{H^\alpha} \leq C \tag{20}$$

with C *not depending on* ϵ.

The estimate (20) follows immediately from (8) and (14). In particular, by Sobolev embedding theorem, (20) gives $\|E^\epsilon(t)\|_{L^\infty} + \|n^\epsilon(t)\|_{L^\infty} \leq C$. Now we give the key estimate for the approximate solution (E^ϵ, n^ϵ).

Proposition 2. *Suppose that* $(E_0, n_0) \in H^2 \times H^{1+\alpha}$ *and* (E^ϵ, n^ϵ) *is the solution obtained in Theorem 2. Then for all* $T > 0$, *there holds that*

$$\|E^\epsilon(t)\|_{H^2} + \|n^\epsilon(t)\|_{H^{1+\alpha}} \leq C(T, \|E_0\|_{H^2}, \|n_0\|_{H^{1+\alpha}}). \tag{21}$$

In particular, the constant C *in* (21) *is independent of* ϵ.

Proof. For the sake of simplicity, we will omit the superscript ϵ in this proof. From the first equation of (7), we can obtain

$$\frac{d}{dt}\|E_{xx}\|_{L^2}^2 = 2\mathrm{Im}\int_{\mathbb{R}} (\mathcal{L}nE)_{xx}\overline{E_{xx}}dx + 2c\mathrm{Im}\int_{\mathbb{R}} (|E|^2E)_{xx}\overline{E_{xx}}dx.$$

By using (20), we can get

$$2c\mathrm{Im}\int_{\mathbb{R}} (|E|^2E)_{xx}\overline{E_{xx}}dx \leq C\|E_{xx}\|_{L^2}^2$$

and

$$\begin{aligned}
2\mathrm{Im}\int_{\mathbb{R}} (\mathcal{L}nE)_{xx}\overline{E_{xx}}dx &= 2\mathrm{Im}\int_{\mathbb{R}} \mathcal{L}n_{xx}E\overline{E_{xx}}dx + 4\mathrm{Im}\int_{\mathbb{R}} \mathcal{L}n_xE_x\overline{E_{xx}}dx \\
&\leq 2\mathrm{Im}\int_{\mathbb{R}} \mathcal{L}n_{xx}E\overline{E_{xx}}dx + C\|n_x\|_{L^\infty}\|E_x\|_{L^2}\|E_{xx}\|_{L^2} \\
&\leq 2\mathrm{Im}\int_{\mathbb{R}} \mathcal{L}n_{xx}E\overline{E_{xx}}dx + C\|n\|_{H^{1+\alpha}}\|E_{xx}\|_{L^2}.
\end{aligned}$$

On the other hand, multiplying (15) by $\Lambda^2 n_t = -n_{xxt}$ and integrating over $\mathbb{R}$, then we can obtain

$$\begin{aligned}\frac{1}{2}\frac{d}{dt}\|\Lambda^{1+\alpha}n\|_{L^2}^2 &= \int_{\mathbb{R}}(\mathcal{L}n)^2\mathcal{L}n_{xxt}dx + \int_{\mathbb{R}}|E|^2\mathcal{L}n_{xxt}dx \\ &= -\frac{d}{dt}\left(\int_{\mathbb{R}}\mathcal{L}n(\mathcal{L}n_x)^2dx + \int_{\mathbb{R}}|E|_x^2\mathcal{L}n_xdx\right) \\ &\quad + \int_{\mathbb{R}}\mathcal{L}n_t(\mathcal{L}n_x)^2dx - \int_{\mathbb{R}}|E|_t^2\mathcal{L}n_{xx}dx,\end{aligned}$$

which gives

$$\begin{aligned}&\frac{d}{dt}\left(\frac{1}{2}\|\Lambda^{1+\alpha}n\|_{L^2}^2 + \int_{\mathbb{R}}\mathcal{L}n(\mathcal{L}n_x)^2dx + \int_{\mathbb{R}}|E|_x^2\mathcal{L}n_xdx\right) \\ &\quad = \int_{\mathbb{R}}\mathcal{L}n_t(\mathcal{L}n_x)^2dx + 2\mathrm{Im}\int_{\mathbb{R}}E_{xx}\overline{E}\mathcal{L}n_{xx}dx,\end{aligned}$$

where we have used the fact $|E|_t^2 = 2\mathrm{Im}(iE_t\overline{E}) = -2\mathrm{Im}(E_{xx}\overline{E})$ in the above equality. Combining the above equalities and noting that

$$\mathrm{Im}\int_{\mathbb{R}}\mathcal{L}n_{xx}E\overline{E_{xx}}dx + \mathrm{Im}\int_{\mathbb{R}}E_{xx}\overline{E}\mathcal{L}n_{xx}dx = 0,$$

then there holds

$$\begin{aligned}&\frac{d}{dt}\left(\|E_{xx}\|_{L^2}^2 + \frac{1}{2}\|\Lambda^{1+\alpha}n\|_{L^2}^2 + \int_{\mathbb{R}}\mathcal{L}n(\mathcal{L}n_x)^2dx + \int_{\mathbb{R}}|E|_x^2\mathcal{L}n_xdx\right) \\ &\qquad \le C(1 + \|\Lambda^{1+\alpha}n\|_{L^2}^2 + \|E_{xx}\|_{L^2}) + \int_{\mathbb{R}}\mathcal{L}n_t(\mathcal{L}n_x)^2dx. \qquad (22)\end{aligned}$$

Now we should estimate the last term in (22). Recall that

$$n_t + \Lambda^{2\alpha}n_x + \mathcal{L}(\mathcal{L}n)_x^2 + \mathcal{L}|E|_x^2 = 0. \qquad (23)$$

Using (23), we have

$$\begin{aligned}\int_{\mathbb{R}}\mathcal{L}n_t(\mathcal{L}n_x)^2dx &= -\int_{\mathbb{R}}\mathcal{L}\Lambda^{2\alpha}n_x(\mathcal{L}n_x)^2dx - \int_{\mathbb{R}}\mathcal{L}^2(\mathcal{L}n)_x^2(\mathcal{L}n_x)^2dx \\ &\quad - \int_{\mathbb{R}}\mathcal{L}^2|E|_x^2(\mathcal{L}n_x)^2dx \\ &=: I_1 + I_2 + I_3. \qquad (24)\end{aligned}$$

By (20) and Sobolev inequalities, I_2 and I_3 can be estimated as follows

$$\begin{aligned}I_3 &\le \||E|_x^2\|_{L^2}\|n_x\|_{L^4}^2 \le C\|n\|_{H^{1+\alpha}}^2 \le C(1 + \|\Lambda^{1+\alpha}n\|_{L^2}^2), \\ I_2 &\le 2\|n\|_{L^\infty}\|n_x\|_{L^2}\|n_x\|_{L^4}^2 \le C\|\Lambda^{\alpha}n\|_{L^2}^{3\alpha-\frac{1}{2}}\|\Lambda^{1+\alpha}n\|_{L^2}^{\frac{7}{2}-3\alpha} \qquad (25) \\ &\le C(1 + \|\Lambda^{1+\alpha}n\|_{L^2}^2),\end{aligned}$$

where the last inequality needs the condition $\alpha \geq 1/2$. For the term I_1, we split it into two terms:

$$I_1 = -\int_{\mathbb{R}} \mathcal{L}\Lambda^\alpha n_x[\Lambda^\alpha(\mathcal{L}n_x)^2 - 2\mathcal{L}n_x\Lambda^\alpha\mathcal{L}n_x]dx - 2\int_{\mathbb{R}}(\mathcal{L}\Lambda^\alpha n_x)^2\mathcal{L}n_x dx.$$

Applying Lemma 5 to the first term in the right-hand side of the above equality, then

$$\begin{aligned}
I_1 &\leq C\|\Lambda^{1+\alpha}n\|_{L^2}\|\Lambda^{\alpha/2}n_x\|_{L^4}^2 - 2\int_{\mathbb{R}}(\mathcal{L}\Lambda^\alpha n_x)^2\mathcal{L}n_x dx\\
&\leq C\|\Lambda^{1+\alpha}n\|_{L^2}\|\Lambda^\alpha n\|_{L^2}^{\alpha-\frac{1}{2}}\|\Lambda^{1+\alpha}n\|_{L^2}^{\frac{3}{2}-\alpha} - 2\int_{\mathbb{R}}(\mathcal{L}\Lambda^\alpha n_x)^2\mathcal{L}n_x dx \qquad (26)\\
&\leq C(1+\|\Lambda^{1+\alpha}n\|_{L^2}^2) - 2\int_{\mathbb{R}}(\mathcal{L}\Lambda^\alpha n_x)^2\mathcal{L}n_x dx.
\end{aligned}$$

Denote

$$\Phi(t) := \|E_{xx}\|_{L^2}^2 + \frac{1}{2}\|\Lambda^{1+\alpha}n\|_{L^2}^2 + \int_{\mathbb{R}}\mathcal{L}n(\mathcal{L}n_x)^2dx + \int_{\mathbb{R}}|E|_x^2\mathcal{L}n_x dx,$$

then it follows from (22) and (24)–(26) that

$$\frac{d}{dt}\Phi(t) \leq C(1+\|\Lambda^{1+\alpha}n\|_{L^2}^2 + \|E_{xx}\|_{L^2}) - 2\int_{\mathbb{R}}(\mathcal{L}\Lambda^\alpha n_x)^2\mathcal{L}n_x dx. \qquad (27)$$

Note that the last term in (27) can not be controlled since the L^∞-norm estimate for n_x is unknown. In order to cancel this term, we introduce the term $\gamma\int_{\mathbb{R}}\mathcal{L}n(\mathcal{L}n_x)^2dx$, where the value of γ will be given later. It is easy to see that

$$\begin{aligned}
\frac{d}{dt}\gamma\int_{\mathbb{R}}\mathcal{L}n(\mathcal{L}n_x)^2dx =& 2\gamma\int_{\mathbb{R}}\mathcal{L}n\mathcal{L}n_x\mathcal{L}n_{xt}dx + \gamma\int_{\mathbb{R}}\mathcal{L}n_t(\mathcal{L}n_x)^2dx\\
=& -\gamma\int_{\mathbb{R}}(\mathcal{L}n)_x^2\mathcal{L}\Lambda^{2\alpha}n_{xx}dx - \gamma\int_{\mathbb{R}}(\mathcal{L}n)_x^2\mathcal{L}^2(\mathcal{L}n)_{xx}^2dx\\
&-\gamma\int_{\mathbb{R}}(\mathcal{L}n)_x^2\mathcal{L}^2|E|_{xx}^2dx + \gamma(I_1+I_2+I_3). \qquad (28)
\end{aligned}$$

From (25) and (26), we have

$$\gamma(I_1+I_2+I_3) \leq C(1+\|\Lambda^{1+\alpha}n\|_{L^2}^2) - 2\gamma\int_{\mathbb{R}}(\mathcal{L}\Lambda^\alpha n_x)^2\mathcal{L}n_x dx. \qquad (29)$$

Note that

$$\begin{aligned}
\left|\gamma\int_{\mathbb{R}}(\mathcal{L}n)_x^2\mathcal{L}^2|E|_{xx}^2dx\right| &\leq C\|n_x\|_{L^2}\|E_{xx}\|_{L^2}\\
&\leq C(1+\|\Lambda^{1+\alpha}n\|_{L^2}^2 + \|E_{xx}\|_{L^2}^2), \qquad (30)
\end{aligned}$$

$$\gamma\int_{\mathbb{R}}(\mathcal{L}n)_x^2\mathcal{L}^2(\mathcal{L}n)_{xx}^2dx = \frac{\gamma}{2}\int_{\mathbb{R}}\left(\mathcal{L}(\mathcal{L}n)_x^2\right)_x^2dx = 0.$$

So it remains to estimate the first term in the right-hand side of (28). In fact, we have

$$\begin{aligned}
-\gamma\int_{\mathbb{R}}(\mathcal{L}n)_x^2\mathcal{L}\Lambda^{2\alpha}n_{xx}dx &= \gamma\int_{\mathbb{R}}\Lambda^{\alpha}(\mathcal{L}n)_{xx}^2\mathcal{L}\Lambda^{\alpha}n_x dx\\
&= \gamma\int_{\mathbb{R}}[\Lambda^{\alpha}(\mathcal{L}n)_{xx}^2 - 2\mathcal{L}n\Lambda^{\alpha}(\mathcal{L}n_{xx}) - 2(\alpha+2)\mathcal{L}n_x\Lambda^{\alpha}(\mathcal{L}n_x)]\mathcal{L}\Lambda^{\alpha}n_x dx\\
&\quad + 2\gamma\int_{\mathbb{R}}\mathcal{L}n\Lambda^{\alpha}\mathcal{L}n_{xx}\mathcal{L}\Lambda^{\alpha}n_x dx + 2(\alpha+2)\gamma\int_{\mathbb{R}}\mathcal{L}n_x\Lambda^{\alpha}\mathcal{L}n_x\mathcal{L}\Lambda^{\alpha}n_x dx\\
&=: I_4 + I_5 + I_6.
\end{aligned} \tag{31}$$

For the term I_4, we use Lemma 6 with $u = v = \mathcal{L}n$, and for the term I_5, we integrate it by parts, then we can obtain

$$\begin{aligned}
I_4 &\le C\|\Lambda^{\frac{2+\alpha}{2}}\mathcal{L}n\|_{L^4}^2\|\mathcal{L}\Lambda^{\alpha}n_x\|_{L^2} \le C\|\Lambda^{\alpha}n\|_{L^2}^{\alpha-\frac{1}{2}}\|\Lambda^{1+\alpha}n\|_{L^2}^{\frac{3}{2}-\alpha}\|\mathcal{L}\Lambda^{\alpha}n_x\|_{L^2}\\
&\le C(1+\|\mathcal{L}\Lambda^{1+\alpha}n\|_{L^2}^2),\\
I_5 &= \gamma\int_{\mathbb{R}}\mathcal{L}n(\Lambda^{\alpha}\mathcal{L}n_x)_x^2 dx = -\gamma\int_{\mathbb{R}}\mathcal{L}n_x(\Lambda^{\alpha}\mathcal{L}n_x)^2 dx.
\end{aligned}$$

Inserting the above two estimates into (31), one sees

$$\begin{aligned}
-\gamma\int_{\mathbb{R}}(\mathcal{L}n)_x^2\mathcal{L}\Lambda^{2\alpha}n_{xx}dx \le& C(1+\|\mathcal{L}\Lambda^{1+\alpha}n\|_{L^2}^2)\\
&+ (2\alpha+3)\gamma\int_{\mathbb{R}}\mathcal{L}n_x(\Lambda^{\alpha}\mathcal{L}n_x)^2 dx.
\end{aligned} \tag{32}$$

So one infers from (28)–(30) and (32) that

$$\begin{aligned}
\frac{d}{dt}\gamma\int_{\mathbb{R}}\mathcal{L}n(\mathcal{L}n_x)^2 dx \le& C(1+\|\Lambda^{1+\alpha}n\|_{L^2}^2 + \|E_{xx}\|_{L^2}^2)\\
&+ (2\alpha+1)\gamma\int_{\mathbb{R}}\mathcal{L}n_x(\Lambda^{\alpha}\mathcal{L}n_x)^2 dx.
\end{aligned} \tag{33}$$

Now we choose $\gamma = \frac{2}{1+2\alpha}$ and set

$$\widetilde{\Phi}(t) := \|E_{xx}\|_{L^2}^2 + \frac{1}{2}\|\Lambda^{1+\alpha}n\|_{L^2}^2 + (1+\gamma)\int_{\mathbb{R}}\mathcal{L}n(\mathcal{L}n_x)^2 dx + \int_{\mathbb{R}}|E|_x^2\mathcal{L}n_x dx,$$

then from (27) and (33), we have

$$\frac{d}{dt}\widetilde{\Phi}(t) \le C(1+\|\Lambda^{1+\alpha}n\|_{L^2}^2 + \|E_{xx}\|_{L^2}). \tag{34}$$

Meanwhile, by interpolation inequalities, it is easy to see that

$$\left|(1+\gamma)\int_{\mathbb{R}}\mathcal{L}n(\mathcal{L}n_x)^2 dx\right| \le C\|\mathcal{L}n\|_{L^\infty}\|\mathcal{L}n_x\|_{L^2}^2 \le \frac{1}{8}\|\Lambda^{1+\alpha}n\|_{L^2}^2 + C \tag{35}$$

and

$$\left|\int_{\mathbb{R}} |E|_x^2 \mathcal{L} n_x dx\right| \le C\|E\|_{L^\infty}\|E_x\|_{L^2}\|\mathcal{L}n_x\|_{L^2}$$
$$\le \frac{1}{8}\|\Lambda^{1+\alpha}n\|_{L^2}^2 + C. \tag{36}$$

From (34)–(36), the desired estimate (21) follows immediately by using Gronwall's inequality. □

By Sobolev embedding theorem, Proposition 3 gives that

$$\|E_x(t)\|_{L^\infty} + \|n_x(t)\|_{L^\infty} \le C,\ t \in [-T, T].$$

Based on this estimate, one can derive the higher order estimates for (E^ϵ, n^ϵ).

Proposition 3. *Assume that $(E_0, n_0) \in H^{k+1} \times H^{k+\alpha}$ with $k \ge 2$ be an integer and (E^ϵ, n^ϵ) is the smooth solution of the regularized system (7). Then for all $T > 0$, there holds that*

$$\|E^\epsilon(t)\|_{H^{k+1}} + \|n^\epsilon(t)\|_{H^{k+\alpha}} \le C(T, \|E_0\|_{H^{k+1}}, \|n_0\|_{H^{k+\alpha}}) \tag{37}$$

with C not depending on ϵ.

Proof. We only prove (37) in the case $k = 2$, since the argument for general k is the same. For the sake of convenience, we again omit the superscript ϵ in the following arguments.

First, the equation for E gives

$$\frac{d}{dt}\|\partial_x^3 E\|_{L^2}^2 = 2\mathrm{Im}\int_{\mathbb{R}} \partial_x^3(\mathcal{L}nE)\overline{\partial_x^3 E}dx + 2c\mathrm{Im}\int_{\mathbb{R}} \partial_x^3(|E|^2E)\overline{\partial_x^3 E}dx.$$

Using (21) and interpolation inequalities in Sobolev space, we can obtain

$$\left|2c\mathrm{Im}\int_{\mathbb{R}} \partial_x^3(|E|^2E)\overline{\partial_x^3 E}dx\right| \le C(1 + \|\partial_x^3 E\|_{L^2}^2)$$

and

$$2\mathrm{Im}\int_{\mathbb{R}} \partial_x^3(\mathcal{L}nE)\overline{\partial_x^3 E}dx$$
$$\le 2\mathrm{Im}\int_{\mathbb{R}} E\partial_x^3(\mathcal{L}n)\overline{\partial_x^3 E}dx + C(1 + \|\Lambda^{2+\alpha}n\|_{L^2}^2 + \|\partial_x^3 E\|_{L^2}^2).$$

Hence, there holds

$$\frac{d}{dt}\|\partial_x^3 E\|_{L^2}^2 \le 2\mathrm{Im}\int_{\mathbb{R}} E\partial_x^3(\mathcal{L}n)\overline{\partial_x^3 E}dx + C(1 + \|\Lambda^{2+\alpha}n\|_{L^2}^2 + \|\partial_x^3 E\|_{L^2}^2). \tag{38}$$

Next, we multiply (15) by $\Lambda^4 n_t = \partial_x^4 n_t$ and obtain

$$\begin{aligned}\frac{1}{2}\frac{d}{dt}\|\Lambda^{2+\alpha}n\|_{L^2}^2 =& -\int_{\mathbb{R}}(\mathcal{L}n)^2\Lambda^4\mathcal{L}n_t dx - \int_{\mathbb{R}}|E|^2\mathcal{L}\partial_x^4 n_t dx\\ =& -\frac{d}{dt}\left(\int_{\mathbb{R}}(\mathcal{L}n)^2\Lambda^4\mathcal{L}n dx - \int_{\mathbb{R}}|E|_x^2\partial_x^3\mathcal{L}n dx\right)\\ &+2\int_{\mathbb{R}}\mathcal{L}n\mathcal{L}n_t\Lambda^4\mathcal{L}n dx - \int_{\mathbb{R}}|E|_{xt}^2\mathcal{L}n_{xxx}dx. \qquad (39)\end{aligned}$$

Using (21) and the fact $|E|_t^2 = -2\mathrm{Im}(E_{xx}\overline{E})$, so there holds

$$\begin{aligned}-\int_{\mathbb{R}}|E|_{xt}^2\mathcal{L}n_{xxx}dx =& 2\mathrm{Im}\int_{\mathbb{R}}E_{xxx}\overline{E}\mathcal{L}n_{xxx}dx - 2\mathrm{Im}\int_{\mathbb{R}}E_{xxx}\overline{E}_x\mathcal{L}n_{xx}dx\\ \le& 2\mathrm{Im}\int_{\mathbb{R}}E_{xxx}\overline{E}\mathcal{L}n_{xxx}dx\\ &+C(1+\|\partial_x^3 E\|_{L^2}^2 + \|\Lambda^{2+\alpha}n\|_{L^2}^2). \qquad (40)\end{aligned}$$

And using (23), we have

$$\begin{aligned}2\int_{\mathbb{R}}\mathcal{L}n\mathcal{L}n_t\Lambda^4\mathcal{L}n dx =& -2\int_{\mathbb{R}}\mathcal{L}n\Lambda^{2\alpha}\mathcal{L}n_x\Lambda^4\mathcal{L}n dx - 2\int_{\mathbb{R}}\mathcal{L}n\mathcal{L}^2(\mathcal{L}n)_x^2\Lambda^4\mathcal{L}n dx\\ &-2\int_{\mathbb{R}}\mathcal{L}n\mathcal{L}^2(|E|_x^2)\Lambda^4\mathcal{L}n dx\\ =:& J_1+J_2+J_3.\end{aligned}$$

For the term J_1, we rewrite it as follows

$$\begin{aligned}J_1 =& -2\int_{\mathbb{R}}\Lambda^{2-\alpha}[\mathcal{L}n\Lambda^{2\alpha}\mathcal{L}n_x]\Lambda^{2+\alpha}\mathcal{L}n dx\\ =& -2\int_{\mathbb{R}}[\Lambda^{2-\alpha}(\mathcal{L}n\Lambda^{2\alpha}\mathcal{L}n_x) - \mathcal{L}n\Lambda^{2+\alpha}\mathcal{L}n_x]\Lambda^{2+\alpha}\mathcal{L}n dx\\ &-2\int_{\mathbb{R}}\mathcal{L}n\Lambda^{2+\alpha}\mathcal{L}n_x\Lambda^{2+\alpha}\mathcal{L}n dx\\ =& -2\int_{\mathbb{R}}[\Lambda^{2-\alpha}(\mathcal{L}n\Lambda^{2\alpha}\mathcal{L}n_x) - \mathcal{L}n\Lambda^{2+\alpha}\mathcal{L}n_x]\Lambda^{2+\alpha}\mathcal{L}n dx\\ &+\int_{\mathbb{R}}\mathcal{L}n_x(\Lambda^{2+\alpha}\mathcal{L}n)^2 dx.\end{aligned}$$

Applying Lemma 4 and the estimate (21), then

$$\begin{aligned}|J_1| \le& C(\|\mathcal{L}n_x\|_{L^\infty}\|\Lambda^{2+\alpha}\mathcal{L}n\|_{L^2} + \|\Lambda^{2-\alpha}\mathcal{L}n\|_{L^p}\|\Lambda^{1+2\alpha}\mathcal{L}n\|_{L^q})\|\Lambda^{2+\alpha}\mathcal{L}n\|_{L^2}\\ &+\|\mathcal{L}n_x\|_{L^\infty}\|\Lambda^{2+\alpha}\mathcal{L}n\|_{L^2}^2\\ \le& C\|\Lambda^{2+\alpha}n\|_{L^2}^2 + C,\end{aligned}$$

where p and q satisfies $\frac{1}{p}+\frac{1}{q}=\frac{1}{2}$ with $p<\infty$ such that

$$\|\Lambda^{2-\alpha}\mathcal{L}n\|_{L^p}\leq C\|n\|_{H^{1+\alpha}},\ \|\Lambda^{1+2\alpha}\mathcal{L}n\|_{L^q}\leq C\|n\|_{H^{2+\alpha}}.$$

For the terms J_2 and J_3, we use (10) and (21), then there holds

$$|J_2|+|J_3|\leq C(\|\Lambda^{2+\alpha}\mathcal{L}n\|_{L^2}^2+\|\partial_x^3E\|_{L^2}^2)+C.$$

Combining the estimates for J_1, J_2 and J_3, then we obtain

$$2\int_{\mathbb{R}}\mathcal{L}n\mathcal{L}n_t\Lambda^4\mathcal{L}ndx\leq C(\|\Lambda^{2+\alpha}n\|_{L^2}^2+\|\partial_x^3E\|_{L^2}^2)+C. \tag{41}$$

Now we collect the estimates (38)–(41) and obtain

$$\frac{d}{dt}\Psi(t)\leq C(\|\Lambda^{2+\alpha}n\|_{L^2}^2+\|\partial_x^3E\|_{L^2}^2)+C, \tag{42}$$

where $\Psi(t)$ is defined as follows

$$\Psi(t)=\|\partial_x^3E\|_{L^2}^2+\frac{1}{2}\|\Lambda^{2+\alpha}n\|_{L^2}^2+\int_{\mathbb{R}}(\mathcal{L}n)^2\Lambda^4\mathcal{L}ndx-\int_{\mathbb{R}}|E|_x^2\partial_x^3\mathcal{L}ndx.$$

Notice that

$$\left|\int_{\mathbb{R}}(\mathcal{L}n)^2\Lambda^4\mathcal{L}ndx\right|\leq\|\Lambda^{2+\alpha}\mathcal{L}n\|_{L^2}\|\Lambda^{2-\alpha}(\mathcal{L}n)^2\|_{L^2}\leq\frac{1}{4}\|\Lambda^{2+\alpha}\mathcal{L}n\|_{L^2}^2+C,$$

$$\left|\int_{\mathbb{R}}|E|_x^2\partial_x^3\mathcal{L}ndx\right|\leq\|\partial_x^3|E|^2\|_{L^2}\|\mathcal{L}n_x\|_{L^2}\leq\frac{1}{2}\|\partial_x^3E\|_{L^2}^2+C.$$

Integrating (42) and using the above two estimates, (37) thus follows in the case $k=2$. When $k\geq 3$, we can prove similarly, and the further details are omitted here. □

4. Existence and Uniqueness of the Solution

In this section, we present the proof of Theorem 1. Based on the estimates obtained in Section 3, letting $\epsilon\to 0$ in (7) and using the standard compact arguments, the existence part of Theorem 1 can be easily proved, and these details are omitted here. In the following, we prove the uniqueness.

Assume $(E^{(1)},n^{(1)})$ and $(E^{(2)},n^{(2)})$ are two solutions of (6) with the same initial data, and we denote $E=E^{(1)}-E^{(2)}$, $n=n^{(1)}-n^{(2)}$. From (6), one can see that (E,n) solves

$$iE_t+E_{xx}=n^{(1)}E+nE^{(2)}+c(|E^{(1)}|^2+|E^{(2)}|^2)E+cE^{(1)}E^{(2)}\overline{E}, \tag{43}$$

$$n_t+\Lambda^{2\alpha}n_x+(nn^{(1)}+nn^{(2)})_x+(E^{(1)}\overline{E}+E\overline{E^{(2)}})_x=0 \tag{44}$$

with initial data $E(0) = 0$ and $n(0) = 0$. Eq. (43) gives the L^2-norm estimate for E:

$$\begin{aligned}\frac{d}{dt}\|E\|_{L^2}^2 &= 2\mathrm{Im}\int_{\mathbb{R}} nE^{(2)}\overline{E}dx + 2c\mathrm{Im}\int_{\mathbb{R}} E^{(1)}E^{(2)}(\overline{E})^2dx \\ &\le C\|E^{(2)}\|_{L^\infty}\|n\|_{L^2}\|\overline{E}\|_{L^2} + C\|E^{(1)}E^{(2)}\|_{L^\infty}\|\overline{E}\|_{L^2}^2 \\ &\le C(\|E\|_{L^2}^2 + \|n\|_{L^2}^2). \end{aligned} \tag{45}$$

And Eq. (44) gives the L^2-norm estimate for n:

$$\frac{d}{dt}\|n\|_{L^2}^2 = -2\int_{\mathbb{R}}(nn^{(1)} + nn^{(2)})_x n dx - 2\int_{\mathbb{R}}(E^{(1)}\overline{E} + E\overline{E^{(2)}})_x n dx.$$

Note that

$$\begin{aligned}\left|2\int_{\mathbb{R}}(nn^{(1)} + nn^{(2)})_x n dx\right| &= \left|\int_{\mathbb{R}}(n^{(1)} + n^{(2)})_x n^2 dx\right| \\ &\le C\|n_x^{(1)} + n_x^{(2)}\|_{L^\infty}\|n\|_{L^2}^2 \le C\|n\|_{L^2}^2,\end{aligned}$$

and

$$\begin{aligned}\left|2\int_{\mathbb{R}}(E^{(1)}\overline{E} + E\overline{E^{(2)}})_x n dx\right| &\le C(\|E^{(1)}\|_{H^1} + \|E^{(2)}\|_{H^1})\|E\|_{H^1}\|n\|_{L^2} \\ &\le C(\|n\|_{L^2}^2 + \|E\|_{H^1}^2).\end{aligned}$$

Hence, the above two estimates give

$$\frac{d}{dt}\|n\|_{L^2}^2 \le C(\|n\|_{L^2}^2 + \|E\|_{H^1}^2). \tag{46}$$

From (45) and (46), we have

$$\|E\|_{L^2}^2 + \|n\|_{L^2}^2 \le C\int_0^t (\|n\|_{L^2}^2 + \|E\|_{H^1}^2)ds. \tag{47}$$

Now we estimate $\|E_x\|_{L^2}$. Since

$$n^{(1)}E^{(1)} - n^{(2)}E^{(2)} = n^{(1)}E + nE^{(2)} = nE^{(1)} + n^{(2)}E,$$

we rewrite (43) as follows:

$$\begin{aligned} iE_t + E_{xx} = f =: &\frac{1}{2}(n^{(1)}E + nE^{(2)}) + \frac{1}{2}(nE^{(1)} + n^{(2)}E) \\ &+ c(|E^{(1)}|^2 + |E^{(2)}|^2)E + cE^{(1)}E^{(2)}\overline{E}. \end{aligned} \tag{48}$$

For simplicity, we let $g := c(|E^{(1)}|^2 + |E^{(2)}|^2)E + cE^{(1)}E^{(2)}\overline{E}$. From (48), one can obtain

$$\begin{aligned}\frac{d}{dt}\|E_x\|_{L^2}^2 =&\mathrm{Im}\int_{\mathbb{R}}(n_x^{(1)}E + n_xE^{(2)} + nE_x^{(2)})\overline{E_x}dx\\&+\mathrm{Im}\int_{\mathbb{R}}(n_xE^{(1)} + nE_x^{(1)} + n_x^{(2)}E)\overline{E_x}dx + 2\mathrm{Im}\int_{\mathbb{R}} g_x\overline{E_x}dx\\\leq&\mathrm{Im}\int_{\mathbb{R}}(E^{(1)}n_x\overline{E_x} + E^{(2)}n_x\overline{E_x})dx + C(\|n\|_{L^2}^2 + \|E\|_{H^1}^2). \quad (49)\end{aligned}$$

In order to cancel the integral term in (49), we should estimate $\|\Lambda^\alpha n\|_{L^2}$. To this end, we first integrate the equation (44) and get (note that the function $E^{(1)}\overline{E} + E\overline{E^{(2)}}$ is real-valued)

$$N_t + \Lambda^{2\alpha}n + n(n^{(1)} + n^{(2)}) + \mathrm{Re}(E^{(1)}\overline{E} + E\overline{E^{(2)}}) = 0,\ N := \int_{-\infty}^{x} n(t,y)dy. \quad (50)$$

Multiplying (50) by n_t, and noticing that $\int_{\mathbb{R}} N_t n_t dx = \int_{\mathbb{R}} N_t N_{xt} dx = 0$, then there holds

$$\frac{1}{2}\frac{d}{dt}\|\Lambda^\alpha n\|_{L^2}^2 = -\int_{\mathbb{R}} n(n^{(1)} + n^{(2)})n_t dx - \mathrm{Re}\int_{\mathbb{R}}(E^{(1)}\overline{E} + E\overline{E^{(2)}})n_t dx,$$

from which we can get

$$\begin{aligned}&\frac{d}{dt}\left(\frac{1}{2}\|\Lambda^\alpha n\|_{L^2}^2 + \frac{1}{2}\int_{\mathbb{R}} n^2(n^{(1)} + n^{(2)})dx + \mathrm{Re}\int_{\mathbb{R}}(E^{(1)}\overline{E} + E\overline{E^{(2)}})ndx\right)\\&= \frac{1}{2}\int_{\mathbb{R}} n^2(n_t^{(1)} + n_t^{(2)})dx + \mathrm{Re}\int_{\mathbb{R}}(E_t^{(1)}\overline{E} + E\overline{E_t^{(2)}})ndx\\&\quad + \mathrm{Re}\int_{\mathbb{R}} E^{(1)}\overline{E_t}ndx + \mathrm{Re}\int_{\mathbb{R}} E_t\overline{E^{(2)}}ndx\\&\leq C(\|n\|_{H^\alpha}^2 + \|E\|_{H^1}^2) + \mathrm{Re}\int_{\mathbb{R}} E^{(1)}\overline{E_t}ndx + \mathrm{Re}\int_{\mathbb{R}} E_t\overline{E^{(2)}}ndx. \quad (51)\end{aligned}$$

So it remains for us to estimate the last two terms in (51). The equation (48) gives that $E_t = iE_{xx} - if$, hence, one has

$$\begin{aligned}\mathrm{Re}\int_{\mathbb{R}} E^{(1)}\overline{E_t}ndx &= \mathrm{Im}\int_{\mathbb{R}} E^{(1)}\overline{E_{xx}}ndx + \mathrm{Re}\int_{\mathbb{R}} E^{(1)}(i\bar{f})ndx\\&= -\mathrm{Im}\int_{\mathbb{R}} E_x^{(1)}\overline{E_x}ndx - \mathrm{Im}\int_{\mathbb{R}} E^{(1)}\overline{E_x}n_xdx\\&\quad + \mathrm{Re}\int_{\mathbb{R}} E^{(1)}(i\bar{f})ndx\\&\leq -\mathrm{Im}\int_{\mathbb{R}} E^{(1)}\overline{E_x}n_xdx + C(\|n\|_{H^\alpha}^2 + \|E\|_{H^1}^2). \quad (52)\end{aligned}$$

With the same argument as above, we have

$$\begin{aligned}\operatorname{Re}\int_{\mathbb{R}} E_t\overline{E^{(2)}}n\,dx &= -\operatorname{Im}\int_{\mathbb{R}} E_{xx}\overline{E^{(2)}}n\,dx - \operatorname{Re}\int_{\mathbb{R}}(if)\overline{E^{(2)}}n\,dx \\ &= \operatorname{Im}\int_{\mathbb{R}} E_x\overline{E^{(2)}_x}n\,dx + \operatorname{Im}\int_{\mathbb{R}} E_x\overline{E^{(2)}}n_x\,dx \\ &\quad - \operatorname{Re}\int_{\mathbb{R}}(if)\overline{E^{(2)}}n\,dx \\ &\le \operatorname{Im}\int_{\mathbb{R}} E_x\overline{E^{(2)}}n_x\,dx + C(\|n\|_{H^\alpha}^2 + \|E\|_{H^1}^2). \end{aligned} \tag{53}$$

Combining (49), (51) and (53), then we can get

$$\frac{d}{dt}\omega(t) \le C(\|n\|_{H^\alpha}^2 + \|E\|_{H^1}^2), \tag{54}$$

where $\omega(t) := \|E_x\|_{L^2}^2 + \frac{1}{2}\|\Lambda^\alpha n\|_{L^2}^2 + \frac{1}{2}\int_{\mathbb{R}} n^2(n^{(1)}+n^{(2)})dx + \operatorname{Re}\int_{\mathbb{R}}(E^{(1)}\overline{E} + E\overline{E^{(2)}})n\,dx$. Note that

$$\left|\frac{1}{2}\int_{\mathbb{R}} n^2(n^{(1)}+n^{(2)})dx\right| \le \|n\|_{L^2}^2\|n^{(1)}+n^{(2)}\|_{L^\infty} \le C\|n\|_{L^2}^2,$$

$$\left|\operatorname{Re}\int_{\mathbb{R}}(E^{(1)}\overline{E} + E\overline{E^{(2)}})n\,dx\right| \le C(\|E\|_{L^2}^2 + \|n\|_{L^2}^2),$$

then (54) gives

$$\|E_x\|_{L^2}^2 + \|\Lambda^\alpha n\|_{L^2}^2 \le C\int_0^t (\|n\|_{H^\alpha}^2 + \|E\|_{H^1}^2)ds + C(\|E\|_{L^2}^2 + \|n\|_{L^2}^2). \tag{55}$$

Collecting the estimates (47) and (55), we thus obtain

$$\|E_x\|_{H^1}^2 + \|n\|_{H^\alpha}^2 \le C\int_0^t (\|n\|_{H^\alpha}^2 + \|E\|_{H^1}^2)ds.$$

And by Gronwall's inequality, one gets $E \equiv 0$ and $n \equiv 0$. Hence, the uniqueness of the solution is proved.

Acknowledgments

B. Ma is supported by the National Science Foundation of China (Grant No. 11271162). J. Zhang is supported by the National Science Foundation of China (Grant No. 11201185).

References

1. D. Bekiranov, T. Ogawa, G. Ponce, Weak solvability and well-posedness of a coupled Schrodinger-Korteweg de Vries equation for capillary-gravity wave interactions, *Proceedings of the AMS.* **125** (1997), no. 10, 2907–2919.
2. A. J. Corcho, F. Linares, Well-posedness for the Schrödinger-Korteweg-de Vries system, *Trans. Amer. Math. Soc.* **359** (2007), 4089–4106.
3. R. Coifman, Y. Meyer, Nonlinear harmonic analysis operator theory and PDE, in: Beijing Lectures in Harmonic Analysis, Princeton University Press, 1986, pp. 3–45.
4. M. Funakoshi, M. Oikawa, The resonant interaction between a long internal gravity wave and a surface gravity wave packet, *J. Phys. Soc. Japan* **52** (1983), 1982–1995.
5. B. Guo, The initial and periodic value problems of one class coupled Schrödinger-Korteweg de Vries equations, *Acta Math. Sinica, Chinese Series.* **26** (1983), no. 5, 513–532.
6. B. Guo, F. Chen, Finite-dimensional behavior of global attractors for weakly damped and forced KdV equations coupling with nonlinear Schrodinger equations, *Acta Math. Sinica, Engl. Series* **29** (1997), no. 5, 569–584.
7. B. Guo, C. Miao, Well-posedness of the Cauchy problem for the coupled system of the Schrödinger-KdV equations, *Acta Math. Sinica, English Series.* **15** (1999), no. 2, 215–224.
8. H. Hojo, H. Ikezi, K. Mima, K. Nishikawa, Coupled nonlinear electron-plasma and ionacoustic waves, *Phys. Rev. Lett.* **33** (1974), 148–151.
9. T. Kato, G. Ponce, Commutator estimates and the Euler and Navier-stokes equations, *Comm. Pure Appl. Math.* **41** (1988), 891–907.
10. C. E. Kenig, G. Ponce, L. Vega, Well-posedness of the initial value problem for the Korteweg-de Vries equation, *J. Amer. Math. Soc.* **4** (1991), 323–347.
11. C. E. Kenig, G. Ponce, L. Vega, Well-posedness and scattering results for the generalized Korteweg-de Vries equation via the contraction principle, *Comm. Pure Appl. Math.* **46** (1993), 527–620.
12. T. Kakutani, T. Kawahara, N. Sugimoto, Nonlinear interaction between short and long capillary-gravity waves, *J. Phys. Soc. Japan* **39** (1975), 1379–1386.
13. J. Satsuma, N. Yajima, Soliton solutions in a diatomic lattice system, *Progr. Theor. Phys.*, **62** (1979), 370–378.
14. M. Tsutsumi, Well-posedness of the Cauchy problem for a coupled Schrödinger-KdV equation, *Math. Sciences Appl.* **2** (1993), 513–528.

Anomaly Cancellation and Modularity

Fei Han

Department of Mathematics, National University of Singapore, Block S17, 10 Lower Kent Ridge Road, Singapore 119076
mathanf@nus.edu.sg

Kefeng Liu

Department of Mathematics, University of California at Los Angeles, Los Angelcs, CA 90095, USA;
Center of Mathematical Sciences, Zhejiang University, Hangzhou 310027, China
liu@math.ucla.edu

Weiping Zhang

Chern Institute of Mathematics & LPMC, Nankai University, Tianjin 300071, China
weiping@nankai.edu.cn

It has been shown that the Alvarez–Gaumé–Witten miraculous anomaly cancellation formula in type IIB superstring theory and its various generalizations can be derived from modularity of certain characteristic forms. In this paper, we show that the Green–Schwarz formula and the Schwarz–Witten formula in type I superstring theory can also be derived from the modularity of those characteristic forms and thus unify the Alvarez–Gaumé-Witten formula, the Green–Schwarz formula as well as the Schwarz–Witten formula in the same framework. Various generalizations of these remarkable formulas are also established.

1. Introduction

Let $Z \to X \to B$ be a fiber bundle with fiber Z being 10 dimensional. Let TZ be the vertical tangent bundle equipped with a metric g^{TZ} and an associated Levi–Civita connection ∇^{TZ} (cf. [3, Proposition 10.2]). Let

$R^{TZ} = (\nabla^{TZ})^2$ be the curvature of ∇^{TZ}. Let $T_{\mathbf{C}}Z$ be the complexification of TZ with the induced Hermitian connection $\nabla^{T_{\mathbf{C}}Z}$. Let $\widehat{A}(TZ, \nabla^{TZ})$, $L(TZ, \nabla^{TZ})$ and $\mathrm{ch}(T_{\mathbf{C}}Z, \nabla^{T_{\mathbf{C}}Z})$ be the Hirzebruch $\widehat{A}$-form, the Hirzebruch L-form and the Chern character form respectively (cf. Refs. 3 and 18).

The Alvarez–Gaumé–Witten "miraculous anomaly cancellation formula"[1] in type IIB superstring theory asserts that

$$\begin{aligned}&\{L(TZ, \nabla^{TZ})\}^{(12)} - 8\{\widehat{A}(TZ, \nabla^{TZ})\mathrm{ch}(T_{\mathbf{C}}Z, \nabla^{T_{\mathbf{C}}Z})\}^{(12)} \\ &+ 16\{\widehat{A}(TZ, \nabla^{TZ})\}^{(12)} = 0,\end{aligned} \tag{1}$$

which assures that the corresponding theory is anomaly-free.

On the other hand, Green and Schwarz[5] (see also Ref. 16) discovered that the anomaly in type I superstring theory with gauge group $SO(32)$ cancels. They found that when the gauge group is $SO(32)$, the anomaly factorizes so that there is a Chern–Simons counterterm making the anomaly cancelled. More precisely, let F be a 32 dimensional Euclidean vector bundle over X with Euclidean connection ∇^F and $F_{\mathbf{C}}$ the complexfication of F with the induced Hermitian connection $\nabla^{F_{\mathbf{C}}}$, then the Green–Schwarz formula reads as follows,[1]

$$\begin{aligned}&\{\widehat{A}(TZ)\mathrm{ch}(\wedge^2 F_{\mathbf{C}})\}^{(12)} + \{\widehat{A}(TZ)\mathrm{ch}(T_{\mathbf{C}}Z)\}^{(12)} - 2\{\widehat{A}(TZ)\}^{(12)} \\ =&(p_1(TZ) - p_1(F)) \cdot \frac{1}{24}\Big(\frac{-3p_1(TZ)^2 + 4p_2(TZ)}{8} \\ &- 2p_1(F)^2 + 4p_2(F) + \frac{1}{2}p_1(TZ)p_1(F)\Big),\end{aligned} \tag{2}$$

where $p_i(TZ)$, $p_i(F), 1 \leq i \leq 2$, are the Pontryagin forms of (TZ, ∇^{TZ}), (F, ∇^F) respectively.

The above formulas of Alvarez–Gaumé–Witten and Green–Schwarz have played crucial roles in the early development of superstring theory.

More recently, Schwarz and Witten[17] analyzed the anomaly in type I theory with additional spacetime-filling D-branes and anti-D-branes pairs and found a similar factorization. More precisely, let F_1 be an m dimensional Euclidean vector bundle over X equipped with a Euclidean connection ∇^{F_1} and F_2 be an n dimensional Euclidean vector bundle over X equipped with a Euclidean connection ∇^{F_2}, when $m = n + 32$, one has

[1]In what follows, we will write characteristic forms without specifying the connections when there is no confusion.

$$\begin{aligned}&\{\widehat{A}(TZ)\mathrm{ch}(\wedge^2 F_{1\mathbf{C}})\}^{(12)} + \{\widehat{A}(TZ)\mathrm{ch}(S^2 F_{2\mathbf{C}})\}^{(12)}\\&- \{\widehat{A}(TZ)\mathrm{ch}(F_{1\mathbf{C}}\otimes F_{2\mathbf{C}})\}^{(12)}\\&+ \{\widehat{A}(TZ)\mathrm{ch}(T_{\mathbf{C}}Z)\}^{(12)} - 2\{\widehat{A}(TZ)\}^{(12)}\\=&\big(p_1(TZ)-p_1(F_1)+p_1(F_2)\big)\\&\cdot\frac{1}{24}\Big(\frac{-3p_1(TZ)^2+4p_2(TZ)}{8} - 2p_1(F_1)^2+4p_2(F_1)+2p_1(F_2)^2\\&-4p_2(F_2)+\frac{1}{2}p_1(TZ)(p_1(F_1)-p_1(F_2))\Big).\end{aligned} \tag{3}$$

Therefore, similarly, there is a Chern–Simons counterterm to make the anomaly cancelled.

In Refs. 6 and 12, it is shown that the Alvarez–Gaumé–Witten "miraculous anomaly cancellation formula" can be derived from the modularity of certain characteristic forms. In fact, let V be a Euclidean vector bundle equipped with a Euclidean connection over X, one can construct two characteristic forms $P_1(TZ,V,\tau)$ and $P_2(TZ,V,\tau)$ such that when $p_1(TZ)=p_1(V)$, $P_1(TZ,V,\tau)$ and $P_2(TZ,V,\tau)$ are level 2 modular forms over $\Gamma_0(2)$ and $\Gamma^0(2)$ respectively. Moreover they are modularly related and form what we call a modular pair (see page 100 for more details). The Alvarez–Gaumé–Witten formula can then be deduced from this modular pair $(P_1(TZ,V,\tau),P_2(TZ,V,\tau))$ if one sets $V=TZ$. This construction is further generalized in[8] to the case where a complex line bundle is involved, in dealing with the Ochanine congruence[15] on spin^c manifolds.

In a recent article,[7] in using the Eisenstein series $E_2(\tau)$, we constructed a pair of modularly related characteristic forms $(\mathcal{P}_1(TZ,V,\xi,\tau)$, $\mathcal{P}_2(TZ,V,\xi,\tau))$ without assuming $p_1(TZ)=p_1(V)$. When $p_1(TZ)=p_1(V)$ and ξ is trivial, $(\mathcal{P}_1(TZ,V,\xi,\tau),\mathcal{P}_2(TZ,V,\xi,\tau))$ degenerates to $(P_1(TZ,V,\tau),P_2(TZ,V,\tau))$.

In the current paper, we will show that the formulas due to Green–Schwarz (2) and Schwarz–Witten (3) can also be deduced from the modularity of the pair $(\mathcal{P}_1(TZ,V,\xi,\tau),\mathcal{P}_2(TZ,V,\xi,\tau))$. Actually, we need only to make use of the modularity of $\mathcal{P}_2(TZ,V,\xi,\tau)$ by replacing V by a super vector bundle F_1-F_2. Our method also generates many generalizations of the Green–Schwarz and Schwarz–Witten formulas. See Theorem 1 and its corollaries for more details.

It is quite amazing that all of the three anomaly cancellation formulas due to Alvarez–Gaumé–Witten, Green–Schwarz, as well as Schwarz–Witten, can be unified through a single modular pair $(\mathcal{P}_1(TZ,V,\xi,\tau),\mathcal{P}_2(TZ,V,\xi,\tau))$. It illustrates one of the deep implications of modularity in physics.

In the rest of this paper, we will first present the Green–Schwarz type factorization formulas in Section 1 and then show how to derive them from modularity in Section 2.

2. Green–Schwarz Type Factorization Formulas

The purpose of this section is to present various generalizations of the Green–Schwarz formula and the Schwarz–Witten formula.

Let $Z \to X \to B$ be a fiber bundle with fiber Z being 10 dimensional. Let TZ be the vertical tangent bundle equipped with a metric g^{TZ} and an associated Levi-Civita connection ∇^{TZ} (cf. [3, Proposition 10.2]). Let $R^{TZ} = (\nabla^{TZ})^2$ be the curvature of ∇^{TZ}. Let $T_{\mathbf{C}}Z$ be the complexification of TZ with the induced Hermitian connection $\nabla^{T_{\mathbf{C}}Z}$.

Let F_1 be an m dimensional Euclidean vector bundle over X equipped with a Euclidean connection ∇^{F_1} and F_2 be an n dimensional Euclidean vector bundle over X equipped with a Euclidean connection ∇^{F_2}.

Let ξ be a rank two real oriented Euclidean vector bundle over X carrying a Euclidean connection ∇^{ξ}. Let $c = e(\xi, \nabla^{\xi})$ be the Euler form canonically associated to ∇^{ξ}.

If E is a real (resp. complex) vector bundle over X, set $\widetilde{E} = E - \dim E \in KO(X)$ (resp. $K(X)$).

If ω is a differential form, denote the degree j-component of ω by $\omega^{(j)}$.

Theorem 1. *The following identity holds,*

$$\begin{aligned}
&\{\widehat{A}(TZ)e^{\frac{c}{2}}\mathrm{ch}(\wedge^2 F_{1\mathbf{C}})\}^{(12)} + \{\widehat{A}(TZ)e^{\frac{c}{2}}\mathrm{ch}(S^2 F_{2\mathbf{C}})\}^{(12)} \\
&- \{\widehat{A}(TZ)e^{\frac{c}{2}}\mathrm{ch}(F_{1\mathbf{C}} \otimes F_{2\mathbf{C}})\}^{(12)} + \{\widehat{A}(TZ)e^{\frac{c}{2}}\mathrm{ch}(T_{\mathbf{C}}Z)\}^{(12)} \\
&+ \left(\frac{(m-n-32)(m-n-31)}{2} - 2\right)\{\widehat{A}(TZ)e^{\frac{c}{2}}\}^{(12)} \\
&- (m-n-32)\{\widehat{A}(TZ)e^{\frac{c}{2}}\mathrm{ch}(F_{1\mathbf{C}} - F_{2\mathbf{C}})\}^{(12)} \\
&+ 5\{\widehat{A}(TZ)e^{\frac{c}{2}}\mathrm{ch}(\widetilde{\xi_{\mathbf{C}}} \otimes \widetilde{\xi_{\mathbf{C}}})\}^{(12)} \\
&+ 3\{\widehat{A}(TZ)e^{\frac{c}{2}}\mathrm{ch}((m-n-31-F_{1\mathbf{C}}+F_{2\mathbf{C}}) \otimes \widetilde{\xi_{\mathbf{C}}})\}^{(12)} \\
=&(p_1(TZ) - p_1(F_1) + p_1(F_2)) \\
&\cdot \Big\{ - \frac{e^{\frac{1}{24}(p_1(TZ)-p_1(F_1)+p_1(F_2))} - 1}{p_1(TZ) - p_1(F_1) + p_1(F_2)} \widehat{A}(TZ)e^{\frac{c}{2}}\mathrm{ch}(\mathfrak{A}) \\
&+ e^{\frac{1}{24}(p_1(TZ)-p_1(F_1)+p_1(F_2))}\widehat{A}(TZ)e^{\frac{c}{2}} \Big\}^{(8)},
\end{aligned} \tag{4}$$

where

$$\begin{aligned}\mathfrak{A} = &\wedge^2 F_{1\mathbf{C}} + S^2 F_{2\mathbf{C}} - F_{1\mathbf{C}} \otimes F_{2\mathbf{C}} + T_{\mathbf{C}} Z \\ &+ \frac{(m-n-32)(m-n-31)}{2} - 2 \\ &- (m-n-32)(F_{1\mathbf{C}} - F_{2\mathbf{C}}) + 5\widetilde{\xi_{\mathbf{C}}} \otimes \widetilde{\xi_{\mathbf{C}}} \\ &+ 3(m-n-31 - F_{1\mathbf{C}} + F_{2\mathbf{C}}) \otimes \widetilde{\xi_{\mathbf{C}}};\end{aligned} \tag{5}$$

if ξ is trivial, the following identity holds,

$$\begin{aligned}&\{\widehat{A}(TZ)\mathrm{ch}(\wedge^2 F_{1\mathbf{C}})\}^{(12)} + \{\widehat{A}(TZ)\mathrm{ch}(S^2 F_{2\mathbf{C}})\}^{(12)} \\ &- \{\widehat{A}(TZ)\mathrm{ch}(F_{1\mathbf{C}} \otimes F_{2\mathbf{C}})\}^{(12)} + \{\widehat{A}(TZ)\mathrm{ch}(T_{\mathbf{C}} Z)\}^{(12)} \\ &+ \left(\frac{(m-n-32)(m-n-31)}{2} - 2\right) \{\widehat{A}(TZ)\}^{(12)} \\ &- (m-n-32)\{\widehat{A}(TZ)\mathrm{ch}(F_{1\mathbf{C}} - F_{2\mathbf{C}})\}^{(12)} \\ =&(p_1(TZ) - p_1(F_1) + p_1(F_2)) \\ &\cdot \Big\{ - \frac{e^{\frac{1}{24}(p_1(TZ)-p_1(F_1)+p_1(F_2))} - 1}{p_1(TZ) - p_1(F_1) + p_1(F_2)} \widehat{A}(TZ)\mathrm{ch}(\mathfrak{B}) \\ &+ e^{\frac{1}{24}(p_1(TZ)-p_1(F_1)+p_1(F_2))} \widehat{A}(TZ) \Big\}^{(8)},\end{aligned} \tag{6}$$

where

$$\begin{aligned}\mathfrak{B} = &\wedge^2 F_{1\mathbf{C}} + S^2 F_{2\mathbf{C}} - F_{1\mathbf{C}} \otimes F_{2\mathbf{C}} + T_{\mathbf{C}} Z \\ &+ \frac{(m-n-32)(m-n-31)}{2} - 2 - (m-n-32)(F_{1\mathbf{C}} - F_{2\mathbf{C}}).\end{aligned} \tag{7}$$

Putting $m = n + 32$ in Theorem 1, we get

Corollary 1. *If* $\dim F_1 - \dim F_2 = 32$, *the following identity holds,*

$$\begin{aligned}&\{\widehat{A}(TZ)e^{\frac{c}{2}}\mathrm{ch}(\wedge^2 F_{1\mathbf{C}})\}^{(12)} + \{\widehat{A}(TZ)e^{\frac{c}{2}}\mathrm{ch}(S^2 F_{2\mathbf{C}})\}^{(12)} \\ &- \{\widehat{A}(TZ)e^{\frac{c}{2}}\mathrm{ch}(F_{1\mathbf{C}} \otimes F_{2\mathbf{C}})\}^{(12)} + \{\widehat{A}(TZ)e^{\frac{c}{2}}\mathrm{ch}(T_{\mathbf{C}} Z)\}^{(12)} \\ &- 2\{\widehat{A}(TZ)e^{\frac{c}{2}}\}^{(12)} + 5\{\widehat{A}(TZ)e^{\frac{c}{2}}\mathrm{ch}(\widetilde{\xi_{\mathbf{C}}} \otimes \widetilde{\xi_{\mathbf{C}}})\}^{(12)} \\ &+ 3\{\widehat{A}(TZ)e^{\frac{c}{2}}\mathrm{ch}((1 - F_{1\mathbf{C}} + F_{2\mathbf{C}}) \otimes \widetilde{\xi_{\mathbf{C}}})\}^{(12)} \\ =&(p_1(TZ) - p_1(F_1) + p_1(F_2)) \\ &\cdot \Big\{ - \frac{e^{\frac{1}{24}(p_1(TZ)-p_1(F_1)+p_1(F_2))} - 1}{p_1(TZ) - p_1(F_1) + p_1(F_2)} \widehat{A}(TZ)e^{\frac{c}{2}}\mathrm{ch}(\mathfrak{C}) \\ &+ e^{\frac{1}{24}(p_1(TZ)-p_1(F_1)+p_1(F_2))} \widehat{A}(TZ)e^{\frac{c}{2}} \Big\}^{(8)},\end{aligned} \tag{8}$$

where

$$\begin{aligned}\mathfrak{C} = & \wedge^2 F_{1\mathbf{C}} + S^2 F_{2\mathbf{C}} - F_{1\mathbf{C}} \otimes F_{2\mathbf{C}} + T_{\mathbf{C}} Z - 2 \\ & + 5\widetilde{\xi_{\mathbf{C}}} \otimes \widetilde{\xi_{\mathbf{C}}} + 3(1 - F_{1\mathbf{C}} + F_{2\mathbf{C}}) \otimes \widetilde{\xi_{\mathbf{C}}};\end{aligned} \tag{9}$$

if ξ is trivial, we obtain the Schwarz–Witten formula (3),

$$\begin{aligned}& \{\widehat{A}(TZ)\mathrm{ch}(\wedge^2 F_{1\mathbf{C}})\}^{(12)} + \{\widehat{A}(TZ)\mathrm{ch}(S^2 F_{2\mathbf{C}})\}^{(12)} \\ & - \{\widehat{A}(TZ)\mathrm{ch}(F_{1\mathbf{C}} \otimes F_{2\mathbf{C}})\}^{(12)} + \{\widehat{A}(TZ)\mathrm{ch}(T_{\mathbf{C}} Z)\}^{(12)} \\ & - 2\{\widehat{A}(TZ)\}^{(12)} \\ = & (p_1(TZ) - p_1(F_1) + p_1(F_2)) \\ & \cdot \Big\{ - \frac{e^{\frac{1}{24}(p_1(TZ) - p_1(F_1) + p_1(F_2))} - 1}{p_1(TZ) - p_1(F_1) + p_1(F_2)} \widehat{A}(TZ)\mathrm{ch}(\mathfrak{D}) \\ & + e^{\frac{1}{24}(p_1(TZ) - p_1(F_1) + p_1(F_2))} \widehat{A}(TZ) \Big\}^{(8)},\end{aligned} \tag{10}$$

where

$$\mathfrak{D} = \wedge^2 F_{1\mathbf{C}} + S^2 F_{2\mathbf{C}} - F_{1\mathbf{C}} \otimes F_{2\mathbf{C}} + T_{\mathbf{C}} Z - 2. \tag{11}$$

Remark 1. It can be checked by direct computation that when $m = n+32$, one indeed has

$$\begin{aligned}& \frac{1}{24}\Big(\frac{-3p_1(TZ)^2 + 4p_2(TZ)}{8} - 2p_1(F_1)^2 + 4p_2(F_1) + 2p_1(F_2)^2 \\ & - 4p_2(F_2) + \frac{1}{2} p_1(TZ)(p_1(F_1) - p_1(F_2))\Big) \\ = & \Big\{ - \frac{e^{\frac{1}{24}(p_1(TZ) - p_1(F_1) + p_1(F_2))} - 1}{p_1(TZ) - p_1(F_1) + p_1(F_2)} \widehat{A}(TZ)\mathrm{ch}(\mathfrak{D}) \\ & + e^{\frac{1}{24}(p_1(TZ) - p_1(F_1) + p_1(F_2))} \widehat{A}(TZ) \Big\}^{(8)}.\end{aligned} \tag{12}$$

If we set $n = 0$ in Theorem 1, we get

Corollary 2. *If* $\dim F = m$, *then the following identity holds,*

$$\begin{aligned}
&\{\widehat{A}(TZ)e^{\frac{c}{2}}\mathrm{ch}(\wedge^2 F_{\mathbf{C}})\}^{(12)} + \{\widehat{A}(TZ)e^{\frac{c}{2}}\mathrm{ch}(T_{\mathbf{C}}Z)\}^{(12)} \\
&+ \left(\frac{(m-32)(m-31)}{2} - 2\right)\{\widehat{A}(TZ)e^{\frac{c}{2}}\}^{(12)} \\
&- (m-32)\{\widehat{A}(TZ)e^{\frac{c}{2}}\mathrm{ch}(F_{\mathbf{C}})\}^{(12)} + 5\{\widehat{A}(TZ)e^{\frac{c}{2}}\mathrm{ch}(\widetilde{\xi_{\mathbf{C}}} \otimes \widetilde{\xi_{\mathbf{C}}})\}^{(12)} \\
&+ 3\{\widehat{A}(TZ)e^{\frac{c}{2}}\mathrm{ch}((m-31-F_{\mathbf{C}}) \otimes \widetilde{\xi_{\mathbf{C}}})\}^{(12)} \\
=&(p_1(TZ) - p_1(F)) \\
&\cdot \Big\{ - \frac{e^{\frac{1}{24}(p_1(TZ)-p_1(F))} - 1}{p_1(TZ) - p_1(F)}\widehat{A}(TZ)e^{\frac{c}{2}}\mathrm{ch}(\mathfrak{E}) \\
&+ e^{\frac{1}{24}(p_1(TZ)-p_1(F))}\widehat{A}(TZ)e^{\frac{c}{2}}\Big\}^{(8)},
\end{aligned} \tag{13}$$

where

$$\begin{aligned}
\mathfrak{E} = \wedge^2 F_{1\mathbf{C}} + T_{\mathbf{C}}Z + \frac{(m-32)(m-31)}{2} - 2 \\
- (m-32)(F_{\mathbf{C}}) + 5\widetilde{\xi_{\mathbf{C}}} \otimes \widetilde{\xi_{\mathbf{C}}} + 3(m-31-F_{\mathbf{C}}) \otimes \widetilde{\xi_{\mathbf{C}}};
\end{aligned} \tag{14}$$

if ξ *is trivial, the following identity holds,*

$$\begin{aligned}
&\{\widehat{A}(TZ)\mathrm{ch}(\wedge^2 F_{\mathbf{C}})\}^{(12)} + \{\widehat{A}(TZ)\mathrm{ch}(T_{\mathbf{C}}Z\}^{(12)} \\
&+ \left(\frac{(m-32)(m-31)}{2} - 2\right)\{\widehat{A}(TZ)\}^{(12)} \\
&- (m-32)\{\widehat{A}(TZ)\mathrm{ch}(F_{\mathbf{C}})\}^{(12)} \\
=&(p_1(TZ) - p_1(F))\Big\{ - \frac{e^{\frac{1}{24}(p_1(TZ)-p_1(F))} - 1}{p_1(TZ) - p_1(F)}\widehat{A}(TZ)\mathrm{ch}(\mathfrak{F}) \\
&+ e^{\frac{1}{24}(p_1(TZ)-p_1(F))}\widehat{A}(TZ)\Big\}^{(8)},
\end{aligned} \tag{15}$$

where

$$\mathfrak{F} = \wedge^2 F_{\mathbf{C}} + T_{\mathbf{C}}Z + \frac{(m-32)(m-31)}{2} - 2 - (m-32)F_{\mathbf{C}}. \tag{16}$$

Putting $m = 32$ in the above corollary, we get

Corollary 3. *If* $\dim F = 32$, *the following identity holds,*

$$\begin{aligned}&\{\widehat{A}(TZ)e^{\frac{c}{2}}\mathrm{ch}(\wedge^2 F_{\mathbf{C}})\}^{(12)} + \{\widehat{A}(TZ)e^{\frac{c}{2}}\mathrm{ch}(T_{\mathbf{C}}Z)\}^{(12)}\\&- 2\{\widehat{A}(TZ)e^{\frac{c}{2}}\}^{(12)} + 5\{\widehat{A}(TZ)e^{\frac{c}{2}}\mathrm{ch}(\widetilde{\xi_{\mathbf{C}}}\otimes\widetilde{\xi_{\mathbf{C}}})\}^{(12)}\\&+ 3\{\widehat{A}(TZ)e^{\frac{c}{2}}\mathrm{ch}((1-F_{\mathbf{C}})\otimes\widetilde{\xi_{\mathbf{C}}})\}^{(12)}\\=&(p_1(TZ)-p_1(F))\\&\cdot\Big\{-\frac{e^{\frac{1}{24}(p_1(TZ)-p_1(F))}-1}{p_1(TZ)-p_1(F)}\widehat{A}(TZ)e^{\frac{c}{2}}\mathrm{ch}(\mathfrak{G})\\&+e^{\frac{1}{24}(p_1(TZ)-p_1(F))}\widehat{A}(TZ)e^{\frac{c}{2}}\Big\}^{(8)},\end{aligned} \tag{17}$$

where

$$\mathfrak{G} = \wedge^2 F_{\mathbf{C}} + T_{\mathbf{C}}Z - 2 + 5\widetilde{\xi_{\mathbf{C}}}\otimes\widetilde{\xi_{\mathbf{C}}} + 3(1-F_{\mathbf{C}})\otimes\widetilde{\xi_{\mathbf{C}}}; \tag{18}$$

if ξ *is trivial, we obtain the Green–Schwarz formula (2),*

$$\begin{aligned}&\{\widehat{A}(TZ)\mathrm{ch}(\wedge^2 F_{\mathbf{C}})\}^{(12)} + \{\widehat{A}(TZ)\mathrm{ch}(T_{\mathbf{C}}Z\}^{(12)} - 2\{\widehat{A}(TZ)\}^{(12)}\\=&(p_1(TZ)-p_1(F))\\&\Big\{-\frac{e^{\frac{1}{24}(p_1(TZ)-p_1(F))}-1}{p_1(TZ)-p_1(F)}\widehat{A}(TZ)\mathrm{ch}(\wedge^2 F_{\mathbf{C}} + T_{\mathbf{C}}Z - 2)\\&+e^{\frac{1}{24}(p_1(TZ)-p_1(F))}\widehat{A}(TZ)\Big\}^{(8)}.\end{aligned} \tag{19}$$

3. Derivation of the Green–Schwarz Type Factorizations from Modularity

In this section, we will derive the Green–Schwarz type factorization formulas presented in Section 2 via the modularity of $\mathcal{P}_2(TZ, F_1 - F_2, \xi, \tau)$.

3.1. *Preliminaries*

In this subsection, we recall some basic knowledge about the Jacobi theta functions, modular forms and Eisenstein series. Although we will not use all the things recalled here, we still put them in this subsection for completeness.

Let $SL_2(\mathbf{Z}) := \left\{\left(\begin{smallmatrix} a & b\\ c & d\end{smallmatrix}\right)\Big| a,b,c,d\in\mathbf{Z},\ ad-bc=1\right\}$ as usual be the modular group. Let $S = \begin{pmatrix} 0 & -1\\ 1 & 0\end{pmatrix}$, $T = \begin{pmatrix} 1 & 1\\ 0 & 1\end{pmatrix}$ be the two generators of $SL_2(\mathbf{Z})$. Their actions on $\mathbf{H}$ are given by $S : \tau \to -\frac{1}{\tau}$, $\quad T : \tau \to \tau + 1$.

The four Jacobi theta functions are defined as follows (cf. Ref. 4):

$$\theta(v,\tau) = 2q^{1/8}\sin(\pi v)\prod_{j=1}^{\infty}\left[(1-q^j)(1-e^{2\pi\sqrt{-1}v}q^j)(1-e^{-2\pi\sqrt{-1}v}q^j)\right],$$

$$\theta_1(v,\tau) = 2q^{1/8}\cos(\pi v)\prod_{j=1}^{\infty}\left[(1-q^j)(1+e^{2\pi\sqrt{-1}v}q^j)(1+e^{-2\pi\sqrt{-1}v}q^j)\right],$$

$$\theta_2(v,\tau) = \prod_{j=1}^{\infty}\left[(1-q^j)(1-e^{2\pi\sqrt{-1}v}q^{j-1/2})(1-e^{-2\pi\sqrt{-1}v}q^{j-1/2})\right],$$

$$\theta_3(v,\tau) = \prod_{j=1}^{\infty}\left[(1-q^j)(1+e^{2\pi\sqrt{-1}v}q^{j-1/2})(1+e^{-2\pi\sqrt{-1}v}q^{j-1/2})\right].$$

They are all holomorphic functions for $(v,\tau) \in \mathbf{C}\times\mathbf{H}$, where $\mathbf{C}$ is the complex plane and $\mathbf{H}$ is the upper half plane.

When acted by S and T, the theta functions obey the following transformation laws (cf. Ref. 4),

$$\begin{aligned}\theta(v,\tau+1) &= e^{\frac{\pi\sqrt{-1}}{4}}\theta(v,\tau)\\ \theta(v,-1/\tau) &= \frac{1}{\sqrt{-1}}\left(\frac{\tau}{\sqrt{-1}}\right)^{1/2} e^{\pi\sqrt{-1}\tau v^2}\theta(\tau v,\tau)\ ;\end{aligned} \tag{20}$$

$$\theta_1(v,\tau+1) = e^{\frac{\pi\sqrt{-1}}{4}}\theta_1(v,\tau),\quad \theta_1(v,-1/\tau) = \left(\frac{\tau}{\sqrt{-1}}\right)^{1/2} e^{\pi\sqrt{-1}\tau v^2}\theta_2(\tau v,\tau)\ ; \tag{21}$$

$$\theta_2(v,\tau+1) = \theta_3(v,\tau),\quad \theta_2(v,-1/\tau) = \left(\frac{\tau}{\sqrt{-1}}\right)^{1/2} e^{\pi\sqrt{-1}\tau v^2}\theta_1(\tau v,\tau)\ ; \tag{22}$$

$$\theta_3(v,\tau+1) = \theta_2(v,\tau),\quad \theta_3(v,-1/\tau) = \left(\frac{\tau}{\sqrt{-1}}\right)^{1/2} e^{\pi\sqrt{-1}\tau v^2}\theta_3(\tau v,\tau)\ . \tag{23}$$

Definition 1. Let Γ be a subgroup of $SL_2(\mathbf{Z})$. A modular form over Γ is a holomorphic function $f(\tau)$ on $\mathbf{H}\cup\{\infty\}$ such that for any $g = \begin{pmatrix} a & b \\ c & d\end{pmatrix} \in \Gamma$, the following property holds,

$$f(g\tau) := f\left(\frac{a\tau+b}{c\tau+d}\right) = \chi(g)(c\tau+d)^l f(\tau),$$

where $\chi : \Gamma \to \mathbf{C}^*$ is a character of Γ and l is called the weight of f.

Let

$$E_{2k} = 1 - \frac{4k}{B_{2k}} \sum_{n=1}^{\infty} \left(\sum_{d|n} d^{2k-1} \right) q^n \tag{24}$$

be the Eisenstein series, where B_{2k} is the $2k$-th Bernoulli number.

When $k > 1$, E_{2k} is a modular form of weight $2k$ over $SL_2(\mathbf{Z})$. However, unlike other Eisenstein series, $E_2(\tau) = 1 - 24\sum_{n=1}^{\infty} \left(\sum_{d|n} d \right) q^n = 1 - 24q - 72q^2 - 96q^3 - \cdots$ is not a modular form over $SL(2, \mathbf{Z})$, instead it is a quasimodular form over $SL(2, \mathbf{Z})$ satisfying

$$E_2\left(\frac{a\tau + b}{c\tau + d} \right) = (c\tau + d)^2 E_2(\tau) - \frac{6\sqrt{-1}c(c\tau + d)}{\pi}. \tag{25}$$

In particular, we have

$$E_2(\tau + 1) = E_2(\tau), \tag{26}$$

$$E_2\left(-\frac{1}{\tau} \right) = \tau^2 E_2(\tau) - \frac{6\sqrt{-1}\tau}{\pi}. \tag{27}$$

For the precise definition of quasimodular forms, see Ref. 11.

Let $\Gamma_0(2) = \left\{ \begin{pmatrix} a & b \\ c & d \end{pmatrix} \in SL_2(\mathbf{Z}) \middle| c \equiv 0 \pmod 2 \right\}$, $\Gamma^0(2) = \left\{ \begin{pmatrix} a & b \\ c & d \end{pmatrix} \in SL_2(\mathbf{Z}) \middle| b \equiv 0 \pmod 2 \right\}$ be the two modular subgroups of $SL_2(\mathbf{Z})$. It is known that the generators of $\Gamma_0(2)$ are T, ST^2ST and the generators of $\Gamma^0(2)$ are STS, T^2STS (cf. Ref. 4).

Consider the q-series:

$$\delta_1(\tau) = \frac{1}{4} + 6\sum_{n=1}^{\infty} \sum_{\substack{d|n \\ d \text{ odd}}} dq^n = \frac{1}{4} + 6q + 6q^2 + \cdots, \tag{28}$$

$$\varepsilon_1(\tau) = \frac{1}{16} + \sum_{n=1}^{\infty} \sum_{d|n} (-1)^d d^3 q^n = \frac{1}{16} - q + 7q^2 + \cdots, \tag{29}$$

$$\delta_2(\tau) = -\frac{1}{8} - 3\sum_{n=1}^{\infty} \sum_{\substack{d|n \\ d \text{ odd}}} dq^{n/2} = -\frac{1}{8} - 3q^{1/2} - 3q - \cdots, \tag{30}$$

$$\varepsilon_2(\tau) = \sum_{n=1}^{\infty} \sum_{\substack{d|n \\ n/d \text{ odd}}} d^3 q^{n/2} = q^{1/2} + 8q + \cdots . \tag{31}$$

Simply writing $\theta_j = \theta_j(0,\tau)$, $1 \le j \le 3$, then we have (cf. Ref. 10 and Ref. 13),

$$\delta_1(\tau) = \frac{1}{8}(\theta_2^4 + \theta_3^4), \qquad \varepsilon_1(\tau) = \frac{1}{16}\theta_2^4\theta_3^4 \,,$$

$$\delta_2(\tau) = -\frac{1}{8}(\theta_1^4 + \theta_3^4), \qquad \varepsilon_2(\tau) = \frac{1}{16}\theta_1^4\theta_3^4 \,.$$

If Γ is a modular subgroup, let $M_{\mathbf{R}}(\Gamma)$ denote the ring of modular forms over Γ with real Fourier coefficients.

Lemma 1 (cf. Ref. 12). *One has that $\delta_1(\tau)$ (resp. $\varepsilon_1(\tau)$) is a modular form of weight* 2 *(resp.* 4*) over $\Gamma_0(2)$, $\delta_2(\tau)$ (resp. $\varepsilon_2(\tau)$) is a modular form of weight* 2 *(resp.* 4*) over $\Gamma^0(2)$ and moreover $M_{\mathbf{R}}(\Gamma^0(2)) = \mathbf{R}[\delta_2(\tau), \varepsilon_2(\tau)]$. Moreover, we have transformation laws*

$$\delta_2\left(-\frac{1}{\tau}\right) = \tau^2\delta_1(\tau), \qquad \varepsilon_2\left(-\frac{1}{\tau}\right) = \tau^4\varepsilon_1(\tau). \tag{32}$$

3.2. *The modular form $\mathcal{P}_2(TZ, F_1 - F_2, \xi, \tau)$*

Let F (resp. G) be a Hermitian vector bundle over X equipped with a Hermitian connection ∇^F (resp. ∇^G). For any complex number t, let

$$\Lambda_t(F) = \mathbf{C}|_X + tF + t^2\Lambda^2(F) + \cdots, \quad S_t(F) = \mathbf{C}|_X + tF + t^2S^2(F) + \cdots$$

denote respectively the total exterior and symmetric powers of F, which live in $K(X)[[t]]$. The following relations between these two operations hold (cf. [2, Chap. 3]),

$$S_t(F) = \frac{1}{\Lambda_{-t}(F)}, \qquad \Lambda_t(F - G) = \frac{\Lambda_t(F)}{\Lambda_t(G)}. \tag{33}$$

The connections ∇^F, ∇^G naturally induce connections on $\Lambda_t(F), S_t(F)$ etc. Moreover, if $\{\omega_i\}$, $\{\omega_j{}'\}$ are formal Chern roots for the Hermitian vector bundles F, G respectively, then (cf. [9, Chap. 1]),

$$\mathrm{ch}\left(\Lambda_t(F), \nabla^{\Lambda_t(F)}\right) = \prod_i (1 + e^{\omega_i} t) \tag{34}$$

and we have the following formulas for Chern character forms,

$$\mathrm{ch}\left(S_t(F),\nabla^{S_t(F)}\right)=\frac{1}{\mathrm{ch}\left(\Lambda_{-t}(F),\nabla^{\Lambda_{-t}(F)}\right)}=\frac{1}{\prod\limits_i(1-e^{\omega_i}t)}\,,\tag{35}$$

$$\mathrm{ch}\left(\Lambda_t(F-G),\nabla^{\Lambda_t(F-G)}\right)=\frac{\mathrm{ch}\left(\Lambda_t(F),\nabla^{\Lambda_t(F)}\right)}{\mathrm{ch}\left(\Lambda_t(G),\nabla^{\Lambda_t(G)}\right)}=\frac{\prod\limits_i(1+e^{\omega_i}t)}{\prod\limits_j(1+e^{\omega_j{}'}t)}\tag{36}$$

Let $q=e^{2\pi\sqrt{-1}\tau}$ with $\tau\in\mathbf{H}$, the upper half complex plane.

Set

$$\begin{aligned}\Theta_2(T_{\mathbf{C}}Z,F_{1\mathbf{C}}-F_{2\mathbf{C}},\xi_{\mathbf{C}})=&\bigotimes_{u=1}^{\infty}S_{q^u}(\widetilde{T_{\mathbf{C}}Z})\otimes\bigotimes_{v=1}^{\infty}\Lambda_{-q^{v-\frac{1}{2}}}(\widetilde{F_{1\mathbf{C}}}-\widetilde{F_{2\mathbf{C}}}-2\widetilde{\xi_{\mathbf{C}}})\\&\otimes\bigotimes_{r=1}^{\infty}\Lambda_{q^{r-\frac{1}{2}}}(\widetilde{\xi_{\mathbf{C}}})\otimes\bigotimes_{s=1}^{\infty}\Lambda_{q^s}(\widetilde{\xi_{\mathbf{C}}}),\end{aligned}\tag{37}$$

which is an element in $K(X)[[q^{\frac{1}{2}}]]$.

Clearly, $\Theta_2(T_{\mathbf{C}}Z,F_{1\mathbf{C}}-F_{2\mathbf{C}},\xi_{\mathbf{C}})$ admits a formal Fourier expansion in $q^{1/2}$ as

$$\Theta_2(T_{\mathbf{C}}Z,F_{1\mathbf{C}}-F_{2\mathbf{C}},\xi_{\mathbf{C}})=B_0+B_1q^{1/2}+B_2q\cdots,\tag{38}$$

where the B_j's are elements in the semi-group formally generated by complex vector bundles over X. Moreover, they carry canonically induced connections denoted by ∇^{B_j} and let ∇^{Θ_2} be the induced connections with $q^{1/2}$-coefficients on Θ_2.

Set

$$\begin{aligned}&\mathcal{P}_2(TZ,F_1-F_2,\xi,\tau)\\:=&\Big\{e^{\frac{1}{24}E_2(\tau)(p_1(TZ)-p_1(F_1)+p_1(F_2))}\widehat{A}(TZ)\cosh\left(\frac{c}{2}\right)\\&\cdot\mathrm{ch}\Big(\Theta_2(T_{\mathbf{C}}Z,F_{1\mathbf{C}}-F_{2\mathbf{C}},\xi_{\mathbf{C}})\Big)\Big\}^{(12)}.\end{aligned}\tag{39}$$

Proposition 1. *$\mathcal{P}_2(TZ,F_1-F_2,\xi,\tau)$ is a modular form of weight 6 over $\Gamma^0(2)$.*

Proof. Let $\{\pm2\pi\sqrt{-1}y_k\}$ (resp. $\{\pm2\pi\sqrt{-1}z_k\}$, $\{\pm2\pi\sqrt{-1}x_j\}$) be the formal Chern roots for $(F_{1\mathbf{C}},\nabla^{F_{1\mathbf{C}}})$ (resp. $(F_{2\mathbf{C}},\nabla^{F_{2\mathbf{C}}})$, $(TZ_{\mathbf{C}},\nabla^{TZ_{\mathbf{C}}})$). Let

$c = 2\pi\sqrt{-1}u$. By the Chern root algorithm, we have

$$\begin{aligned}&\mathcal{P}_2(TZ, F_1 - F_2, \xi, \tau)\\&=\Big\{e^{\frac{1}{24}E_2(\tau)(p_1(TZ)-p_1(F_1)+p_1(F_2))}\widehat{A}(TZ)\cosh\left(\frac{c}{2}\right)\\&\quad\cdot \mathrm{ch}\Big(\Theta_2(T_{\mathbf{C}}Z, F_{1\mathbf{C}} - F_{2\mathbf{C}}, \xi_{\mathbf{C}})\Big)\Big\}^{(12)}\\&=\left\{e^{\frac{1}{24}E_2(\tau)(p_1(TZ)-p_1(F_1)+p_1(F_2))}\left(\prod_{j=1}^{5}\left(x_j\frac{\theta'(0,\tau)}{\theta(x_j,\tau)}\right)\right)\left(\prod_{j=1}^{[\frac{m}{2}]}\frac{\theta_2(y_j,\tau)}{\theta_2(0,\tau)}\right)\right.\\&\quad\left.\cdot\left(\prod_{k=1}^{[\frac{n}{2}]}\frac{\theta_2(0,\tau)}{\theta_2(z_k,\tau)}\right)\cdot\frac{\theta_2^2(0,\tau)}{\theta_2^2(u,\tau)}\frac{\theta_3(u,\tau)}{\theta_3(0,\tau)}\frac{\theta_1(u,\tau)}{\theta_1(0,\tau)}\right\}^{(12)}.\end{aligned} \tag{40}$$

Then we can apply the transformation laws (20)-(23) for theta functions as well as the transformation laws (26), (27) to (40) to get the desired results. □

In addition to the above modular form, we have also constructed in Ref. 7 the modular form

$$\begin{aligned}\mathcal{P}_1(TZ, V, \xi, \tau) := \Big\{&e^{\frac{1}{24}E_2(\tau)(p_1(TZ)-p_1(V))}\\&\cdot\frac{\widehat{A}(TZ)\det^{1/2}\left(2\cosh\left(\frac{\sqrt{-1}}{4\pi}R^V\right)\right)}{\cosh^2\left(\frac{c}{2}\right)}\mathrm{ch}\left(\Theta_1(T_{\mathbf{C}}Z, V_{\mathbf{C}}, \xi_{\mathbf{C}})\right)\Big\}^{(12)},\end{aligned} \tag{41}$$

where

$$\begin{aligned}\Theta_1(T_{\mathbf{C}}Z, V_{\mathbf{C}}, \xi_{\mathbf{C}}) = &\bigotimes_{u=1}^{\infty}S_{q^u}(\widetilde{T_{\mathbf{C}}Z})\otimes\bigotimes_{v=1}^{\infty}\Lambda_{q^v}(\widetilde{V_{\mathbf{C}}}-2\widetilde{\xi_{\mathbf{C}}})\\&\otimes\bigotimes_{r=1}^{\infty}\Lambda_{q^{r-1/2}}(\widetilde{\xi_{\mathbf{C}}})\otimes\bigotimes_{s=1}^{\infty}\Lambda_{-q^{s-1/2}}(\widetilde{\xi_{\mathbf{C}}}).\end{aligned} \tag{42}$$

We showed in Ref. 7 that $\mathcal{P}_1(TZ, V, \xi, \tau)$ is a modular form of weight 6 over $\Gamma_0(2)$ while $\mathcal{P}_2(TZ, V, \xi, \tau)$ is a modular form of weight 6 over $\Gamma^0(2)$ and moreover they are modularly related in the sense that

$$\mathcal{P}_1\left(TZ, V, \xi, -\frac{1}{\tau}\right) = 2^{[\frac{\dim V}{2}]}\tau^6\mathcal{P}_2(TZ, V, \xi, \tau).$$

We call such a pair of modular forms a modular pair (see Ref. 7 for the cases of general dimensions).

One can use this modular pair $(\mathcal{P}_1(TZ,V,\xi,\tau),\mathcal{P}_2(TZ,V,\xi,\tau))$ to derive the Alvarez–Gaumé–Witten miraculous anomaly cancellation formula by setting $V=TZ$, $\xi=\mathbf{C}$ and obtain its various generalizations (see Refs. 6–8, 12–14 for details).

In the following subsection, we will use the modularity of $\mathcal{P}_2(TZ,F_1-F_2,\xi,\tau)$ to derive the Green–Schwarz type factorization formulas. It's amazing to see that all these anomaly cancellations due to Alvarez–Gaumé–Witten, Green–Schwarz as well as Schwarz–Witten can be derived from the modular pair $(\mathcal{P}_1(TZ,V,\xi,\tau),\mathcal{P}_2(TZ,V,\xi,\tau))$.

3.3. *Derivation of Green–Schwarz type factorizations from modularity*

From Proposition 1, we see that $\mathcal{P}_2(TZ,F_1-F_2,\xi,\tau)$ is a modular form of weight 6 over $\Gamma^0(2)$. Therefore, by Lemma 1, there exist $h_1,h_2\in\Omega^{12}(X)$ such that

$$\mathcal{P}_2(TZ,F_1-F_2,\xi,\tau)=h_0(8\delta_2)^3+h_1(8\delta_2)\varepsilon_2. \tag{43}$$

Therefore

$$\begin{aligned}&\left\{e^{\frac{1}{24}(1-24q+\cdots)(p_1(TZ)-p_1(F_1)+p_1(F_2))}\widehat{A}(TZ)\cosh\left(\frac{c}{2}\right)\right.\\&\left.\cdot\,\mathrm{ch}\Big(B_0+B_1q^{\frac{1}{2}}+B_2q+\cdots\Big)\right\}^{(12)}\\=&h_0(8\delta_2)^3+h_1(8\delta_2)\varepsilon_2\\=&h_0(-1-24q^{\frac{1}{2}}-24q-\cdots)^3\\&+h_1(-1-24q^{\frac{1}{2}}-24q-\cdots)(q^{\frac{1}{2}}+8q+\cdots).\end{aligned}\tag{44}$$

Comparing the coefficients of 1, $q^{\frac{1}{2}}$ and q in both sides of (44), we have

$$\left\{e^{\frac{1}{24}(p_1(TZ)-p_1(F_1)+p_1(F_2))}\widehat{A}(TZ)\cosh\left(\frac{c}{2}\right)\mathrm{ch}(B_0)\right\}^{(12)}=-h_0, \tag{45}$$

$$\left\{e^{\frac{1}{24}(p_1(TZ)-p_1(F_1)+p_1(F_2))}\widehat{A}(TZ)\cosh\left(\frac{c}{2}\right)\mathrm{ch}(B_1)\right\}^{(12)}=-h_1-72h_0, \tag{46}$$

$$\begin{aligned}&\Big\{e^{\frac{1}{24}(p_1(TZ)-p_1(F_1)+p_1(F_2))}(-(p_1(TZ)+p_1(F_1)-p_1(F_2)))\\&\quad\cdot\widehat{A}(TZ)\cosh\left(\frac{c}{2}\right)\mathrm{ch}(B_0)\\&\quad+e^{\frac{1}{24}(p_1(TZ)-p_1(F_1)+p_1(F_2))}\widehat{A}(TZ)\cosh\left(\frac{c}{2}\right)\mathrm{ch}(B_2)\Big\}^{(12)}\\&=-32h_1-1800h_0.\end{aligned}\tag{47}$$

By (45)-(47), we see that

$$\begin{aligned}&\Big\{e^{\frac{1}{24}(p_1(TZ)-p_1(F_1)+p_1(F_2))}(-(p_1(TZ)+p_1(F_1)-p_1(F_2)))\\&\quad\cdot\widehat{A}(TZ)\cosh\left(\frac{c}{2}\right)\mathrm{ch}(B_0)\\&\quad+e^{\frac{1}{24}(p_1(TZ)-p_1(F_1)+p_1(F_2))}\widehat{A}(TZ)\cosh\left(\frac{c}{2}\right)\mathrm{ch}(B_2)\Big\}^{(12)}\\&=\Big\{e^{\frac{1}{24}(p_1(TZ)-p_1(F_1)+p_1(F_2))}\widehat{A}(TZ)\cosh\left(\frac{c}{2}\right)\mathrm{ch}(32B_1-504B_0)\Big\}^{(12)}.\end{aligned}\tag{48}$$

In the following, let's expand $\Theta_2(T_{\mathbf{C}}Z,F_{1\mathbf{C}}-F_{2\mathbf{C}},\xi_{\mathbf{C}})$ to find B_0, B_1, B_2. In fact, we have

$$\begin{aligned}&\Theta_2(T_{\mathbf{C}}Z,F_{1\mathbf{C}}-F_{2\mathbf{C}},\xi_{\mathbf{C}})\\&=\bigotimes_{u=1}^{\infty}S_{q^u}(\widetilde{T_{\mathbf{C}}Z})\otimes\bigotimes_{v=1}^{\infty}\Lambda_{-q^{v-\frac{1}{2}}}(\widetilde{F_{1\mathbf{C}}}-\widetilde{F_{2\mathbf{C}}}-2\widetilde{\xi_{\mathbf{C}}})\\&\quad\otimes\bigotimes_{r=1}^{\infty}\Lambda_{q^{r-\frac{1}{2}}}(\widetilde{\xi_{\mathbf{C}}})\otimes\bigotimes_{s=1}^{\infty}\Lambda_{q^s}(\widetilde{\xi_{\mathbf{C}}})\\&=\bigotimes_{u=1}^{\infty}S_{q^u}(\widetilde{T_{\mathbf{C}}Z})\otimes\bigotimes_{v=1}^{\infty}\Lambda_{-q^{v-\frac{1}{2}}}(\widetilde{F_{1\mathbf{C}}})\\&\quad\otimes\frac{1}{\bigotimes_{v=1}^{\infty}\Lambda_{-q^{v-\frac{1}{2}}}(\widetilde{F_{2\mathbf{C}}})}\otimes\frac{1}{\bigotimes_{v=1}^{\infty}(\Lambda_{-q^{v-\frac{1}{2}}}(\widetilde{\xi_{\mathbf{C}}}))^2}\\&\quad\otimes\bigotimes_{r=1}^{\infty}\Lambda_{q^{r-\frac{1}{2}}}(\widetilde{\xi_{\mathbf{C}}})\otimes\bigotimes_{s=1}^{\infty}\Lambda_{q^s}(\widetilde{\xi_{\mathbf{C}}})\\&=(1+(T_{\mathbf{C}}Z-10)q+O(q^2))\\&\quad\otimes\left(1+(m-F_{1\mathbf{C}})q^{\frac{1}{2}}+\left(\wedge^2F_{1\mathbf{C}}-mF_{1\mathbf{C}}+\frac{m(m+1)}{2}\right)q+O(q^{\frac{3}{2}})\right)\\&\quad\otimes\left(1+(F_{2\mathbf{C}}-n)q^{\frac{1}{2}}+\left(S^2F_{2\mathbf{C}}-nF_{2\mathbf{C}}+\frac{n(n-1)}{2}\right)q+O(q^{\frac{3}{2}})\right)\end{aligned}$$

$$\begin{aligned}
&\otimes\left(1+2\widetilde{\xi_{\mathbf{C}}}q^{\frac{1}{2}}+\left(3\widetilde{\xi_{\mathbf{C}}}\otimes\widetilde{\xi_{\mathbf{C}}}+4\widetilde{\xi_{\mathbf{C}}}\right)q+O(q^{\frac{3}{2}})\right)\\
&\otimes\left(1+\widetilde{\xi_{\mathbf{C}}}q^{\frac{1}{2}}-2\widetilde{\xi_{\mathbf{C}}}q+O(q^{\frac{3}{2}})\right)\\
&\otimes(1+\widetilde{\xi_{\mathbf{C}}}q+O(q^2))\\
=&1+(m-F_{1\mathbf{C}}+F_{2\mathbf{C}}-n+3\widetilde{\xi_{\mathbf{C}}})q^{\frac{1}{2}}\\
&+\Big(\wedge^2F_{1\mathbf{C}}+S^2F_{2\mathbf{C}}-F_{1\mathbf{C}}\otimes F_{2\mathbf{C}}+T_{\mathbf{C}}Z+\frac{(m-n)^2+(m-n)}{2}\\
&\qquad-10-(m-n)(F_{1\mathbf{C}}-F_{2\mathbf{C}})+5\widetilde{\xi_{\mathbf{C}}}\otimes\widetilde{\xi_{\mathbf{C}}}\\
&\qquad+3(m-F_{1\mathbf{C}}+F_{2\mathbf{C}}-n+1)\otimes\widetilde{\xi_{\mathbf{C}}}\Big)q+O(q^{\frac{3}{2}}).
\end{aligned}\tag{49}$$

Therefore, we have

$$\begin{aligned}
B_0=&1,\\
B_1=&m-F_{1\mathbf{C}}+F_{2\mathbf{C}}-n+3\widetilde{\xi_{\mathbf{C}}},\\
B_2=&\wedge^2F_{1\mathbf{C}}+S^2F_{2\mathbf{C}}-F_{1\mathbf{C}}\otimes F_{2\mathbf{C}}+T_{\mathbf{C}}Z+\frac{(m-n)^2+(m-n)}{2}\\
&-10-(m-n)(F_{1\mathbf{C}}-F_{2\mathbf{C}})+5\widetilde{\xi_{\mathbf{C}}}\otimes\widetilde{\xi_{\mathbf{C}}}\\
&+3(m-F_{1\mathbf{C}}+F_{2\mathbf{C}}-n+1)\otimes\widetilde{\xi_{\mathbf{C}}}.
\end{aligned}\tag{50}$$

From (48), we see that

$$\begin{aligned}
&\left\{\widehat{A}(TZ)\cosh\left(\frac{c}{2}\right)\mathrm{ch}(B_2-32B_1+504B_0)\right\}^{(12)}\\
=&(p_1(TZ)-p_1(F_1)+p_1(F_2))\\
&\cdot\left\{-\frac{e^{\frac{1}{24}(p_1(TZ)-p_1(F_1)+p_1(F_2))}-1}{p_1(TZ)-p_1(F_1)+p_1(F_2)}\widehat{A}(TZ)e^{\frac{c}{2}}\mathrm{ch}(B_2-32B_1+504B_0)\right.\\
&\qquad\left.+e^{\frac{1}{24}(p_1(TZ)-p_1(F_1)+p_1(F_2))}\widehat{A}(TZ)e^{\frac{c}{2}}\mathrm{ch}(B_0)\right\}^{(8)}.
\end{aligned}\tag{51}$$

However, from (50), we have

$$\begin{aligned}
&B_2-32B_1+504B_0\\
=&\wedge^2F_{1\mathbf{C}}+S^2F_{2\mathbf{C}}-F_{1\mathbf{C}}\otimes F_{2\mathbf{C}}+T_{\mathbf{C}}Z\\
&+\frac{(m-n-32)(m-n-31)}{2}-2-(m-n-32)(F_{1\mathbf{C}}-F_{2\mathbf{C}})\\
&+5\widetilde{\xi_{\mathbf{C}}}\otimes\widetilde{\xi_{\mathbf{C}}}+3(m-n-31-F_{1\mathbf{C}}+F_{2\mathbf{C}})\otimes\widetilde{\xi_{\mathbf{C}}}.
\end{aligned}\tag{52}$$

Theorem 1 follows from (51) and (52).

Acknowledgements

The first author is partially supported by a start-up grant from National University of Singapore. The second author is partially supported by NSF. The third author is partially supported by MOE and NNSFC.

References

1. L. Alvarez-Gaumé and E. Witten, Gravitational anomalies, *Nucl. Physics*, B234, 269-330, (1983).
2. M. F. Atiyah, $K-theory$. Benjamin, New York, 1967.
3. N. Berline, E. Getzler and M. Vergne, *Heat Kernels and Dirac Operators*. Springer, 2002.
4. K. Chandrasekharan, *Elliptic Functions*. Springer-Verlag, 1985.
5. M. B. Green, J. H. Schwarz, Anomaly cancellations in supersymmetric d=10 gauge theory and superstring theory, *Physics Letters* B149 (1984) pp. 11722.
6. F. Han and K. Liu, Gravitational anomaly cancellation and modular invariance, *Arxiv:1007.5295.*
7. F. Han, K. Liu and W. Zhang, Modular forms and generalized anomaly cancellation formulas, *J. Geom. Phys.*, 62 (2012), pp. 1038-1053.
8. F. Han and W. Zhang, Modular invariance, characteristic numbers and η invariants. *J. Diff. Geom.* 67 (2004), 257-288.
9. F. Hirzebruch, *Topological Methods in Algebraic Geometry*. Springer-Verlag, 1966.
10. F. Hirzebruch, T. Berger and R. Jung, *Manifolds and Modular Forms*. Aspects of Mathematics, vol. E20, Vieweg, Braunschweig 1992.
11. M. Kaneko and D. Zagier, A generalized Jacobi theta function and quasi-modular forms. in *The moduli space of curves,* Ed. R. Dijkgraaf, C. Faber and G. van der Geer, Boston: Birkhäuser, (1995), 165-172.
12. K. Liu, Modular invariance and characteristic numbers, *Comm. Math. Phys*, 174, 29-42 (1995).
13. K. Liu, Modular forms and topology. *Moonshine, the Monster, and related topics (South Hadley, MA, 1994), 237–262, Contemp. Math.*, 193, Amer. Math. Soc., Providence, RI, 1996.
14. K. Liu and Y. Wang, A note on modular forms and generalized anomaly cancellation formulas, *Arxiv:1109.5794*.
15. S. Ochanine, Signature modulo 16, invariants de Kervaire géneralisés et nombre caractéristiques dans la K-théorie reelle. *Mémoire Soc. Math. France*, Tom. 109 (1987), 1-141.
16. J. H. Schwarz, Anomaly cancellation: a retrospective from a modern perspective, *2001: A Spacetime Odyssey, Proceedings of the Inaugural Conference of the Michigan Center for Theoretical Physics*, Ed. M.J. Duff and J.T. Liu, 2002. [*Arxiv: hep-th/0107059*].
17. J. H. Schwarz and E. Witten, Anomaly analysis and brane-antibrane system, *JHEP*, 0103, 032 (2001).

18. W. Zhang, *Lectures on Chern-Weil Theory and Witten Deformations.* Nankai Tracts in Mathematics Vol. 4, World Scientific, Singapore, 2001.

On Interior Estimates for Mean Curvature of Convex Surfaces in R^3 and Its Applications

Jiaxing Hong*

School of Mathematics Sciences, LMNS, Fudan University, Shanghai, China, 200433

jxhong@fudan.edu.cn

In the present paper a direct proof for the interior estimates for mean curvature of convex surfaces in R^3 is given. Meanwhile, when the Gaussian curvature of such a C^2- convex surface is degenerate on the boundary at degree m where m is some integer, a global lower bound away from zero for mean curvature is also obtained. Therefore, global smooth isometric embedding for smooth positive disks in R^3 can be constructed by means of the techniques in Ref. 12.

1. Main Results

In 1916, Weyl[16] stated and partially solved the problem of whether there exists a convex surface in R^3 with the given positively curved metric on S^2. The existence in R^3 of such an isometric embedding is called Weyl problem. In 1950s, Nirenberg[14] and Pogorelov[15] independently and completely solved Weyl problem. Later Guan-Li,[3] and Hong-Zuily[13] independently generalized to the case of nonnegatively curved metric. Also Hong[10] proved the existence of isometric embedding for any complete smooth nonnegatively curved metric defined on R^2. In 1970, Pogorelov [15] studied a kind of boudnary value problem for a given positively curved metric g defined on the closed unit disk $D \subset R^2$, sometimes denoted by $(\bar{D}, g)$ and called positive disk. In Ref. 18 it is called Dirichlet problem. Such an isometric embedding in R^3 of $(\bar{D}, g)$ leads to solve a nonlinear partial differential equation of Monge Ampere type

$$\det(\nabla^2 z) = K \det(g_{ij})(1 - g^{ij} z_i z_j) \ \text{ in } D \subset R^2, \tag{1}$$

*Supported by NNSF(11121101 and 11131005).

where $\nabla^2 z = (z_{ij} - \Gamma_{ij}^k z_k)$ denotes the Hessian of z with respect to the given smooth metric $g = g_{ij} du^i du^j$ defined on $\bar{D}$, g^{ij} the inverse of the metric tensor and K the curvature of the metric g. Write the Gauss equations of the desired isometric embedding $\vec{r} = (x, y, z)$,

$$\vec{r}_{ij} = \Gamma_{ij}^l \vec{r}_l + h_{ij} \vec{n}, \qquad i, j = 1, 2,$$

where h_{ij} are the coefficients of its second fundamental form and $\vec{n}$ is its normal. Then computing the inner products of the last expressions and the unit vector $\vec{k}$ of the z axis we have

$$\begin{aligned} \nabla_{ij} z &= h_{ij}(\vec{r}, \vec{n}) \\ \text{and hence,} \quad \det(\nabla^2 z) &= \det(h_{ij})(\vec{n}, \vec{k})^2. \end{aligned} \tag{2}$$

Notice that $\det(h_{ij}) = K \det(g_{ij})$ and

$$(\vec{n}, \vec{k})^2 = 1 - \left(\frac{\vec{r}_1 X \vec{r}_2}{|\vec{r}_1 X \vec{r}_2|} X \vec{k} \right)^2 = 1 - g^{ij} z_i z_j. \tag{3}$$

It turns out that z satisfies (1) which is called Darboux equation. Ones are very interested in the existence of the smooth convex caps, i.e., a convex surface whose boundary is on some plane and the projection of this surface on this plane is a one to one mapping. This leads to solve (1) with

$$z = 0 \quad \text{on } \partial\Omega \tag{4}$$

and a subsidiary condition

$$|\nabla z|^2 = g^{ij} z_i z_j < 1 \quad \text{on } \bar{\Omega}. \tag{5}$$

We say z is a global smooth solution to the problem (1) if $z \in C^\infty(\bar{D})$ is a solution to (1). If z is a global smooth solution to (1) with (4) and (5), it follows that $g - dz^2$ is a smooth flat metric defined on $\bar{D}$, for details, refer to Refs. 1 and 8, p.36 and we can find two functions x and y in $C^\infty(\bar{D})$ such that $dx^2 + dy^2 = g - dz^2$, i.e., $\vec{r} = (x, y, z)$ is the desired convex cap. Therefore the existence of smooth isometric embedding of $(\bar{D}, g)$ is equivalent to solving a global smooth solution to (1) with (4) and (5). As is well known, the problem (1) with (4) is not always solvable. There are some counter examples, for example, Ref. 4, Appendix 3 for an analytic positively curved metric and Ref. 11 for a positive disk with the geodesic curvature on ∂D, k_g changing its sign. Therefore we always assume that

$$\partial D \text{ has positive geodesic curvature }, i.e., k_g > 0 \text{ on } \partial D. \tag{6}$$

and

$$\text{the Gaussian curvature } K > 0 \text{ on } \bar{D}. \tag{7}$$

Under the assumptions (6) and (7), a local smooth solution (in $C^\infty(D) \cap C^{0.1}(\bar{D})$) to the problem (1) with (4) and (5) is obtained in Ref. 15, p. 104. Instead of (7) for the case of curvature $K > 0$ in D and $K = 0 \neq |\nabla K|$ on ∂D, a global smooth solution in $C^\infty(\bar{D})$ is also given in Ref. 9. In study of Darboux equation (1) we have to control the lower bound of the factor $(1 - |\nabla z|^2)$ in its right hand side away from null, otherwise this factor might destroy its ellipticity. This is a difficulty we ran across even when the curvature K is strictly positive on $\bar{D}$. The paper[9] gives such a lower bound for solutions to (1) with (4) by means of an interior estimates in Refs. 6 and 7 related to the theory of 2×2 elliptic system of second order equations. Ones wish there is a direct proof for such interior estimates of mean curvature of convex caps. This is one motivation of the present paper. The Main result is as follows.

Theorem 1. *Let g be a smooth metric defined $\bar{D}$ with curvature $K > 0$ in D and let $\vec{r} = (x, y, z) \in C^\infty(\bar{D})$ its isometric embedding satisfying (4). Then its mean curvature*

$$H(p) \leq C \exp \frac{8l}{|z(p)|} \quad \forall p \in D$$

where the constant C depends only on the intrinsic diameter l of $(\bar{D}, g)$ and $|g|_4$.

Section 2 shows that $-Cz(p) \geq dis(p, \partial D), \forall p \in \partial D$ for some constant C under control and hence, combining it with the above theorem soon gives the desire interior estimates for the mean curvature H. Therefore, by means of Theorem 12.4 in Ref. 5 without difficulty we can get the interior Holder estimates for the second fundamental form of the given positive convex cap. To author's knowledge it seems that the usual way of multiplying a cut off function does not work for the above interior estimates.

Up to now it is unknown whether Problem (1) with (4) and (5) still admits a global smooth solution under the assumptions (6) and

$$K = \varphi^m \tilde{K} \text{ for some smooth strictly positive function } \tilde{K} \text{ on } \bar{D} \qquad (8)$$

where m is nonnegative integer, φ a smooth defining function of ∂D. (8) means the curvature K degenerate at finite degree at ∂D. This is another motivation of the present paper. As we know, since $K = 0$ on ∂D, it follows that there is at least one eigenvalue of Hessian z tending to zero while going to ∂D. In order to obtain smooth solution, as a key step, we must separate the eigenvalues of Hessian z. (8) makes such arguments much more difficult than the case in Ref. 9 where K is degenerate at first degree. Following the technique in Ref. 12 and generalizing it to 2-dimensional Riemannian manifolds, we have

Theorem 2. *Let $(\bar{D}, g)$ be smooth and (6)(8) be fulfilled. Then for any solution $z \in C^2(\bar{D})$ to (1) with (4) and (5) there holds*

$$\Delta_g z \geq \frac{1}{C} \text{ for some constant } C$$

depending only on the intrinsic diameter l of $(\bar{D}, g)$ and $|g|_4$, the maxima of $1/\tilde{K}$ and $1/k_g$ on $\bar{D}$ and ∂D respectively.

Theorem 1 and Theorem 2 present all C^2- estimates for the isometric embedding of the given positive disks. Higher order estimates as well as the existence of its global smooth isometric embedding can be constructed by means of the techniques in Refs. 9 and 12.

2. Interior Estimates for Mean Curvature of Convex Cap

This section is concerned with some geometric lemmas related to smooth disks with positive or nonnegative curvature.

The proof for Theorem 1. Let g be a smooth metric in $\bar{D}$ with curvature $K > 0$ in D and let $\vec{r} = (x, y, z) \in C^\infty(\bar{D})$ be its isometric embedding in $\mathbb{R}^3$ satisfying (4). We first recall a differential equation satisfied by its mean curvature H. For any $p \in D$, in the sequel we often take the orthonormal coordinates centered at p and then have

$$Kh^{ij}H_{ij} = (h_{12,l}h_{12,l} - h_{11,l}h_{22,l}) + 2KH^2 + \frac{1}{2}(\Delta K - 4K^2) \text{ at } p \tag{9}$$

where $h_{ij}, i, j = 1, 2$ denote the coefficients of the second fundamental form of $\vec{r}$ and (h^{ij}) is the inverse matrix of (h_{ij}). See Refs. 14 and 17 or 8, p. 182 for its derivation. In what follows we shall derive its interior estimates.

Set $w = He^\phi$ where ϕ is a function to be specified. Then a direct differentiation of w yields

$$\begin{aligned} w_i &= (H_i + \phi_i H)e^\phi \\ w_{ij} &= \Big(H_{ij} + H_i\phi_j + H_j\phi_i + H(\phi_{ij} + \phi_i\phi_j)\Big)e^\phi. \end{aligned} \tag{10}$$

Then w solves,

$$e^{-\phi}Kh^{ij}w_{ij} = (h_{12,l}h_{12,l} - h_{11,l}h_{22,l}) \\ + 2KH^2 + \frac{1}{2}(\Delta_g K - 4K^2) + 2Kh^{ij}H_i\phi_j + HKh^{ij}(\phi_{ij} + \phi_i\phi_j). \tag{11}$$

If $\phi \longrightarrow -\infty$ as $p \to \partial D$, it follow that w attends its maximum at some point $p \in D$ where orthonormal coordinates are taken. Then it is easy to see $H(p) > 0$. From

$$w_i = 0, \left(w_{ij}\right) \leq 0 \text{ at } p,$$

and (10) it follows that $H_i(p) = -\phi_i H(p)$ and

$$0 \geq e^{-\phi}Kh^{ij}w_{ij} = (h_{12,l}h_{12,l} - h_{11,l}h_{22,l}) \\ + 2KH^2 + \frac{1}{2}(\Delta_g K - 4K^2) - \frac{K}{H}h^{ij}H_iH_j + HKh^{ij}\phi_{ij} \text{ at } p. \tag{12}$$

Without loss of generality we may assume $z_{12}(p) = 0$ and hence, $h_{12}(p) = 0$. Fix

$$\phi = \frac{1}{2l^2}|\vec{r}|^2 + a\frac{1}{z}, \text{ and obviously, } \phi \to -\infty \text{ as } p \to \partial D$$

where a is a positive constant to be fixed and l is the intrinsic diameter of $(\bar{D}, g)$. Hence,

$$\begin{aligned} \phi_i &= \frac{1}{l^2}(\vec{r}, \vec{r}_i) - \frac{a}{z^2}z_i \\ Kh^{ij}\phi_{ij} &= Kh^{ii}\phi_{ii} \\ &= Kh^{ii}\Big(\frac{1}{l^2}(\vec{r}, \vec{r}_{ii}) + \frac{1}{l^2}\delta_{ii} + \frac{2a}{z^3}z_i^2 - \frac{a}{z^2}z_{ii}\Big) \\ &= Kh^{ii}\Big(\frac{1}{l^2}h_{ii}(\vec{r}, \vec{n}) + \frac{1}{l^2}\delta_{ii} + \frac{2a}{z^3}z_i^2 - \frac{a}{z^2}z_{ii}\Big) \\ &= \frac{2}{l^2}K(\vec{r}, \vec{n}) + \frac{2}{l^2}H + I. \end{aligned} \tag{13}$$

By means of the fact that $z_{ij} = h_{ij}\sqrt{1 - |\nabla z|^2}$ and $H_i = -\phi_i H$ at p, it follows that

$$\begin{aligned}I &= -\frac{a}{z^2}Kh^{ii}z_{ii} + \frac{2a}{z^3}Kh^{ii}z_i^2\\&= -\frac{2a}{z^2}K\sqrt{1-|\nabla z|^2} + \frac{2z}{a}Kh^{ii}\Big(\frac{H_i}{H} + \frac{1}{l^2}(\vec{r},\vec{r}_i)\Big)^2\\&= \frac{2z}{aH^2}Kh^{ii}H_i^2 + \frac{4z}{aHl^2}Kh^{ii}H_i(\vec{r},\vec{r}_i) + Kh^{ii}\frac{2z}{al^4}(\vec{r},\vec{r}_i)^2 - \frac{2a}{z^2}K\sqrt{1-|\nabla z|^2}\\&\geq -\frac{4l}{a}\Big(\frac{Kh^{ii}H_i^2}{H^2} + \frac{2H}{l^2}\Big) - \frac{C_a}{z^2}.\end{aligned} \tag{14}$$

Consequently, combining (13) with (14) gives

$$\begin{aligned}Kh^{ij}\phi_{ii} &\geq \big(2 - \frac{8l}{a}\big)\frac{H}{l^2} - \frac{4l}{a}\frac{Kh^{ii}H_i^2}{H^2} - \frac{C_a}{z^2}\\&\geq \frac{H}{l^2} - \frac{1}{2}\frac{Kh^{ii}H_i^2}{H^2} - \frac{C_a}{z^2} \quad \text{if we take } a = 8l.\end{aligned} \tag{15}$$

We must compute the first term in the right hand side of (12). Indeed, distinguish two cases:

Case 1. $H^2(p) \leq 2\sup_D K$. Then the present lemma is readily proved.

Case 2. $H^2(p) > 2\sup_D K$. Under the present circumstance, there holds

$$(h_{11} - h_{22})^2 = 4(H^2 - K) \geq 2H^2 \text{ at } p. \tag{16}$$

Since at p,

$$\begin{aligned}h_{11,i} + h_{22,i} &= 2H_i\\h_{22}h_{11,i} + h_{11}h_{22,i} &= K_i\end{aligned} \tag{17}$$

by a differentiation of the equation: $\det(h_{ij}) = K\det(g_{ij})$, we have

$$h_{11,i} = \frac{2h_{11}H_i - K_i}{h_{11} - h_{22}}, \quad h_{22,i} = \frac{2h_{22}H_i - K_i}{h_{22} - h_{11}}. \tag{18}$$

Hence

$$\begin{aligned}&h_{12,i}h_{12,i} - h_{11,i}h_{22,i}\\&= (h_{12,1})^2 + (h_{12,2})^2 - h_{11,1}h_{22,1} - h_{11,2}h_{22,2}\\&= \frac{1}{4(H^2-K)}\Big(8HKh^{ii}H_i^2 - 4Kh^{ij}H_iK_i - 4HH_iK_i + 2|\nabla K|^2\Big).\end{aligned} \tag{19}$$

Inserting (15), (19) into (12) we have, at p,

$$0 \geq e^{-\phi} K h^{ij} w_{ij} \geq \frac{3H^2+K}{2H(H^2-K)} K h^{ii} H_i^2 + \frac{HKh^{ii}+H^2\delta_{ii}}{H^2-K}\Big(\frac{1}{l^2}(\vec{r},\vec{r}_i) - \frac{az_i}{z^2}\Big)K_i + \frac{|\nabla K|^2}{2(H^2-K)} + \frac{H^2}{l^2} - \frac{C}{z^2}(H+1).$$

In view of (16) one can soon obtain

$$0 \geq e^{-\phi} K h^{ij} w_{ij} \geq \frac{H^2}{l^2} - \frac{C}{z^2}(H+1) \text{ at } p.$$

Multiplying the both sides of the above inequality by $e^{2\phi}$ gives at once

$$0 \geq w^2 - C(w+1) \text{ at } p \tag{20}$$

for some constant C depending only on the quantities mentioned in the present lemma since $z^{-2}e^{\phi}$ bounded above by some universal constant. Thus so far we have proved, at any point $q \in D$

$$w(q) \leq \max_{\bar{D}} w = w(p) = C$$

for another constant C under control. Therefore

$$H(q) \leq C \exp\{-\frac{1}{2l^2}|\vec{r}|^2\} \exp\{\frac{8l}{|z|}\} \leq C \exp\{\frac{8l}{|z|}\} \quad \forall q \in D.$$

This ends the proof for Theorem 1.

Next we shall first consider the following boundary value problem:

$$\Delta_g \tilde{\psi} = 2\sqrt{K}, \text{ in } D \text{ with } \tilde{\psi} = 0 \text{ on } \partial D. \tag{21}$$

By the standard theory on elliptic boundary value problem the above Dirichlet problem always admits a solution $\tilde{\psi} \in W^{2,p}(D) \cap C^1(\bar{D})$ for any $p \in (2,\infty)$ since $\sqrt{K} \in L^\infty(D)$. From the Hopf lemma it is easy to see $\partial_n \tilde{\psi} > 0$ on ∂D as long as K is not identically zero. Define

$$\psi^+ = \frac{\theta \tilde{\psi}}{\sqrt{1 + \sup_D |\nabla \tilde{\psi}|^2}}$$

for some positive constant $\theta \in (0,1)$ to be specified. Obviously,

$$\min_{\partial D} \frac{\partial \psi^+}{\partial n} > 0 \quad \text{on } \partial D \quad \text{and } |\nabla \psi^+| < 1 \text{ in } D \quad \forall \theta \in (0,1).$$

And a direct computation at once provides

$$\begin{aligned}\Delta_g\psi^+ &= 2\frac{\theta\sqrt{K}}{\sqrt{1+|\nabla\tilde{\psi}|^2}}\\ &= \frac{\theta\sqrt{1+|\nabla\tilde{\psi}|^2}}{\sqrt{1+(1-\theta^2)\sup_D|\nabla\tilde{\psi}|^2}}2\sqrt{K(1-|\nabla\psi^+|^2)}\\ &\leq 2\sqrt{K(1-|\nabla\psi^+|^2)} \text{ in } D\end{aligned}$$

if we take $\theta = 1/\sqrt{2}$. It turns out

$$\Delta_g\psi^+ \leq 2\sqrt{K(1-|\nabla\psi^+|^2)} \quad \text{in } D \text{ with } \psi^+ = 0, \frac{\partial\psi^+}{\partial n} > 0 \text{ on } \partial D. \tag{22}$$

Lemma 1. *Let g be a smooth metric defined on $\bar{D}$ with curvature $K > 0$ in D. Suppose that z is a global smooth solution to (1) with (4) and (5). Then the following inequalities,*

$$z(p) \leq -\frac{1}{C}dis_g(p,\partial D) \text{ and } |\nabla z(p)| \leq \max_{\partial D}|\nabla z| \leq 1 \quad \forall p\in D \tag{23}$$

hold for a constant C depending only on the intrinsic diameter l of $(\bar{D}, g)$, $|g|_4$ and the maximum of the reciprocal of $\partial\psi^+/\partial n$ over ∂D.

Proof. By the geometric-arithematic mean theorem (1) implies

$$\Delta_g z \geq 2\sqrt{K(1-|\nabla z|^2)} \quad \text{in } D \text{ with } z = 0, \text{ on } \partial D. \tag{24}$$

Combining (24) with (22) gives

$$\Delta_g(z-\psi^+) + 2\sqrt{K}\frac{(\nabla\psi^+ + \nabla z)\nabla(z-\psi^+)}{\sqrt{1-|\nabla z|^2}+\sqrt{1-|\nabla\psi^+|^2}} \geq 0 \text{ in } D$$
$$\text{with } z-\psi^+ = 0 \text{ on } \partial D.$$

By the maximum principle we know $z \leq \psi^+$ on $\bar{D}$. In view of the last part of (22), we have

$$z(p) \leq -\frac{1}{2}\min_{\partial D}\frac{\partial\psi}{\partial n}dis_g(p,\partial D)$$

when p is in a neighbourhood of ∂D. Thus the first part of (23) is true for another large constant depending only on the quantities mentioned in the present lemma. As for the second part, it is the immediate consequence of the geodesic convexity of z. □

Remark 1. Let the assumptions in Lemma 1 and (8) be fulfilled. Then the constant C mentioned in (23) depends only on the intrinsic diameter l of $(\bar{D}, g)$, $|g|_4$ and the maximum of $1/\tilde{K}$ on $\bar{D}$.

Indeed, replacing K by $\varphi^m \inf_D \tilde{K}$ in (21), by the same arguments we can show that the lower bound of $\partial\psi^+/\partial_n$ on ∂D, is completely determined by $\inf_D \tilde{K}$ and hence the present remark is proved.

In what follows we need the geodesic coordinates based on ∂D in which the given metric can be written in the form:

$$\begin{aligned} g = B^2 ds^2 + dt^2 \text{ for some function } B(s,t) \in C^\infty([0, 2\pi] \times [-2\delta, 0]) \\ \text{where} \quad B(s, 0) = 1 \text{ and } B_t(s, 0) = k_g(s) > 0, \end{aligned} \tag{25}$$

for some positive constant δ completely determined by the given metric. Here $t = 0$ corresponds to ∂D. Also we assume that

$$B_t \geq \frac{1}{2} \min_{\partial D} k_g, \quad \forall t \in [-2\delta, 0] \tag{26}$$

Lemma 2. *Let $g \in C^\infty(\bar{D})$ be a metric with curvature $K > 0$ in D and let (6) be fulfilled. Then the following inequality*

$$|\nabla z| \leq C\Big(l, |g|_4, \frac{1}{\min_{\partial D} k_g}, \frac{1}{\min_{\partial D} \frac{\partial \psi^+}{\partial n}}, \frac{1}{\min_{D_\delta} K}, \frac{1}{\delta}\Big) < 1$$

holds.

Proof. Under the geodesic coordinates (25)(26) we have

$$\sup_{-2\delta \leq t \leq 0} \frac{BK}{\partial_t B} \leq M(l, |K|_0, |g|_2, \frac{1}{\min k_g})$$

for some constant M. Set

$$\phi = -\int_t^0 \sqrt{1 - \exp\{-M(R + \tau)\}} d\tau, -2\delta \leq t \leq 0.$$

A simple computation yields

$$\det(\nabla^2 \phi) \geq KB^2(1 - |\nabla\phi|^2), -2\delta \leq t \leq 0 \tag{27}$$

for arbitrary $R > 2\delta$.

Next we choose R so large that ϕ is a subbarrier as $-2\delta \leq t \leq 0$ to (1) with (4) in the region $-2\delta \leq t \leq 0$. Since $\vec{r} = (x, y, z)$ is a global smooth convex cap, Theorem 1 and Lemma 1 provide us with an above bound for its mean curvature H on $\bar{D}_\delta$, which depends only on $1/\delta$, l, $|g|_4$

and the maximum of the reciprocal of $\partial\psi^+/\partial n$ over ∂D. So their principal curvatures k_i, $i = 1, 2$ satisfy

$$C \geq k_2 \geq k_1 \geq \frac{K}{C} \geq \frac{1}{C_\delta} \text{ on } \bar{D}_\delta$$

for some constant C_δ under control. Now let us study the image of the Gauss map G from $\vec{r}(D_\delta) \to S^2$. Denote the angle α between $\vec{n}(q_1) \in S^2$ and $\vec{n}(q_2) \in S^2$ by $\alpha < \vec{n}(q_1), \vec{n}(q_2) >$. From the third fundamental form

$$\begin{aligned} d\vec{n}^2 &= 2Hh_{ij}du^i du^j - Kg_{ij}du^i du^j \\ &\geq k_1^2 g_{ij}du^i du^j \geq \frac{1}{C_\delta} g_{ij}du^i du^j \quad \forall t \in [-2\delta, -\delta]. \end{aligned}$$

Suppose that the distance between $G(\vec{r}(\partial D_\delta))$ and $G(\vec{r}(\partial D_{2\delta}))$ is attended by the curve $\vec{\gamma}$ with the terminals $\vec{\gamma}(q_1) \in G(\vec{r}(\partial D_\delta))$ and $\vec{\gamma}(q_2) \in G(\vec{r}(\partial D_{2\delta)})$. Then we have

$$\alpha < \vec{\gamma}(q_1), \vec{\gamma}(q_2) >= \int_\gamma \sqrt{d\vec{n}^2} \geq \frac{1}{\sqrt{C_\delta}} \int_{G^{-1}(\gamma)} \sqrt{g_{ij}du^i du^j} \geq \frac{1}{\sqrt{C_\delta}}\delta.$$

Therefore the distance of the image of the spherical mapping $G(\vec{r}(\partial D_{2\delta}))$ from the equator $\geq 1/C_\delta$ for another bigger constant C_δ under control since $\vec{r}$ is a convex cap and the spherical mapping for convex surfaces is homeomorphic. Consequently

$$|\nabla z|^2 = 1- < \vec{n}, \vec{k} >^2 \leq 1 - \sin^2\left(\frac{1}{C_\delta}\right) = \Theta^2 < 1 \text{ on } \partial D_{2\delta}.$$

Now let us fix the constant R so large that

$$\sqrt{1 - \exp\{-M(R - 2\delta)\}} > \Theta$$

which implies

$$\phi_t > \partial_t z \quad \text{on } t = -2\delta. \tag{28}$$

With the aid of (27), (1), (28) and the fact that $\phi = z = 0$ on ∂D, an application of the maximum principle to $\phi - z_\lambda$ on the region $-2\delta \leq t \leq 0$ provides at once

$$\phi \leq z, \text{ as } -2\delta \leq t \leq 0$$

since $(\varphi - z)$ attains its maximum at $t = 0$. Thus, we have

$$1 > \phi_t(0) = \sqrt{1 - e^{-MR}} \geq \sup_{\partial D} \partial_t z = \max_{\partial D} |\nabla z|.$$

This completes the proof of the present lemma since $|\nabla z|$ attends its maximum on ∂D. □

The following *a priori* estimates are very useful and independent of Theorem 1.

Lemma 3. *Let g be a smooth metric defined on $\bar{D}$ and let (6)(8) be fulfilled. Suppose that z is a global smooth solution to (1) with (4) with (5). Then*

$$\max_{\bar{D}} |D^2 z| \le C\Big(l, |g|_4, \quad \frac{1}{\min_{\partial D} k_g}, \quad \frac{1}{\min_{\partial D} \frac{\partial \psi^+}{\partial n}}\Big). \tag{29}$$

It is worth pointing out that the bound given in (29) is independent of the lower bound of the Gauss curvature over $\bar{D}$. We can take this advantage to the nonnegative curvature case. The proof for (29) can be obtained by a standard procedure, for details, see the paper [9].

3. Estimates for the Lower Bound of $\Delta_g z$

Since for the given metric in Theorem 2, its Gaussian curvature $K(p)$ tends to zero while p is going to the boundary ∂D, it follows that at least there is an eigenvalue of $(\nabla^2 z)$ going to zero. In order to get smooth solutions we must separate the eigenvalues of the Hessian $(\nabla^2 z)$. In two dimensional case it suffices to estimate the lower bound of $\Delta_g z$ away from zero.

On ∂D it is easy to get the lower bound of $\Delta_g z$. Indeed, for any point $p \in \partial D$, under the coordinates (25) from Lemma 1 and the fact that $z = 0$ on ∂D it turns out

$$\nabla_{11} z = z_{ss} - \frac{B_s}{B} z_s + B B_t z_t = k_g z_t \ge \frac{1}{C} \tag{30}$$

for some positive constant C under control. Hence, on ∂D, i.e., at $t = 0$,

$$\Delta_g z = \frac{1}{B^2} \nabla_{11} z + \nabla_{22} z \ge \nabla_{11} z \ge \frac{1}{C}.$$

Such a positive lower bound over $\bar{D}$ is the key point of finding global smooth solutions for (1) in the degenerate case. Unfortunately, the way to directly estimate the lower bound of $\Delta_g z$ does not work, we have to find an auxiliary function to replace $\Delta_g z$. Suppose that $X = b^k \partial_k$ is a smooth vector field defined on $\bar{D}$. Denote by (a^{ij}) the cofactor matrix of $(\nabla_{ij} z)$. Throughout the present section, unless otherwise stated, the normal coordinates centered at the point p considered are always used and moreover, without loss of generality we assume $z_{12}(p) = 0$. Then a direct differentiation of the equation in (1) yields

$$a^{ij} z_{ijl} + 2K z_\lambda z_{\lambda l} = K_l (1 - |\nabla z|^2). \tag{31}$$

By Ricci identity $z_{ijl} = z_{lij} + z_m R_{mijl}$ where R_{mijl} are the components of Riemann tensor, multiplying the above identity by b^l and summing up we have

$$\begin{aligned} &a^{ij}((Xz)_{ij} - G_{ij}(z)) + 2Kz_\lambda(Xz)_\lambda \\ &= XK(1-|\nabla z|^2) + 2K(1-|\nabla z|^2)div(X) - a^{ij}z_m b^l R_{mijl} + 2Kb^l_\lambda z_l z_\lambda \end{aligned} \quad (32)$$

where $G_{ij}(u) = b^l{}_{,ij}\, z_l$. Suppose that $X = b^l\partial_l = z_2\partial_1 - z_1\partial_2 = T^{lm}z_m\partial_l$ where

$$T = (T^{lm}) = \begin{pmatrix} 0 & 1 \\ -1 & 0 \end{pmatrix}.$$

Define

$$\begin{aligned} \Phi &= \nabla z_{22}z_1^2 - 2\nabla z_{12}z_1z_2 + \nabla z_{11}z_2^2 \\ &= z_{22}z_1^2 - 2z_{12}z_1z_2 + z_{11}z_2^2 \quad \text{at } p \\ &= T^{\lambda m}T^{\mu l}z_m z_l z_{\lambda\mu} = a^{ij}z_iz_j. \end{aligned} \quad (33)$$

Lemma 4. *Then there hold*

$$X(z) = 0 \text{ and } div(X) = 0 \quad (34)$$

$$G_{ij}(z) = T^{\lambda m}z_{mij}z_\lambda \quad (35)$$

$$T^{\lambda m}T^{\mu l}z_m z_l z_{\lambda\mu i} = \Phi_i - 2z_i\det(z_{ij}), i = 1, 2. \quad (36)$$

Proof. Since $X(z) = z_2\partial_1 z - z_1\partial_2 z = 0$ and $div(X) = \partial_1 z_2 + \partial_2(-z_1) = 0$, (34) is readily proved. According to the definition of $G_{ij}(z)$ we have

$$G_{ij}(z) = (T^{\lambda m}z_m)_{ij}z_\lambda = T^{\lambda m}z_{mij}z_\lambda.$$

This is just (35). By the definition of Φ we have

$$\Phi_i = T^{\lambda m}T^{\mu l}z_m z_l z_{\lambda\mu i} + \xi_i \text{ where}$$

$$\xi_1 = 2z_{22}z_{11}z_1 - 2z_{12}^2z_1 - 2z_{12}z_{11}z_2 + 2z_{11}z_{21}z_2 = 2z_1\det(z_{ij})$$

and $\xi_2 = 2z_2\det(z_{ij})$. Hence, (36) follows immediately and the present lemma is proved. □

As is pointed out in [12], in Euclidean space Φ defined in (33) is the product of $|Dz|^3$ and the curvature of the level set of $z = const$.

Lemma 5. *With $D_0 = a^{ij} z_m b^l R_{mijl} - XK(1-|\nabla z|^2) - 2Kb^l_\lambda z_\lambda z_l = 0(K)$ and $\sigma = K(4-5|\nabla z|^2)$ there hold*

$$\begin{aligned}
\Phi G_{11}(z) &= -a^{22}(\Phi_1 z_2 + \Phi_2 z_1) + D_0 z_1^2 + a^{22} z_1 z_2 \sigma \\
\Phi G_{12}(z) &= (a^{11}\Phi_1 z_1 - a^{22}\Phi_2 z_2) + D_0 z_1 z_2 \\
&\quad + 2K(1-\nabla z|^2)(a^{22} z_2^2 - a^{11} z_1^2) + a^{11} z_1^2 K|\nabla z|^2 \\
\Phi G_{22}(z) &= a^{11}(\Phi_1 z_2 + \Phi_2 z_1) + D_0 z_2^2 - a^{11} z_1 z_2 \sigma \\
G_{21} &= G_{12} - K|\nabla z|^2.
\end{aligned}$$

Proof. By means of Ricci identity it is easy to see

$$\begin{aligned}
G_{21} &= T^{\lambda m} z_m z_{\lambda 21} \\
&= T^{\lambda m} z_m (z_{\lambda 12} + z_r R_{r\lambda 21}) = G_{12} - K|\nabla z|^2
\end{aligned}$$

which implies the last part of the present lemma. Combining (35) with (36) gives

$$\begin{aligned}
T^{il} z_l G_{ij}(z) &= T^{il} T^{\lambda m} z_{mij} z_\lambda z_l = -T^{il} T^{m\lambda} z_{mij} z_\lambda z_l \\
&= -\Phi_j + 2z_j K(1-|\nabla z|^2) = D_j, j = 1, 2. \qquad (37)
\end{aligned}$$

On the other hand, from (32) and (34) it follows that

$$a^{ij} G_{ij}(z) = D_0. \qquad (38)$$

(37) and (38) form a algebraic system for G_{ij}:

$$\begin{pmatrix} z_2 & -z_1 & 0 \\ 0 & z_2 & -z_1 \\ z_{22} & 0 & z_{11} \end{pmatrix} \begin{pmatrix} G_{11} \\ G_{12} \\ G_{22} \end{pmatrix} = \begin{pmatrix} D_1 - z_1|\nabla z|^2 K \\ D_2 \\ D_0 \end{pmatrix}. \qquad (39)$$

Therefore solving G_{ij} from (39) one can get

$$\begin{pmatrix} G_{11} \\ G_{12} \\ G_{22} \end{pmatrix} = \frac{1}{\Phi} \begin{pmatrix} z_2 z_{11} & z_1 z_{11} & z_1^2 \\ -z_1 z_{22} & z_2 z_{11} & z_1 z_2 \\ -z_2 z_{22} & -z_1 z_{22} & z_2^2 \end{pmatrix} \begin{pmatrix} D_1 - z_1|\nabla z|^2 K \\ D_2 \\ D_0 \end{pmatrix}.$$

Consequently,

$$G_{11}(z) = \frac{1}{\Phi}[D_0 z_1^2 + a^{22}(D_1 z_2 + D_2 z_1) - a^{22} z_1 z_2 K|\nabla z|^2] \qquad (40)$$

$$G_{12}(z) = \frac{1}{\Phi}[D_0 z_1 z_2 - a^{11} D_1 z_1 + a^{22} D_2 z_2 + a^{11} z_1^2 K|\nabla z|^2] \qquad (41)$$

$$G_{22}(z) = \frac{1}{\Phi}[D_0 z_2^2 - a^{11}(D_2 z_1 + D_1 z_2) + a^{11} z_1 z_2 K|\nabla z|^2]. \qquad (42)$$

Thus Lemma 5 follows immediately by inserting the expressions of D_1 and D_2 into (40)-(42). □

Lemma 6.

$$G_{12}^2 - G_{11}G_{22} = \frac{a^{ij}\Phi_i\Phi_j}{\Phi} + \frac{a^{ij}\Phi_i Q_j^1 + \Phi_i Q_i^2}{\Phi} + \tilde{E} \tag{43}$$

where $Q_j^i, i, j = 1, 2, E$ depend only on Dz, D^2z and K, DK and furthermore

$$|Q_j^i| \le C(K + |X(K)|) \text{ and } \tilde{E} = 0(K) \tag{44}$$

for some constant $C = C(|z|_{C^2})$.

Proof. Since (43) is invariant under the transformation of variables, it suffices to locally prove (43) where $z_{12} = 0$. From Lemma 5 direct computations yield

$$\begin{aligned}
& \Phi^2\left(G_{12}^2(z) - G_{11}(z)G_{22}(z)\right) \\
= & \left((a^{11}\Phi_1 z_1 - a^{22}\Phi_2 z_2)^2 + a^{11}a^{22}(\Phi_1 z_2 + \Phi_2 z_1)^2\right) \\
+ & \ 2(a^{11}\Phi_1 z_1 - a^{22}\Phi_2 z_2)\left(2K(a^{22}z_2^2 - a^{11}z_1^2) + D_0 z_1 z_2\right) \\
+ & \ a^{22}(\Phi_1 z_2 + \Phi_2 z_1)(-\sigma z_2 z_1 a^{11} + D_0 z_2^2) \\
- & \ a^{11}(\Phi_1 z_2 + \Phi_2 z_1)(\sigma z_1 z_2 a^{22} + D_0 z_1^2) + I \\
= & \ I_1 + I_2 + I_3 + I_4 + I
\end{aligned} \tag{45}$$

where

$$\begin{aligned}
I = & \left(2K(1 - |\nabla z|^2)(a^{22}z_2^2 - a^{11}z_1^2) + D_0 z_1 z_2 + a^{11}z_1^2 K|\nabla z|^2\right)^2 \\
& -(\sigma z_1 z_2 a^{22} + D_0 z_1^2)(D_0 z_2^2 - \sigma z_2 z_1 a^{11})
\end{aligned}$$

It is easy to see

$$\begin{aligned}
I_1 &= a^{11}\Phi_1^2(a^{11}z_1^2 + a^{22}z_2^2) + a^{22}\Phi_2^2(a^{11}z_1^2 + a^{22}z_2^2) \\
&= \Phi(a^{11}\Phi_1^2 + a^{22}\Phi_2^2) = \Phi a^{ij}\Phi_i\Phi_j
\end{aligned} \tag{46}$$

which just gives the first term of right hand side of (43). Evidently,

$$I_2 + I_3 + I_4 = a^{11}\Phi_1\tilde{Q}_1^1 + a^{22}\Phi_2\tilde{Q}_2^1 + \Phi_1\tilde{Q}_1^2 + \Phi_2\tilde{Q}_2^2$$

where

$$\tilde{Q}_j^1 = \Phi K O(1) \text{ and } \tilde{Q}_j^2 = \Phi D_0 O(1)$$

where the factors $O(1)$ denote ones depending only on Dz, D^2z. Consequently, (44) follows immediately. As for I, there holds

$$I = \Phi^2\tilde{E} = \Phi K^2 0(1).$$

Therefore the second part of (44) is verified and the proof for the present lemma is ended. □

Now we are in a position to derive an equation satisfied by Φ. A direct computation gives

$$\begin{aligned} a^{ij}\Phi_{ij} &= T^{\lambda m}T^{\mu l}a^{ij}\left(z_m z_l z_{\lambda\mu ij} + z_m z_{lj} z_{\lambda\mu i} + z_{mj} z_l z_{\lambda\mu i}\right) \\ &+ T^{\lambda m}T^{\mu l}a^{ij}\left(z_m z_{li} z_{\lambda\mu j} + z_{mi} z_l z_{\lambda\mu j} + z_m z_{lij} z_{\lambda\mu} + z_{mij} z_l z_{\lambda\mu}\right) \\ &+ T^{\lambda m}T^{\mu l}a^{ij}\left(z_{mj} z_{li} z_{\lambda\mu} + z_{mi} z_{lj} z_{\lambda\mu}\right) \\ &= T^{\lambda m}T^{\mu l}a^{ij} z_m z_l z_{\lambda\mu ij} \\ &+ K(1-|\nabla z|^2)\left(T^{\lambda m}T^{\mu j} z_{mj} z_{\lambda\mu} + T^{\lambda j}T^{\mu l} z_{lj} z_{\lambda\mu}\right) + I_6 + I_7 \quad (47) \end{aligned}$$

where

$$\begin{aligned} I_6 &= K(1-|\nabla z|^2)\big(T^{\lambda m}T^{\mu i} z_m z_{\lambda\mu i} + T^{\lambda i}T^{\mu l} z_l z_{\lambda\mu i}\big) \\ &\quad + K(1-|\nabla z|^2)\big(T^{\lambda m}T^{\mu j} z_m z_{\lambda\mu j} + T^{\lambda j}T^{\mu l} z_l z_{\lambda\mu j}\big) \\ I_7 &= a^{ij}\left(T^{\lambda m}T^{\mu l} z_m z_{lij} z_{\lambda\mu} + T^{\lambda m}T^{\mu l} z_{mij} z_l z_{\lambda\mu}\right). \end{aligned}$$

By means of the Ricci identity and $T^{\mu i} = -T^{i\mu}$ it follows that

$$\begin{aligned} &T^{\lambda m}T^{\mu i} z_m z_{\lambda\mu i} K(1-|\nabla z|^2) \\ &= K(1-|\nabla z|^2) T^{\lambda m}T^{\mu i} z_m (z_{\mu i\lambda} + z_r R_{r\lambda\mu i}) = 0(K^2) \end{aligned} \quad (48)$$

and hence, $I_6 = 0(K^2)$. Since $z_{12} = 0$ at the point considered, there holds

$$\begin{aligned} &a^{ij}\left(T^{\lambda m}T^{\mu l} z_m z_{lij} z_{\lambda\mu}\right) \\ &= a^{ii} z_i\big(K(1-|\nabla z|^2)\big)_i + 2(T^{il})^2 a^{ii} z_{ii} z_l^2 R_{lili} \\ &= a^{ii} z_i\big(K(1-|\nabla z|^2)\big)_i + 0(K^2) \end{aligned} \quad (49)$$

which implies $I_7 = 2a^{ij} z_i\big(K(1-|\nabla z|^2)\big)_j + 0(K^2)$.

$$T^{\lambda m}T^{\mu j} z_{mj} z_{\lambda\mu} K(1-|\nabla z|^2) = 2(z_{11}z_{22} - z_{12}^2)K(1-|\nabla z|^2) = 0(K^2). \quad (50)$$

Hence it turns out

$$a^{ij}\Phi_{ij} = T^{\lambda m}T^{\mu l} z_m z_l a^{ij} z_{\lambda\mu ij} + 2a^{ij} z_i K_j + 0(K^2). \quad (51)$$

It remains to compute the first term in the right hand side of (51). By differentiations of (31) we have

$$\begin{aligned} &a^{ij} z_{ij\lambda\mu} + 2K z_r z_{r\lambda\mu} + 2K z_{r\mu} z_{r\lambda} + 2K_\mu z_r z_{r\lambda} \\ &= K_{\lambda\mu}(1-|\nabla z|^2) - (a^{ij})_\mu z_{ij\lambda} = K_{\lambda\mu}(1-|\nabla z|^2) - \left(K(1-|\nabla z|^2) z^{ij}\right)_\mu z_{ij\lambda} \\ &= K_{\lambda\mu}(1-|\nabla z|^2) - K(1-|\nabla z|^2)(z^{ij}z^{lm} - z^{im}z^{jl}) z_{ml\mu} z_{ij\lambda}. \end{aligned} \quad (52)$$

From (36) it is easy to see

$$T^{\lambda m}T^{\mu l} z_m z_l 2K z_r z_{r\lambda\mu} = 2K z_r \Phi_r + 0(K^2)$$

$$
\begin{aligned}
&T^{\lambda m}T^{\mu l}z_m z_l 2K_\mu z_r z_{r\lambda} = T^{\lambda m}z_m(|\nabla z|^2)_\lambda T^{\mu l}z_l K_\mu \\
&= X(|\nabla z|^2)X(K) = 0(K) \text{ and } T^{\lambda m}T^{\mu l}z_m z_l 2K z_{r\mu}z_{r\lambda} = 0(K)
\end{aligned}
$$

Therefore

$$
\begin{aligned}
&T^{\lambda m}T^{\mu l}z_m z_l a^{ij}z_{ij\lambda\mu} + 2Kz_r\Phi_r \\
&= T^{\lambda m}T^{\mu l}z_m z_l K_{\lambda\mu}(1-|\nabla z|^2) \\
&+ (-T^{\lambda m}T^{\mu l}z_m z_l K(1-|\nabla z|^2)(z^{ij}z^{rs}-z^{ir}z^{js})z_{rs\mu}z_{ij\lambda} + 0(K)) \\
&= R(K)(1-|\nabla z|^2) + I_8.
\end{aligned}
\tag{53}
$$

Notice that

$$
z_{rs\mu} = z_{\mu rs} + z_\theta R_{\theta rs\mu}, z_{ij\lambda} = z_{\lambda ij} + z_\theta R_{\theta ij\lambda}
$$

$$
G_{ij} = G_{ji} + 0(K) \text{ and } R_{\theta\lambda ij} = -R_{\theta\lambda ji}
$$

with the aid of (35) it turns out

$$
\begin{aligned}
I_8 &= -K(1-|\nabla z|^2)(z^{ij}z^{rs}-z^{ir}z^{js})G_{rs}G_{ij} + 0(K) \\
&= (G_{21}^2 + G_{12}^2 - 2G_{11}G_{22}) + 0(K).
\end{aligned}
\tag{54}
$$

By means of Lemma 5

$$
\begin{aligned}
G_{21}^2 &= G_{12}^2 - 2K|\nabla z|^2 G_{12} + 0(K^2) \\
&= G_{12}^2 + \frac{a^{ij}\Phi_i Q_j^1 + \Phi_i Q_i^2}{\Phi} + 0(K)
\end{aligned}
$$

for some $Q_i^j, i,j = 1,2$ satisfying (44). It is easy to see

$$
\begin{aligned}
T^{\lambda m}T^{\mu l}z_m z_l a^{ij}z_{ij\lambda\mu} &= 2(G_{12}^2 - G_{11}G_{22}) + R(K)(1-|\nabla z|^2) - 2Kz_r\Phi_r \\
&+ \frac{a^{ij}\Phi_i Q_j^1 + \Phi_i Q_i^2}{\Phi} + 0(K).
\end{aligned}
\tag{55}
$$

Note that

$$
\begin{aligned}
z_{\lambda\mu ij} =& z_{ij\lambda\mu} + z_{ri}R_{r\mu j\lambda} + z_{rj}R_{r\lambda\mu i} + z_{r\lambda}R_{r\mu ji} \\
&+ z_{r\mu}R_{ri\lambda j} + z_r R_{r\lambda\mu i,j} + z_r R_{ri\lambda j,\mu}
\end{aligned}
\tag{56}
$$

and that

$$
\begin{aligned}
&T^{\lambda m}T^{\mu l}z_m z_l a^{ij}z_{ri}R_{r\mu j\lambda} = 0(K^2) \\
&T^{\lambda m}T^{\mu l}z_m z_l a^{ij}z_{r\mu}R_{ri\lambda j} = (T^{\lambda i})^2 a^{ii}z_i^2 z_{\lambda\lambda}R_{\lambda i\lambda i} = 0(1)\Phi K \\
&T^{\lambda m}T^{\mu l}z_m z_l z_r a^{ij}R_{r\lambda\mu i,i} = 0(1)\sqrt{a^{ii}z_i^2}|DK|.
\end{aligned}
$$

Hence

$$\begin{aligned}a^{ij}\Phi_{ij}+2Kz_r\Phi_r &= T^{\lambda m}T^{\mu l}z_m z_l a^{ij}z_{ij\lambda\mu}+0(K)+0(1)\sqrt{\Phi}|DK|\\ &= 2(G_{12}^2-G_{11}G_{22})+R(K)(1-|\nabla z|^2)+0(1)\sqrt{\Phi}|DK|\\ &+\frac{a^{ij}\Phi_i Q_j^1+\Phi_i Q_i^2}{\Phi}+0(K).\end{aligned}\tag{57}$$

Finally by means of Lemma 6, (47), (51) and (57) it follows that

$$\begin{aligned}L(\Phi) &= a^{ij}\Phi_{ij}+2Kz_r\Phi_r\\ &= 2\frac{a^{ij}\Phi_i\Phi_j}{\Phi}+\frac{a^{ij}\Phi_i Q_j^1+D_0Q_j^2\Phi_j}{\Phi}+E+R(K)(1-|\nabla z|^2)\end{aligned}\tag{58}$$

where $R(K)=K_{11}z_2^2-2K_{12}z_1z_2+K_{22}z_1^2$ and

$$E=2a^{ij}z_iK_j+0(K)+0(1)\sqrt{\Phi}|DK|,\quad |Q_j^l|\le CK.\tag{59}$$

Lemma 7. *Let (6)(8) be fulfilled. Then we have*

$$X(K)=0(K),\quad R(K)(1-|\nabla z|^2)+E\le-\frac{1}{C}\varphi^{m-1}\big(1-C\sqrt{\Phi}-C^2\varphi\big)\ \textit{in } N_\delta\tag{60}$$

where C depends only on $|Dz|_1, |g|_4$ and the maximum of $1/|Dz|$ and $1/k_g$ on ∂D.

Proof. Still write $b^1=z_2, b^2=-z_1$ and $X=b^l\partial_l$. Since ∂D is convex with respect to metric g, without loss of generality we may assume φ geodesic-convex with respect to g and hence, under any coordinates

$$-\nabla_{ij}\varphi\xi_i\xi_j\ge\theta|\xi|^2\forall\text{ any vector }\xi=(\xi_1,\xi_2)$$

for some universal positive constant θ independent of ξ. Then by the definition,

$$\begin{aligned}X(K) &= z_1K_2-z_2K_1\\ &= m\varphi^{m-1}\tilde{K}(z_1\phi_2-z_2\varphi_1)+\frac{1}{m}\varphi^m X(\tilde{K})=0(\varphi^m)=0(K)\end{aligned}$$

since $z_1\phi_2-z_2\varphi_1=0$ on ∂D. Obviously, the first part of (60) follows at once. Now let us deal with the second part. Evidently,

$$\begin{aligned}R(K) &= \nabla_{ij}Kb^ib^j\\ &= \varphi^{m-1}\Big(\tilde{K}m\nabla_{ij}\varphi b^ib^j+\frac{m(m-1)}{\varphi}\tilde{K}(\varphi_ib^i)^2+2m(\varphi_ib^i\tilde{K}_jb^j)+\varphi\nabla_{ij}\tilde{K}b^ib^j\Big)\\ &\le \varphi^{m-1}(-m\theta\min\{|Dz|^2\tilde{K}\}+C\varphi)\le-\frac{1}{C}\varphi^{m-1}(1-C^2\varphi)\end{aligned}$$

since $|b^i\varphi_i| \le C\varphi$ for some constant C depending the quantities mentioned in the present lemma. Finally

$$|E| = |2a^{ij}z_iK_j + 0(K)| \le CK + 2\sqrt{a^{ij}z_iz_j}\sqrt{a^{ij}K_iK_j}$$
$$\le C(K + \sqrt{\Phi}\varphi^{m-1}).$$

Thus by Lemma 2 we have

$$E + R(K)(1 - |\nabla z|^2) \le C\big(\sqrt{\Phi}\varphi^{m-1} + \varphi^m\big) - \frac{1}{C}\varphi^{m-1}.$$

This proves the present lemma. □

The proof for Theorem 2. Suppose that z is a global smooth solution to (1) with (4) (5) and that (6) (8) hold. Then it follows that under the coordinates (25)(26), (30) tells us

$$\partial_{ss}z - \frac{B_s}{B} + BB_tz_t \ge \frac{1}{C} \text{ as } t = 0$$

for some constant C under control. Suppose that

$$\min_{-\delta_1 \le t \le 0} \Phi = \Phi(p), p \in \bar{N}_{\delta_1}$$

where $N_{\delta_1} = \bar{D}\backslash\bar{D}_{\delta_1}$ and δ_1 to be specified. Then if $p \in \partial D$, there holds

$$\Phi = z_t^2\big(\partial_{ss}z - \frac{B_s}{B} + BB_tz_t\big) \ge \frac{\min_{\partial D} z_t^2}{C} \ge \frac{1}{C_1} \tag{61}$$

for another constant C_1 under control from (23). If $p \in D \cap N_{\delta_1}$ it is easy to see that at p with $p = (s_p, t_p)$, $\Phi_i = 0, i = 1, 2$ and (Φ_{ij}) nonnegative and hence by (58) and (60) it turns out that at p.

$$0 \le L(\Phi) \le -\frac{1}{C}\varphi^{m-1}(1 - C\sqrt{\Phi} - C\varphi) \le -\frac{1}{C}\varphi^{m-1}(1 - C\sqrt{\Phi} - C_1t_p)$$

which implies

$$\Phi(p) \ge (\frac{1}{C} - \frac{C_1t_p}{C})^2 \ge \frac{1}{4C^2} \text{ if } p \in N_{\delta_1} \text{ where } \delta_1 = \min\{\delta, \quad \frac{1}{2C_1}\}. \tag{62}$$

If $p \in \partial N_{\delta_1} \cap D$, (23) tells us: $|\nabla z| \ge \frac{1}{C}$ as $\delta_1 \le t \le 0$ for some constants C and some smaller δ_1 under control and hence we have

$$\min_{t=\delta_1}\Phi \ge \min_{t=\delta_1}|Dz|^2 \min_{t=\delta_1}\{\text{ eigenvalues of } (\nabla_{ij}z)\}$$
$$\ge \min_{t=\delta_1}|Dz|^2\frac{\min_{t=\delta_1}K(1 - |\nabla z|^2)}{\max_{\bar{D}}(\Delta_g z)} \ge \frac{\delta_1^m \inf_D \tilde{K}}{C_1}. \tag{63}$$

Finally, combining (63) with (61) and (62) it is easily seen that $\Phi \geq 1/C$ as $-\delta_1 \leq t \leq 0$. Consequently

$$\partial_{ss}z - \frac{B_s}{B} + BB_t z_t = \frac{1}{z_t^2}(\Phi - z_s z_t(z_{zt} - \frac{B_t}{B}z_s) - z_{tt}z_s^2)$$
$$\geq (\frac{1}{C} - Ct) \geq \frac{1}{2C} \quad \text{as } 0 \leq -t \leq \delta_2 = \min\{\delta_1, \frac{1}{2C^2}\}.$$

Similarly,

$$\Delta_g z \geq \frac{1}{|Dz|^2}\Phi \geq \frac{1}{C} \text{ on } \bar{N}_{\delta_2}$$

for another larger constant C under control. Now Theorem 2 follows immediately from the interior estimates for (1) and the proof for Theorem 2 is ended.

References

1. Amano, K., Global isometric embedding of a Riemannian 2-manifold with nonnegative curvature into a Euclidean 3-space, J. Diff. Geometry, 35(1991), 49-83.
2. Delanoe, Ph., Relations globalement regulieres de disques strictement convexes dans les espaces d'Euclide et de Minkowski par la methode de Weingarten, Ann. Sci. Ec Norm. Super., 21(1988), 637–652.
3. Guan, P., Li, Y., The Weyl problem with nonnegative Gauss curvature, J. Diff. Geometry, 39(1994), 331-342.
4. Gromov, M., Rokhlin, V.A., Embeddings and immersions in Riemannian geometry, Russian Mathematical Surveys, 25(1970), 1-57.
5. Gilbarg, D., Trudinger, N.S, Elliptic partial differential equations of second order, Reprint of the 1998 edition, Springer (2001).
6. Heinz, E., Neue a priori Abschatsungen fur den ortsvektor iener Flache positiver Gaussscher Krummung durch ihr Linienelment, Math. Z., 74(1960), 1-52.
7. Heinz, E., On elliptic Monge-Ampere equations and Weyl's embedding problem, Journal D'Analyse Mathematique, 7(1959–1960), 1–52.
8. Hang, Q., Hong, J.X., Isometric embedding of Riemannian manifolds, in Euclidean spaces, Mathematical Survey and Monographs Vol. 130 American Mathematical Society.
9. Hong, J.X, Darboux equations and isometric embedding of Riemannian manifolds with nonnegative curvature in R^3, Chin. Ann. of Math., 20B(1999), 123-136.
10. Hong, J. X., Isometric embedding in R^3 of complete noncompact Riemannian nonnegatively curved surfaces, Manuscripta Math., 94(1997), 271-286.
11. Hong, J.X., Recent developments of realization of surfaces in R^3, First International Conference of Chinese Mathematicians (Beijing, 1998), 47-62, AMS/IP Stud. Adv. Math., 20, Amer. Math. Soc., Providence, RI, 2001.

12. Hong, J.X., Huang, G.G., Wang, W.Y., Existence of Global smooth solutions to Dirichlet problem for degenerate elliptic Monge-Ampère equations, Comm. in Partial Diff. Equations, 36(4)(2011), 635-656.
13. Hong, J.X., Zuily, C., Isometric embedding of the 2-sphere with nonnegative curvature in R^3, Math. Z., 219(1995), 323-334.
14. Nirenberg, L., The Weyl and Minkowski problems in differential geometry in the large, Comm. Pure Appl. Math., 6(1953), 337-394.
15. Pogorelov, A.V., Extinsic geometry of convex surfaces, Amer. Math. Soc., Providence RI. (1973).
16. Weyl, H., Über die Bestimmheit einer geschlossenen konvex Fläche durch ihr Linienelement, Vierteljahresschrift der nat.-Forsch. Ges. Zürich, 61(1916), 40-72.
17. Yau, S.T., Lecture on Differential Geometry, in Berkeley, (1977).
18. Yau, S.T., Survey on Geometry and Analysis, Asian J. Math., 4(2000), 235-278.

Geometric Invariant Theory of the Space — A Modern Approach to Solid Geometry

Wu-yi Hsiang

Department of Mathematics, University of California,
Berkeley, CA94720
mahsiang@ust.hk

In this survey article we shall outline a systematic approach to the study of solid geometry in the setting of geometric invariant theory, namely the invariant theory of the isometric group of the physical space often referred as the Euclidean group. As a testing of the powerfulness of such a theory, we shall discuss a much simplified recent proof of Kepler's conjecture on sphere packings and the least action principle of crystal formation of the dense type as an application of such an approach to solid geometry.

1. Introduction

The *space* is the totality of all possible locations of the universe in which we and everything else are located, while all kinds of natural phenomena are happening within. In his formulation of the philosophical belief which is, nowadays, referred to as the *least action principle*, Leibniz stated that our world is the *best* among all possible worlds. First of all, this must, of course, include that our space is the *best* among all possible spaces, a perfect creation that makes the embodiments of all the other wonderful aspects of our world, at all, possible. Following the foresight of Pythagoras, a good understanding of natural sciences must be based upon an indepth understanding of the space (i.e. solid geometry). The foundational study of quantitative geometry of Pythagoreans pioneered the long history of geometric studies beginning with *plane geometry* as a basic stepping stone for the study of *solid geometry*.

However, a comparison study of the current status of our knowledges on plane geometry and that of solid geometry will clearly show that the level and depth of the latter is far below that of the formers. For example, various kinds of geometric problems of corresponding types had already been successfully solved in the case of plane geometry, while that of the case of solid geometry remain to be disappointing. Of course, this is just a manifestation of the reality that solid geometry is far more difficult than plane geometry. At here, one might ask what are the fundamental reasons that make solid geometry so much harder to master than that of plane geometry. Of course, the dimension of the space is 3 while that of the plane is 2, namely, there is a difference of 1 higher in dimension. Besides such a major difference, the other basic properties, in fact, appear to be quite similar, namely

(i) *the continuity property*: it has already been characterized by the continuity of a line,

(ii) *the flatness property*: it has already been characterized by that the sum of inner angles of triangles is always equal to a flat angle, thus again the same for both,

(iii) *symmetry properties*: the *plane* is reflectionally symmetric with respect to any given line, while the *space* is reflectionally symmetric with respect to any given plane, thus they are seemingly again of the same kind of symmetries. However, a deeper analysis will reveal that it is the symmetry aspect that lies the *major difference between plane geometry* and *solid geometry.*

Note that, the *continuity* property and the *flatness* are essentially local, while the symmetry property is global which should be comprehended and used in the setting of the *totality* of those reflections with respect to all lines (resp. all planes), namely, the *isometry groups generated by those reflections* of the plane (resp. in the space), say denoted by $G(\pi)$ (resp. $G(V)$). It is, nowadays, well known that $G(\pi)$ (resp. $G(V)$) contain a *translation subgroup* $\mathbb{R}^2$ (resp. $\mathbb{R}^3$) as their *normal* subgroups and

$$G(\pi)/\mathbb{R}^2 \cong O(2) \supset SO(2)$$
$$G(V)/\mathbb{R}^3 \cong O(3) \supset SO(3).$$

Algebraically, $G(\pi)$ (resp. $G(V)$) is a split extension of $\mathbb{R}^2$(resp. $\mathbb{R}^3$) with $O(2)$ (resp. $O(3)$) as the quotient, while their connected component of identity $G_0(\pi)$ (resp. $G_0(V)$) are that of $SO(2)$ (resp. $SO(3)$). Geometrically, $G(\pi)$ ($G(V)$) act simple transitively on the totality of orthonormal frames of π (resp. V), while $G_0(\pi)$ (resp. $G(V)$) act simple transitively on that of *oriented* orthonormal frames. All *geometric quantities* of plane (resp. solid) geometry are invariants of $G_0(\pi)$ (resp. $G_0(V)$) which are, nowadays, simply referred to as *geometric invariants* of $G_0(\pi)$ (resp. $G(V)$), and hence, the quantitative study of plane (resp. solid) geometry can be regarded as the *geometric invariant theory* of $G_0(\pi)$ (resp. $G_0(V)$), while the $SO(3)$-factor of $G_0(V)$ makes its geometric invariant theory much more difficult than that of $G_0(\pi)$. Anyhow, this is the fundamental reason why solid geometry is much harder than plane geometry.

Algebraically, the *translation group* $\mathbb{R}^3$ is a *normal subgroup* of $G_0(V)$ and hence the inner automorphisms of $G_0(V)$ naturally provides an $SO(3)$ action on $\mathbb{R}^3$ which is just the usual $SO(3)$-vector space with inner product. The *inner product* and the *determinant* already constitute a *complete system* of *basic geometric invariants* such that all the other geometric invariants are theoretically expressible in terms of them, thus making the vector algebra an advantageous, powerful tool for the study of solid geometry. Specifically, the *basic theorems* of quantitative solid geometry are, now, systematically transformed into the *law of operations* of the vector algebra of $(SO(3),\ \mathbb{R}^3)$. Thus, the *basic part* of geometric invariant theory of the space is exactly the *vector algebra* of $(SO(3),\ \mathbb{R}^3)$, but the *major part* of geometric invariant theory of the space actually lies in those *profound relations* among various geometric invariants, such as that of spherical triangles, tetrahedra, sphere packings etc. (cf. §2, 3 and 4). In particular, spherical geometry, itself, is the geometric invariant theory of $SO(3)$ (or $O(3)$), which should be regarded as an important part of the geometric invariant theory of the space.

In retrospect, the first fundamental formula of solid geometry should be the volume formula of a cone, while the first fundamental theorem of the geometric invariant theory of the space is the Archimedes Theorem that the total amount of solid angle (i.e. the area of unit sphere) is equal to 4π. Historically, the proof of the volume formula of cones by Eudoxus heralded the beginning of integration theory and the proof of the area of sphere by Archimedes using Eudoxian principle of exhaustion represents one of the glory of Greek geometry. Moreover, motivated by the need of quantitative analysis of astronomical observations, basic elements of *spherical trigonometry* had already been developed way ahead of *plane trigonometry.* (cf. [Pto])

2. Spherical Trigonometry, the Basic Geometric Invariant Theory of Spherical Triangles

2.1. *Basic Theorems*

Theorem 1. *(Archimedes) The area of unit sphere in the space is equal to* 4π.

Corollary 1. *The area of a spherical triangle* $\sigma(ABC)$ *is equal to its angular excess, namely*

$$|\sigma| = A + B + C - \pi. \tag{1}$$

Proof. It is essentially a straightforward localization of Archimedes Theorem, first to the area formula of a lune, namely

$$A(L(\theta)) = 2\theta.$$

Set A', B', C' to be the antipodal point of A, B, C. Then

$$|\sigma(ABC)| + |\sigma(A'BC)| = 2A, \text{ etc.}$$

Summing up the above three equations and using $|\sigma(ABC')| = |\sigma(A'B'C)|$, one gets

$$\begin{aligned} &2|\sigma(ABC)| + 2\pi = 2A + 2B + 2C \\ &\Rightarrow \sigma(ABC) = A + B + C - \pi. \end{aligned}$$

□

As usual, the angles (resp. side lengths) of $\sigma(ABC)$ will simply be denoted by $\{A, B, C\}$ (resp. $\{a, b, c\}$). Set $\{\mathbb{a}, \mathbb{b}, \mathbb{c}\}$ to be the position (unit) vectors of $\{A, B, C\}$, $D = \mathbb{a} \cdot (\mathbb{b} \times \mathbb{c})$ and $R =$ the circumradius of $\sigma(ABC)$. Then, one has

(i) $\cos a = \mathbb{b} \cdot \mathbb{c}$ etc; (ii) $\sin a = |\mathbb{b} \times \mathbb{c}|$ etc and

$$D^2 = \begin{vmatrix} \mathbb{a} \cdot \mathbb{a}, & \mathbb{b} \cdot \mathbb{a}, & \mathbb{c} \cdot \mathbb{a} \\ \mathbb{a} \cdot \mathbb{b}, & \mathbb{b} \cdot \mathbb{b}, & \mathbb{c} \cdot \mathbb{b} \\ \mathbb{a} \cdot \mathbb{c}, & \mathbb{b} \cdot \mathbb{c}, & \mathbb{c} \cdot \mathbb{c} \end{vmatrix} = 1 + 2\cos a \cos b \cos c - \cos^2 a - \cos^2 b - \cos^2 c.$$

Theorem 2. *(Spherical sine law):*

$$\frac{\sin A}{\sin a} = \frac{\sin B}{\sin b} = \frac{\sin C}{\sin c} = \frac{D}{\sin a \sin b \sin c}. \tag{2}$$

Proof. Set $\mathbb{b}' = \mathbb{b} - (\mathbb{b} \cdot \mathbb{a})\mathbb{a}$, $\mathbb{c}' = \mathbb{c} - (\mathbb{c} \cdot \mathbb{a})\mathbb{a}$, Thus $\mathbb{b}'$ and $\mathbb{c}'$ are both orthogonal to $\mathbb{a}$ and $|\mathbb{b}'| = \sin c$, $|\mathbb{c}'| = \sin b$. Therefore

$$D = \mathbb{a} \cdot (\mathbb{b} \times \mathbb{c}) = \mathbb{a} \cdot (\mathbb{b}' \times \mathbb{c}') = \sin c \sin b \sin A$$

and hence (2) follows readily. □

Theorem 3. *(Spherical cosine law):*

$$\sin c \sin b \sin A = \cos a - \cos c \cos b \quad etc. \tag{3}$$

Proof. Again by direct application of vector algebra, one has

$$\begin{aligned}
&\sin c \sin b \cos A = \mathbb{b}' \cdot \mathbb{c}' = (\mathbb{b} - (\mathbb{b} \cdot \mathbb{a})\mathbb{a}) \cdot (\mathbb{c} - (\mathbb{c} \cdot \mathbb{a})\mathbb{a}) \\
&= \mathbb{b} \cdot \mathbb{c} - (\mathbb{b} \cdot \mathbb{a})(\mathbb{a} \cdot \mathbb{c}) - (\mathbb{c} \cdot \mathbb{a})(\mathbb{b} \cdot \mathbb{a}) + (\mathbb{b} \cdot \mathbb{a})(\mathbb{c} \cdot \mathbb{a})\mathbb{a} \cdot \mathbb{a} \\
&= \cos a - \cos c \cos b.
\end{aligned}$$

□

Remark 1. The above two basic laws of spherical trigonometry have quite a few different kinds of proofs. However, the above direct simple proofs via vectors algebra not only fully reveal their natural close relationships, but also fitting nicely in the framework of geometric invariant theory. In fact, the neat area formula of (1) still needs another formulation that fits well with the basic invariants of inner products.

Theorem 4. *(Another area formula):*

$$\tan \frac{|\sigma|}{2} = \frac{D}{u}, \quad u = 1 + \cos a + \cos b + \cos c. \tag{4}$$

Proof. The sine law and cosine law assert that

$$\sin A = \frac{D}{\sin b \sin c}, \quad \cos A = \frac{\cos a - \cos b \cos c}{\sin b \sin c}, \quad \text{etc.}$$

Therefore, straightforward substitution and simplification will show that

$$\begin{aligned}
\sin |\sigma| &= \sin(A + B + C - \pi) = \frac{2uD}{u^2 + D^2} \\
\cos |\sigma| &= \cos(A + B + C - \pi) = \frac{u^2 - D^2}{u^2 + D^2}
\end{aligned}$$

and hence

$$\tan \frac{|\sigma|}{2} = \frac{\sin |\sigma|}{1 + \cos |\sigma|} = \frac{D}{u}.$$

□

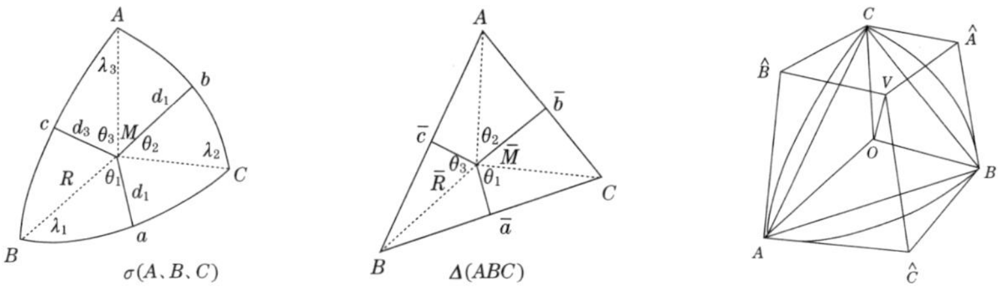

Figure 1

The special case of right-angle triangles: In the special case that $C = \frac{\pi}{2}$, the above set of basic formulas simplified into the following, which are much easier to use and hence quite useful:

$$\begin{cases} \cos c = \cos a \cos b;\ \sin A = \frac{\sin a}{\sin c},\ \sin B = \frac{\sin b}{\sin c}; \\ \cos A = \frac{\tan b}{\tan c},\ \cos B = \frac{\tan a}{\tan c};\ \tan A = \frac{\tan a}{\sin b},\ \tan B = \frac{\tan b}{\sin a} \\ \tan A \cdot \tan B = \frac{1}{\cos c},\quad \cos|\sigma| = \frac{\cos a + \cos b}{1 + \cos c} \\ \sin|\sigma| = \frac{\sin a \sin b}{1 + \cos c}. \end{cases} \tag{5}$$

2.2. *Geometric invariants of circumcentric subdivision*

To a given spherical triangle $\sigma(ABC)$, the Euclidean triangle $\triangle ABC$, the 1-isosceles tetrahedron $\tau(\sigma, 1) = \tau(OABC)$ and the portion of the solid angle cone $\Gamma(\sigma)$ bounded by the tangent planes at A, B, C, namely

$$T(\sigma) := \{X \in \Gamma(\sigma);\ \overrightarrow{OX} \cdot \mathbb{a},\ \overrightarrow{OX} \cdot \mathbb{b} \textit{ and } \overrightarrow{OX} \cdot \mathbb{c} \leq 1\}$$

are a triple of canonically associated solid geometric objects, while the circumcentric subdivision of both $\sigma(ABC)$ and $\triangle ABC$ naturally correspond to each other via radial projection. As indicated in Figure 1, set M (resp. $\overline{M}$) to be the circumcenters of $\sigma(ABC)$ (resp. $\triangle ABC$), $\{\theta_1, \theta_2, \theta_3\}$ to the central angles at both M and $\overline{M}$, $\{d_1, d_2, d_3\}$ to be the distances of M towards the sides of σ and $\{\lambda_1, \lambda_2, \lambda_3\}$ to be the oriented side-angles of that of σ. Moreover, set $\{\sigma_A, \sigma_B, \sigma_C\}$ to be the spherical triangles of $\tau(\sigma, 1)$ at $\{A, B, C\}$, $\{\bar{a},\ \bar{b},\ \bar{c}\}$ (resp. $|\triangle|$, $\overline{R}$) to be the side-lengths (resp. area, circumradius) of $\triangle ABC$. Then, one has the following system of basic formulas of the above set of basic geometric invariants canonically associated to $\sigma(ABC)$:

$$\begin{cases} \bar{a} = 2\sin\frac{a}{2}, \text{ etc. } \overline{R} = \sin R, \ \bar{a}\bar{b}\bar{c} = 4\overline{R}\cdot|\triangle| \\ D = 6\text{vol}\,\tau(\sigma,1) = 2|\triangle|\cos R = \frac{1}{2}\bar{a}\bar{b}\bar{c}\cot R \\ \tan^2 R = \frac{2}{D^2}(1-\cos a)(1-\cos b)(1-\cos c). \end{cases} \tag{6}$$

Moreover, it is easy to see that the S.S.S. (resp. A.A.A. and S.A.S.) data of σ_A are given by $\{\frac{1}{2}(\pi-b), \frac{1}{2}(\pi-c), \theta_1\}$ (resp. $\{A, \frac{\pi}{2}-d_2, \frac{\pi}{2}-d_3\}$ and $\{\frac{1}{2}(\pi-b), \frac{1}{2}(\pi-c), A\}$). Therefore,

$$\begin{cases} \cos\theta_1 = \sin\frac{b}{2}\sin\frac{c}{2} + \cos\frac{b}{2}\cos\frac{c}{2}\cos A, \quad \text{etc.} \\ 2\lambda_1 = B + C - A = |\sigma| + \pi - 2A, \ \lambda_1 = \frac{|\sigma|}{2} + (\frac{\pi}{2} - A), \quad \text{etc.} \\ \tan\lambda_1 = \frac{1}{D}(1+\cos a - \cos b - \cos c), \quad \text{etc.} \\ \tan d_1 = \tan\lambda_1 \sin\frac{a}{2} = \cos\theta_1 \tan R, \quad \text{etc.} \\ \nu(\sigma) := |\sigma| + |\sigma_A| + |\sigma_B| + |\sigma_C| \\ \qquad = \pi + 2|\sigma| - 2(d_1+d_2+d_3) \\ \omega(\sigma) := \text{vol of } T(\sigma) = \frac{1}{6}\sin|\sigma|\{\frac{8}{u} - \tan^2 R\} \text{ (cf. p.141 of [Hs-1])} \end{cases} \tag{7}$$

2.3. *Area preserving deformations of spherical triangles*

In general, the congruence classes of spherical triangles with a given area constitutes a two dimensional level-subset in the three dimensional moduli space. In view of the paramount importance of area and the often crucial role of area-wise estimate of various kinds of pertinent geometric invariants such as $\omega(\sigma)$, etc., it is useful to have indepth understanding on *area preserving deformations* of spherical triangles. Let us begin with Lexell's Theorem and Lexell's deformation.

Theorem 5. *(Lexell) Let $\sigma(A_1BC)$ and $\sigma(A_2BC)$ be two spherical triangles with the same base $\overline{BC}$ and the same oriented area. Then A_1, A_2 and the antipodal points B', C' (of B, C) are cocircular.*

Proof. Let O be their common circumcenter. Then

$$\angle C'B'A_1 + \angle A_1C'B' - \angle B'A_1C' = \angle C'B'A_2 + \angle A_2C'B' - \angle B'A_2C'$$

because they both equal to the sum of the pair of base angles of $\sigma(OB'C')$. As indicated in Figure 2, one has

$$\angle C'B'A_i = \pi - \angle CBA_i, \quad \angle A_iC'B' = \pi - \angle A_iCB \quad (i=1,2).$$

Therefore

$$\angle C'B'A_i + \angle A_iC'B' - \angle B'A_iC' = 2\pi - (\angle CBA_i + \angle A_iCB + \angle BA_iC)$$

and hence both equal to $\pi - |\sigma|$. This implies that $\{B',\ C', A_1, A_2\}$ are cocircular. □

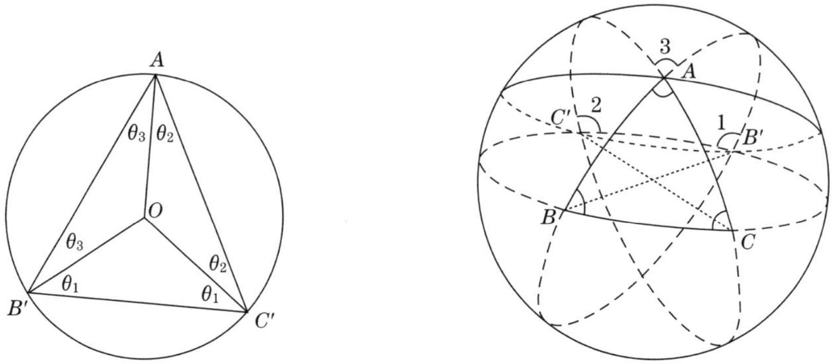

Figure 2

Lexell's deformations: As indicated in Figure 3, $\square A_1A_2BC$ is an (a, b)–"rectangle" with $2|\sigma|$ as its area, O is the center of the Lexell's circle (i.e., passing A_1, A_2 and B', C'). Set its radius to be $r(a, |\sigma|)$ and

$$\theta = \frac{1}{2}\angle A_2OA_1,\ \angle BOA = \theta + \phi,\ \angle AOC = \theta - \phi.$$

Then

$$\begin{cases} \tan^2 \frac{|\sigma|}{2} = \frac{(1-\cos a)(1-\cos b)}{2(\cos a+\cos b)} \ (\text{cf.}(9)) \\ \cos b = \dfrac{1-\cos a(2\tan^2 \frac{|\sigma|}{2}+1)}{2\tan^2 \frac{|\sigma|}{2}+1-\cos a} \\ \sin r(a, |\sigma|) = \cos \frac{b}{2}, \sin \theta = \sin \frac{a}{2} \sec \frac{b}{2} \\ \cos \frac{1}{2}\overline{AB} = \cos \frac{b}{2} \cos \frac{\theta+\varphi}{2}, \cos \frac{1}{2}\overline{AC} = \cos \frac{b}{2} \cos \frac{\theta-\varphi}{2}. \end{cases} \tag{8}$$

The above Lexell's deformations of $\sigma(ABC)$ keep the area and the base-length $a = \overline{BC}$ fixed. The following is a corollary of Theorem 4 which provides a neat answer of those deformations that keep the area and the top angle A fixed.

Corollary 1 of Theorem 4: In terms of the S.A.S data of $\sigma(ABC)$, $\{b, A, c\}$, one has the area formula

$$\tan \frac{|\sigma|}{2} = \frac{\sin A}{\cot \frac{b}{2} \cot \frac{c}{2} + \cos A}. \tag{4'}$$

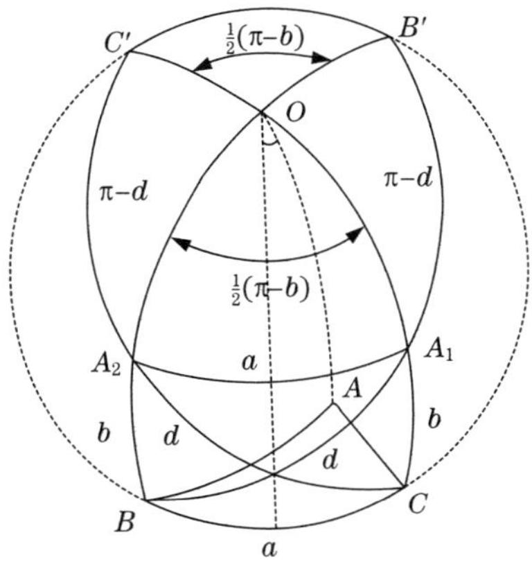

Figure 3

Proof. By Theorems 2 and 3,

$$D = \sin b \sin c \sin A, \;\; \cos c = \cos b \cos c + \sin b \sin c \cos A$$

Therefore

$$\tan \frac{|\sigma|}{2} = \frac{D}{u} = \frac{\sin b \sin c \sin A}{(1+\cos b)(1+\cos c) + \sin b \sin c \cos A}$$
$$= \frac{\sin A}{\cot \frac{a}{2} \cot \frac{b}{2} + \cos A}.$$

□

Thus those deformations of $\sigma(ABC)$ that keep both $|\sigma|$ and A fixed are exactly those keep $\cot \frac{a}{2} \cos \frac{b}{2}$ constant which will be henceforth referred to as δ-deformation. Set

$$k = \cot \frac{b}{2} \cot \frac{c}{2}, \;\; \delta = \cot \frac{b}{2} - \cot \frac{c}{2} \geq 0 \;(i.e., b \leq c).$$

Then $\{A, k, \delta\}$ also constitutes a complete set of S.A.S. type congruence invariants of σ. Set $t = \tan \frac{A}{2}$. Then, by $(4')$

$$(k-1)t^2 - 2 \cot \frac{|\sigma|}{2} t + (k+1) = 0.$$

Therefore, the real-ness of t implies that

$$\cot^2 \frac{|\sigma|}{2} \geq k^2 - 1 = \frac{2(\cos b + \cos c)}{(1-\cos b)(1-\cos c)}$$

which provides the upper bound of $|\sigma|$ for $\sigma(ABC)$ with given $\{b, c\}$ as its pair of side-length, namely

$$|\sigma| \leq 2\arctan\left\{\frac{(1-\cos b)(1-\cos c)}{2(\cos b+\cos c)}\right\}^{\frac{1}{2}} = \frac{1}{2}\,\Box(b,c) \tag{9}$$

where $\Box(b,c)$ denotes the area of spherical rectangle with (b,c) as its pair of side-lengths whose quadruple angles are equal to

$$A_0 = \frac{\pi}{2} + \arctan\left\{\frac{(1-\cos b)(1-\cos c)}{2(\cos b+\cos c)}\right\}^{\frac{1}{2}}.$$

In the general case of $|\sigma| < \frac{1}{2}\,\Box(b,c)$, the above quadratic equation has a pair of roots $t_i = \tan\frac{A_i}{2}, i = 1, 2$, while $\{A_1, A_2\}$ are exactly the pair of angles of the "sphericla //-gram" with (b,c) as its side-lengths and $2|\sigma|$ as its area. However, if we specify that $\sigma(ABC)$ should contain its circumcenter, then A should be the smaller one of $\{A_1, A_2\}$ and hence again unique, thus making $(|\sigma|, k, \delta)$ also constitutes a complete set of basic congruence invariants for such spherical triangles.
Note that

$$A = \arcsin(k\sin\frac{|\sigma|}{2}) + \frac{|\sigma|}{2}. \tag{10}$$

3. Tetranometry, the Basic Geometric Invariant Theory of Tetrahedra

Just as the roles of triangles and circles in the study of plane geometry, one naturally expects that *tetrahedra* and *spheres* will, of course, of central importance in the study of solid geometry. Therefore, *tetranometry*, namely, the basic geometric invariant theory of tetrahedra, must be a chapter of fundamental importance in any approach to the study of solid geometry, just as that of plane trigonometry in the study of plane geometry. However, a tetrahedron has many more basic geometric invariants than that of a triangle. For examples, it has six edge lengths, four triangular faces, six dihedral angles, four solid angles and its volume, circumradius, inradius, etc. Moreover, the interlocking relationships among them are also much more complicated and challenging to deal with. Anyhow, tetranometry is much more difficult to master than trigonometry and this is another reason that makes solid geometry so much harder than plane geometry. However, our basic attitude towards understanding solid geometry can only be that *understanding we must and understanding we will*, simply because the space is where we live.

A basic geometric idea of reduction via canonical decomposition:

In retrospect, the geometry of isosceles triangles is much easier than that of general triangles, while an isosceles triangle can be further subdivided into a pair of matching right-angle triangles. In fact, the *trigonometric functions* are defined in terms of the simple geometry of right-angle triangles and it is where the whole theory of trigonometry is based upon. Therefore, it is quite natural to look for proper subfamily of simple type of tetrahedra that will play the same kind of roles in the study of tetranometry. First of all, the *isosceles tetrahedra* such as the $\tau(\sigma, 1)$ we already encounter in §2 would be a natural generalization of isosceles triangles, and moreover, the circumcentric subdivision of $\triangle ABC$ also provides a canonical decomposition of $\tau(\sigma, 1)$ into a triple of pairs of doubly orthogonal tetrahedra, namely a tetrahedra $\tau(ABCD)$ satisfying

$$\overline{AB} \perp \triangle BCD \text{ and } \triangle ABC \perp \overline{CD}.$$

3.1. *Geometric invariant theory of doubly orthogonal tetrahedra*

As indicated in Figure 4, the geometric invariants of a doubly orthogonal tetrahedra $\tau(ABCD)$ are as follows:

$$\begin{aligned} &l_1 = \overline{AB},\ l_2 = \overline{BC},\ l_3 = \overline{CD}, \\ &\overline{AC}^2 = l_1^2 + l_2^2,\ \overline{BD}^2 = l_2^2 + l_3^2,\ \overline{AD}^2 = l_1^2 + l_2^2 + l_3^2; \end{aligned}$$

Set the dihedral angles at $\overline{AB}, \overline{CD}$ and $\overline{AD}$ to be λ_1, λ_3 and λ, while the other triple of dihedral angles are all equal to $\frac{\pi}{2}$. Then the solid angles (i.e. spherical triangles) at the vertices are given as follows:

(i) σ_B (resp. σ_C) are $\dfrac{\pi}{2}$-isosceles spherical triangles with λ_1 (resp. λ_3) as their top angles

(ii) σ_A (resp. σ_D) are right-angle spherical triangles with $\{\lambda, \lambda_1\}$ (resp. $\{\lambda, \lambda_3\}$) as the other pair of angles, and moreover, the S.S.S data of σ_A (resp. σ_D) are given by $\{\frac{\pi}{2} - \lambda_3, \frac{\pi}{2} - \delta, \alpha\}$ (resp. $\{\frac{\pi}{2} - \lambda_1, \frac{\pi}{2} - \alpha, \delta\}$), while

$$\begin{aligned} &\tan\lambda_1 \cdot \tan\lambda = \sec\alpha,\ \tan\lambda_3 \cdot \tan\lambda = \sec\delta \\ &\sin\lambda_3 \cdot \sin\delta = \cos\alpha,\ \sin\lambda_1 \cdot \sin\alpha = \cos\delta. \end{aligned} \tag{11}$$

Theorem 6. *(Laws of doubly orthogonal tetrahedra): Let $\Omega(ABCD)$ and $\Omega(A'B'C'D')$ be a pair of doubly orthogonal tetrahedra. Then they are similar (i.e. congruent up to scaling) to each other provide one of the following conditions holds, namely*

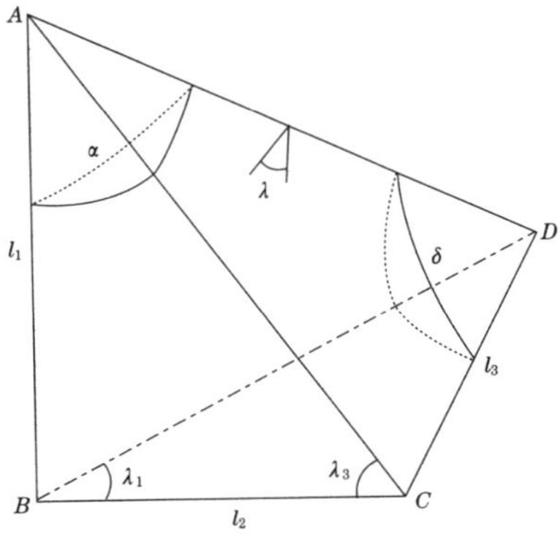

Figure 4

(i) $\overline{AB} : \overline{BC} : \overline{CD} = \overline{A'B'} : \overline{B'C'} : \overline{C'D'}$
(ii) $\lambda_1 = \lambda_1'$ *and* $\lambda_3 = \lambda_3'$
(iii) $\lambda_i = \lambda_i'$ *(*$i = 1$ *or* 2*) and* $\lambda = \lambda'$
(iv) $\lambda_1 = \lambda_1'$ *and* $\alpha = \alpha'$ *(or* $\lambda_3 = \lambda_3'$ *and* $\delta = \delta'$*)*
(v) $\lambda = \lambda'$ *and* $\alpha = \alpha'$ *(resp.* $\delta = \delta'$*)*

Proof. Follows readily from the doubly orthogonal condition and (11). □

Remark 2. The above theorem and the tight relationships among those angular invariants of doubly orthogonal tetrahedra should be regarded as the generalization of that of the right-angle triangles.

3.2. *Geometric invariant theory of isosceles tetrahedra*

Up to a scaling factor, an *isosceles* tetrahedra is uniquely determined by the congruence class of its top spherical triangle σ_A. We should denote the isosceles tetrahedra with $\sigma_A \cong \sigma$ and $\overline{AB} = \overline{AC} = \overline{AD} = l$ simply by $\tau(\sigma, l)$. For the convenience of later usage, we shall from now on denote the top vertices of an isosceles tetrahedra by O and σ_O by σ.

As indicated in Figure 5, set $\overline{M}$, M to be the circumcenters of $\triangle ABC$ and $\sigma(ABC)$. Then the *circumcentric subdivision* of $\triangle ABC$ subdivides it into three pairs of matching right-angle triangles, while their cones with O as the common vertex provide a *canonical decomposition* of $\tau(\sigma, 1)$ into three pairs of matching *doubly orthogonal tetrahedra* whose *spherical triangle at*

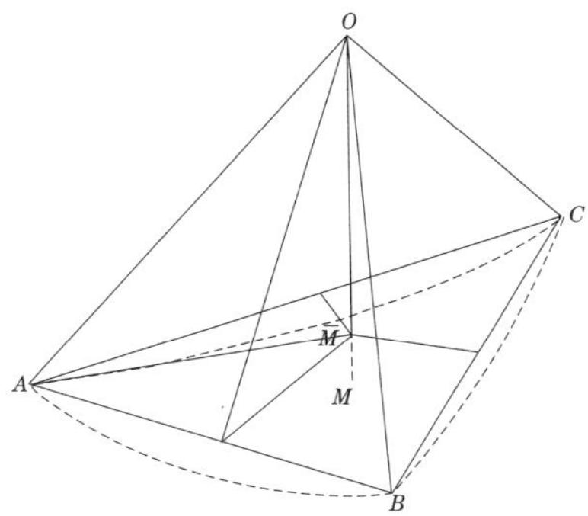

Figure 5

O are exactly that of the circumcentric subdivision of $\sigma(ABC)$ (cf. §(2.2)), namely, right-angle spherical triangle with $\{\theta_i, \lambda_i\}$ as the other two angles respectively. Therefore, the geometric invariants as well as their correlated formulas of such a canonical decomposition of $\tau(\sigma, 1)$ can be deduced simply by that of the *circumcentric subdivision of* σ (cf. §(2.2))

3.3. *Circumcentric subdivision of tetrahedra and the geometric invariant theory of tetranometry*

Up to a scaling factor, the geometric invariant theory of tetrahedra can be reduced to that of tetrahedra with unit circumradii, while the circumcentric subdivision of such a tetrahedra provides a *canonical decomposition* into quadruple 1-isosceles tetrahedra $\{\tau(\sigma_i, 1)\}$ with $\{\sigma_i\}$ being the solid angles at the center O spanned by its four face-triangles. In solid geometry, the totality of solid angle at O is naturally represented by the unit sphere centered at O. Thus the above decomposition of the total solid at O into $\{\sigma_i\}$ constitutes a spherical configuration canonically associated to the geometry of tetrahedra, which will be henceforth referred to as the *associated spherical configuration of a tetrahedra.* Such a canonical association enables us to reduce the study of geometric invariant theory of tetrahedra (i.e. tetranometry) neatly and concisely to that of *spherical configurations with four spherical triangles* together with their circumcentric subdivisions. Anyhow, this will provide an advantageous way of studying tetranometry.

4. An Outstanding Example of Problems of Solid Geometry: The Sphere Packing Problem

There are many natural as well as challenging problems in solid geometry, however, the *sphere packing problem* is certainly one of the most outstanding problem suggested by the crystal structure of dense type in solid state physics and the Kepler's conjecture of 1611 [Kep]. It certainly deserves to have a clean-cut, elementary proof revealing in depth understanding of solid geometry, and in the opinion of the authors, it also provides a natural testing problem for the depth of understanding of solid geometry. Anyhow, it is exactly for the purpose of achieving a clean-cut, elementary proof of Kepler's conjecture as well as the least action principle of crystal formation that motivates and leads to the development of such an approach of solid geometry (cf. [Hs-2]). We shall summarize the main results of [Hs-2] in this section and then provide a concise outline of such a proof via geometric invariant theory in the next section.

4.1. *Three kinds of sphere packings, various kinds of densities and problems of their optima*

Basically, there are three different kinds of packings of spheres of identical size, namely

(i) packing with container; $\mathcal{P} \subset \Gamma$(container)

(ii) finite packings without container, e.g. crystals;

(iii) infinite packings with the whole space as the container.

For examples, putting marbles into a jar, say beans into a silo are daily life examples of the first kind; while a piece of crystals of gold, silver etc. are Nature-created examples of the second kind. On the other hand, those infinite packings such as the f.c.c packing, the hexagonal close packings and lattice packings are, in fact, just some mathematical models served as the limiting situations of the other two kinds as their sizes tends to infinity.

The central problems in the study of sphere packings are the problems of optimalities of various kinds of densities. For example, the density of a packing $\mathcal{P}$ into a container Γ is defined by setting

$$\rho(\mathcal{P} \subset \Gamma) := \operatorname{vol}\mathcal{P}/\operatorname{vol}\Gamma$$

while $\rho(r,\Gamma) = \text{l.u.b}\{\rho(\mathcal{P},\Gamma)\}$ as $\mathcal{P}$ runs through all possible packings of r-spheres into Γ is defined to be *optimal density* of such packings. Note that $\rho(r,\Gamma)$ will certainly depend on the shape of Γ and the relative size between Γ and r-spheres. Anyhow, this motivates us to study

$$\hat{\rho}(\Gamma) := \limsup_{k\to\infty} \rho(1, k\Gamma)$$

where $k\Gamma$ denotes the k-time magnification of Γ, and the following formulation of Kepler's conjecture, namely

Kepler's conjecture (2nd version): For a large class of container Γ, such as those with piecewise smooth boundary $\partial\Gamma$, $\hat{\rho}(\Gamma)$ should always be equal to $\pi/\sqrt{18}$.

Problem of least action principle on the crystal formation of dense type: How to *properly define* the density of packings of the second kind so that the geometric regularity of crystal of dense type (i.e. hexagonal close packing type) is actually the consequence of density optimality.

4.2. *A simple local invariant and the definition of global densities of packings of the second and the third kinds*

In his remarkable booklet of 1611 [Kep], Kepler had already introduce the concept of local cell, nowadays often referred to as Voronoi cell, which assigns a surrounding polyhedron to each S_i in $\mathcal{P}$ consisting of those points that are as close to the center of S_i, say O_i, as to that of the others, say $\{O_j, j \neq i\}$. Let C_i be the local cell of S_i in $\mathcal{P}$. Then $\{C_i, i \in I\}$ constitutes a decomposition of the space into non-overlapping union of convex polyhedra, each containing a single sphere of $\mathcal{P}$, which will be, henceforth, referred to as the local cell decomposition associated to the packing $\mathcal{P}$, or simply as the *L-decomposition of* $\mathcal{P}$. Moreover, the L-decomposition has a natural dual decomposition into convex polyhedra, often referred to as the Delaunay decomposition, or simply as the *D-decomposition of* $\mathcal{P}$. The duality between such a pair of decompositions associated to a given packing $\mathcal{P}$ can be summarized as follows:

(i) The set of vertices of the D-decomposition are just those centers of $S_i \in \mathcal{P}$, thus $\{O_i\} \leftrightarrow \{C_i\}$.

(ii) The set of edges of the D-decomposition are those $\{\overline{O_iO_j}\}$ linking the centres of neighbouring pairs $\{S_i, S_j\} \leftrightarrow$ those common faces of $\{C_i, C_j\}$ situated on the perpendicular bisector of $\overline{O_iO_j}$.

(iii) The set of faces of the D-decomposition are those polygons spanned by the centres of those local cells with a *common edge* $\leftrightarrow$ {their common edge, perpendicular to the face at its circumcenter}.

(iv) The set of convex polyhedra spanned by the centres of those local cells with a *common vertex* $\leftrightarrow$ {their common vertex which is exactly the circumcenter of its corresponding D-cell}.

A new kind of locally averaged density
(cf.§1.4 of [Hs-1] for another kind.)

Set I (resp. J) to be the indices sets of the L-cells (resp. D-cells) of the above dual pair of decomposition of $\mathcal{P}$, and set

$$w_i = \operatorname{vol} C_i,\ w^j = \operatorname{vol} w_j,\ w_i^j = \operatorname{vol} C_i \cap \Omega_j.$$

Definition. To each D-cells Ω_j, set

$$\rho(\Omega_j) := \operatorname{vol}(\Omega_j \cap \mathcal{P})/\operatorname{vol}\Omega_j \tag{12}$$

and call it the density of the D-cell Ω_j. To each $S_i \in \mathcal{P}$, the locally averaged density of $\mathcal{P}$ at S_i is defined to be

$$\bar{\rho}(S_i, \mathcal{P}) := \sum_{j \in J} w_i^j \rho(\Omega_j) / \sum_{j \in J} w_i^j. \tag{13}$$

Note that, for each given i, there are only a rather small number of j with non-zero w_i^j and $\sum_j w_i^j = w_i$

Remarks:

(i) In retrospect, the cluster of spheres centred at the vertices of a given Ω_j can be regarded as the sub-cluster of $\mathcal{P}$ of the most localized kind. Thus, $\rho(\Omega_j)$ can be regarded as a kind of *ultimate localization* of the concept of densities associated to a given $\mathcal{P}$.

(ii) Note that $\bar{\rho}(S_i, \mathcal{P})$ is, itself, a weighted averaged of $\{\rho(\Omega_j)\}$ that makes use of the dual pair of decompositions. Of course, its usefulness will only be determined by the ultimate test whether it can provide a better result in the study of global optimal of sphere packings, (cf. §(4.3)).

Relative density and global densities of the second and the third kind
Let $\mathcal{P} = \{S_i : i \in I\}$ be an infinite packing and $\mathcal{P}' = \{S_i : i \in I'\}$ be one of its finite sub-packings.

Definition (relative density). The *relative density* of $\mathcal{P}'$ in $\mathcal{P}$, denoted by $\bar{\rho}(\mathcal{P}', \mathcal{P})$, is defined to be

$$\bar{\rho}(\mathcal{P}', \mathcal{P}) = \frac{\sum_{i \in I'} w_i \bar{\rho}(S_i, \mathcal{P})}{\sum_{i \in I'} w_i},\ \mathcal{P}' = \{S_i, i \in I'\} \tag{14}$$

Definition (intrinsic density). Let $\mathcal{P}'$ be a given finite packing without container. The intrinsic density of $\mathcal{P}'$, denoted by $\bar{\rho}(\mathcal{P}')$, is defined to be

$$\bar{\rho}(\mathcal{P}') := \text{l.u.b}\,\{\bar{\rho}(\mathcal{P}', \overline{\mathcal{P}})\} \tag{15}$$

where $\overline{\mathcal{P}}$ runs through all possible extension of $\mathcal{P}'$.

Definition (Global density of infinite packings).

$$\rho(\mathcal{P}) := \text{l.u.b}\,\{\limsup_{n\to\infty} \bar{\rho}(\mathcal{P}_n, \mathcal{P})\} \tag{16}$$

where $\{\mathcal{P}_n\}$ runs through all possible exhaustion sequence of $\mathcal{P}$.

Example 1.1. Let $\mathcal{P}$ be a hexagonal close packing. Then

$$\bar{\rho}(S_i, \mathcal{P}) = \pi/\sqrt{18} \quad \forall S_i \in \mathcal{P} \tag{17}$$

and hence

$$\bar{\rho}(\mathcal{P}', \mathcal{P}) = \pi/\sqrt{18} \tag{18}$$

for all finite sub-packings $\mathcal{P}'$ in $\mathcal{P}$, and $\rho(\mathcal{P})$ is also equal to $\pi/\sqrt{18}$.

Proof. The local cluster of D-cells that occur in (1.8) consists of octuple regular 2-tetrahedron and sextuple regular 2-octahedron. Therefore their volumes (resp. total solid angles) are given by

$$\begin{cases} \sqrt{8}/3 \\ 8\sqrt{2}/3 \end{cases} \left(resp. \begin{cases} 4(3\arccos\dfrac{1}{3} - \pi) \\ 6(4\arccos(-\dfrac{1}{3}) - 2\pi) \end{cases} \right)$$

and hence

$$\rho(\Omega_j) = \begin{cases} \sqrt{2}(3\alpha_0 - \pi) & \sim 0.7796356 \\ \frac{1}{\sqrt{8}}(3\pi - 6\alpha_0) & \sim 0.7209029. \end{cases}$$

Thus

$$\begin{aligned} \rho(S_i, \mathcal{P}) &= \frac{1}{4\sqrt{2}} \left\{ \frac{4}{3}\sqrt{2} \cdot \sqrt{2}(3\alpha_0 - \pi) + \frac{8}{3}\sqrt{2} \cdot \frac{1}{\sqrt{8}}(3\pi - 6\alpha_0) \right\} \\ &= \pi/\sqrt{18}. \end{aligned}$$

□

4.3. *Fundamental problem and fundamental theorem of sphere packings*

Note that a pair of spheres with their center distance less than $2\sqrt{2}r$ are automatically neighbours of each other (i.e. their local cells must have a common face). Thus, it is quite natural to introduce the following definition of cluster of spheres.

Definition. A finite packing of r-spheres is called a *cluster* if any pair of them can always be linked by a chain with consecutive center distance less than $2\sqrt{2}r$.

Remark: A single sphere is, of course, regarded as a special case of cluster.

Fundamental Problem of Sphere Packings:
Set ρ_N to be the optimal intrinsic density for all possible N-cluster, namely

$$\rho_N := l.u.b.\{\rho(\mathcal{C})\}$$

where $\mathcal{C}$ runs through all possible clusters of N spheres. What is ρ_N equal to? And what are the geometric structures of those N-clusters together with their tightest surrounding with $\bar{\rho}(\mathcal{C}, \mathcal{C}^*) = \rho_N$?

In the beginning case of $N = 1$, ρ_1 is just the *optimal locally averaged density.* The above fundamental problem naturally leads to the proof of the following fundamental theorem, namely

Theorem I. *The optimally locally averaged density is equal to $\pi/\sqrt{18}$ and $\bar{\rho}(S_0, \mathcal{L}) = \pi/\sqrt{18}$ when and only when the local packing $\mathcal{L}(S_0)$ is isometric to either that of the f.c.c. or that of the h.c.p. packing, which surrounds S_0 with twelve touching neighbours with their touching points as indicated in Figure 6.*

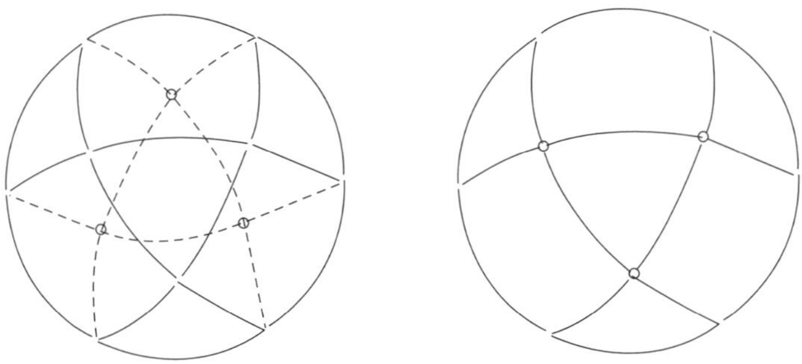

f.c.c. *h.c.p.*

Figure 6

It is remarkable that the following major theorems on the optimalities of global densities of sphere packing can readily be deduced from Theorem I, namely

Theorem II (Kepler's conjecture, 1st version). *The optimal global density of infinite sphere packings is equal to $\pi/\sqrt{18}$, namely*

$$\rho(\mathcal{P}) \leq \pi/\sqrt{18} = \rho \text{ (hexagonal close packings)}.$$

Theorem III (Least action principle of crystal formation of dense type).

$$\rho_N = \pi/\sqrt{18} \quad \forall N$$

and $\bar{\rho}(\mathcal{C}, \mathcal{C}^*) = \rho_N$ when and only when the N-cluster $\mathcal{C}$ together with its tightest surrounding $\mathcal{C}^*$ can be embedded into a certain hexagonal close packing.

Theorem IV (Kepler's conjecture, 2nd version). For all kinds of containers Γ with piecewise smooth boundaries $\partial\Gamma$,

$$\hat{\rho} = \pi/\sqrt{18}.$$

In retrospect, the fact that the *optimalities* of three kinds of sphere packings as stated in Theorems II, III and IV can all be deduced from Theorem I on the optimality of a single local invariant $\bar{\rho}(S_0, L)$ can be regarded as the kinds of far-reaching proofs via *proper ways of localizations* which are, at all, feasible only when *the global optimalities are actually the consequence of local optimalities everywhere.* In fact, the very *existence* of those crystals of dense type already reveals that this must be the way how Nature created such wonderful optimalities, while the above proof amounts to demonstrate the *uniqueness aspect of global optimalities on sphere packings.* Naturally, one certainly needs in depth understanding on solid geometry in order to achieve the proof of Theorem I. We refer to [Hs-2] for a detail exposition of such a proof, thus only providing a concise outline on some crucial geometric ideas and pertinent lemmas of [Hs-2] in §5.

5. Two Stages of Localizations and Two Stages of Area-wise Estimates

In this section, we shall first outline the basic geometric ideas and some crucial geometric insights on the proof of Theorem I in [Hs-2], and then discuss a few key lemmas on pertinent area-wise estimates of the specific weighted average of densities.

5.1. *The geometry of local packings and the analysis of density of D-cells*

The (new kind) *locally averaged density* $\bar{\rho}(S_0, L)$ is defined to be a specific weighted average of the densities of the *cluster of D-cells* with the center of S_0 as their common vertex. Thus $\bar{\rho}(S_0, L)$ is geometrically rooted in the geometry of local packing L, while it is analytically rooted in density function of D-cells and weight function $w_0^j = \operatorname{vol} C_0 \cap \Omega_j$.
Remark: In comparison with the locally averaged density of [Hs-1] which depends on the geometry of a doubly layer local packing $L^2(S_0)$, the latter is much, much simpler to deal with both geometrically and analytically.

The geometry of local packing $L(S_0)$, consisting of S_0 together with its surrounding neighbors, say S_j, can be concisely represented by the set of position vectors $\{\overrightarrow{OO_j}\}$, or equivalently, the set of $\{A_j\}$ and $\{2(1+h_j)\}$ recording the directions and the lengths of $\overrightarrow{OO_j}$ respectively. We shall call $\Sigma = \{A_j\}$ the *directional profile* of L and $\{h_j\}$ the set of *buckling heights* of L, which are recording the individual non-touchingness between S_0 and S_j. One of the, nowadays, well-known basic geometric fact on local packings is that there are *at most twelve touching neighbors* [S-W] (i.e. at most twelve zero buckling heights). Intuitively, one naturally expects that the most critical case in the proof of Theorem I should be that of this type of local packings which will be, henceforth, referred as *Type-I local packings* (i.e. with exactly twelve touching neighbors).

5.2. *Some basic geometric ideas and insights*

1) The proof of Theorem I for the case of Type-I local packings is not only the most critical case, but it will also provide generalizable geometric insights as well as analytical techniques for the proof of non-Type-I case. In fact, it is possible to find a clean-cut, elementary proof for the Type I case which is naturally generalizable to the case of non-Type-I local packings (cf. [Hs-2]).

2) The geometry of a Type-I local packing is essentially determined by that of the twelve touching points $\Sigma \subseteq S_0$, which can be concisely organized and advantageous represented by the associated spherical configuration of Σ, henceforth denoted by $\mathcal{S}(\Sigma)$. In the generic case, such an $\mathcal{S}(\Sigma)$ consists of twenty triangular faces which are exactly those central solid angles of the cluster of D-cells with O as their common vertex, while $\bar{\rho}(S_0, L)$ is the weighted average of the densities of these twenty *isosceles tetrahedra*,

namely

$$\bar{\rho}(S_0, L) = \frac{1}{\operatorname{vol} T(\Sigma)} \sum \left\{ w(\sigma_j)\rho(\tau(\sigma_j, 2)); \sigma_j \in \mathcal{S}(\Sigma) \right\} \tag{19}$$

where (cf. (7)),

$$\begin{aligned} w(\sigma_j) &= \text{vol of}\, T(\sigma_j) = \tfrac{1}{6} \sin |\sigma_j| \{ \tfrac{8}{u(\sigma_j)} - \tan^2 R(\sigma_j) \}, \\ \rho(\tau(\sigma_j, 2)) &= \tfrac{1}{4D(\sigma_j)} \nu(\sigma_j) \end{aligned} \tag{20}$$

3) Analytically, the proof of Theorem I for such a generic geometric situation amounts to provide an optimal upper bound estimate of $\pi/\sqrt{18}$ for the RHS of (19) together with the uniqueness of the optimality being the f.c.c. and the h.c.p of Figure 6. Anyhow, such an optimal estimate for such a weighted average defined on the *moduli space of Type-I configurations* of dimension 21 is still a very formidable task to achieve. The pertinent geometric insight for achieving such a formidable task in the realm of spherical geometry is a kind of two stages localization by means of two stages of area-wise bound estimates.

5.3. *Two stages of area-wise estimates*

1) As a direct application of the geometric invariant theory of spherical triangles and isosceles tetrahedron, it is not difficult to analyze the restriction of $\rho(\tau(\sigma, 2))$ on a $|\sigma|$-level surface satisfying further geometric restrictions on σ which are *automatically* satisfied for those triangular faces belonging to a *Type-I configurations.* Anyhow, the area preserving deformations such as the Lexell's deformations and the δ-deformations provide a set of powerful tools for such an analysis which enable us to prove the optimal area-wise estimate of $\rho(\tau(\sigma, 2))$ for the first stage of localization, namely

The first stage optimal estimate:

First of all, for the purpose of proving Theorem I in the case of Type I local packings, it suffices to study the up bound areawise estimate of $\rho(\tau(\sigma, 2))$ for those σ satisfying the following conditions, namely, of edges lengths of at least $\pi/3$ and containing its own circumcenter. Let σ be such a spherical triangle with its area $|\sigma|$ at most equal to 0.97, it is natural to study the first stage *areawise optimal estimate* of $\rho(\tau(\sigma, 2))$, namely

What are such σ's that $\rho(\tau(\sigma, 2))$ achieve the maximal among those σ's of the same given areas, lying between $\sigma_0 = 2 \arctan \frac{\sqrt{2}}{5}$ and 0.97?

The (k, δ)-representation of $|\sigma|$-level surface:

Set $c \geq b \geq a \geq \frac{\pi}{3}$ to be the edge-lengths of σ and

$$k = \cot\frac{a}{2}\cot\frac{b}{2}, \quad \delta = \cot\frac{a}{2} - \cot\frac{b}{2} \geq 0.$$

Then $\{C, k, \delta\}$ also constitutes a complete set of S.A.S.-type of congruence invariants of σ, and moreover, $\{|\sigma|, k, \delta\}$ also constitutes a complete set of congruence invariants for those σ containing their own circumcenters (cf. Cor 1 of Thm 4, §2), namely

$$C = \arcsin(k \sin\frac{|\sigma|}{2}) + \frac{|\sigma|}{2}. \tag{21}$$

For a given value of $|\sigma| \in [2\arctan\sqrt{2}/5, 0.97]$, it is convenient to parametrize the $|\sigma|$-level surface by (k, δ), thus representing it as a domain in the (k, δ)-plane. It is natural to subdivide into two cases, namely

Case 1: $|\sigma| \leq 2\arcsin\frac{1}{3} = \frac{1}{2}\Box_{\pi/3}$,

Case 2: $|\sigma| > 2\arcsin\frac{1}{3}$ and at most equal to 0.97

Case 1: $|\sigma| \in [2\arctan\frac{\sqrt{2}}{5}, \frac{1}{3}]$:

Note that the special case of $|\sigma| = 2\arctan\frac{\sqrt{2}}{5}$ only consists of a single point (i.e. (3,0)). Thus, we shall assume that $|\sigma| > 2\arctan\frac{\sqrt{2}}{5}$ and at most equal to $2\arcsin\frac{1}{3} = \frac{1}{2}\Box_{\pi/3}$. For such a given $|\sigma|$, there exist a unique $\pi/3$-isosceles (resp. equilateral) triangle with their areas equal to $|\sigma|$, and moreover, a continuous family of isosceles triangles of area $|\sigma|$ linking them in between, namely, with $\{(k, 0), 3 \geq k \geq k_0\}$ as their (k, δ)-coordinates where

$$k_0 = \frac{1}{1 - 2\cos\frac{1}{3}(\pi + |\sigma|)}.$$

Now, beginning with such an isosceles triangle, say denoted by $\bar{\sigma}(a, C)$, with $a = 2\arctan 1/\sqrt{k}$ and C given by (21), one has the area-preserving deformation keeping the (k, C) fixed, while increases δ up until either its shortest side-length already reaches the lower bound of $\pi/3$, or it becomes an isosceles triangle with C as its base angles. Therefore, the domain of (k, δ) representation of such a $|\sigma|$-level surface is as indicated in Figure 7-(i), where (k_1, δ_1) represents the unique c_1-isosceles triangle with the given area $|\sigma|$ and $\pi/3$-base, namely

$$c_1 = \arccos\left\{\frac{\cos\frac{|\sigma|}{2} - \frac{\sqrt{3}}{2}}{\frac{2}{\sqrt{3}} - \cos\frac{|\sigma|}{2}}\right\} \left(i.e. \cos\frac{|\sigma|}{2} = \frac{\frac{\sqrt{3}}{2} + \frac{2}{\sqrt{3}}\cos c_1}{1+\cos c_1}\right)$$

$$k_1 = \sqrt{3}\cot\frac{c_1}{2} = \sqrt{3}(7 - 4\sqrt{3}\cos\frac{|\sigma|}{2})^{-\frac{1}{2}},\ \sigma_1 = \sqrt{3} - \frac{k_1}{\sqrt{3}}.$$

Case 2: $2\arcsin\frac{1}{3} < |\sigma| \leq 0.97$

Set $\hat{a}$ to be the side-length of the spherical square with its area $2|\sigma| > \Box_{\pi/3}$, namely

$$\tan\frac{|\sigma|}{2} = \frac{1-\cos\hat{a}}{2\sqrt{\cos\hat{a}}},\ \hat{k} = \cot^2\frac{\hat{a}}{2} = \csc\frac{|\sigma|}{2}.$$

Therefore, the only difference between Case 2 and Case 1 is that k is bounded above by $\hat{k} < 3$. Thus, the domain of (k, δ) representation for Case 2 is as indicated in Figure 7-(ii).

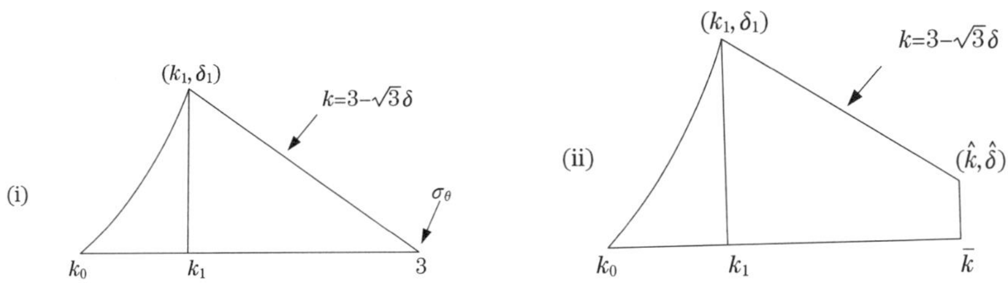

Figure 7

Note that the boundary of the above (k, δ)-domain consists of the following segments, namely

(i) The horizontal segment: $\{(k, 0), k_0 \leq k \leq \hat{k}\}$ representing those a-isosceles $\bar{\sigma}(a, C)$ with $a = 2\arctan\frac{1}{\sqrt{k}}$ and C given by (21).

(ii) The slant segment: $\{(k, \sqrt{3} - \frac{k}{\sqrt{3}}), k_1 \leq k \leq \hat{k}\}$ representing those σ with $\pi/3$ as their shortest side-length, the deformation along it is exactly the Lexell's deformation fixing the $\pi/3$ side.

(iii) The curved segment: Representing those isosceles with their base angles larger than their top angles.

(iv) The vertical segment: In the case of $|\sigma| > \frac{1}{2}\Box_{\pi/3}$, $\hat{k} < 3$ and one has as an additional vertical segment: $\{(\hat{k}, \delta), 0 \leq \delta \leq \sqrt{3} - \hat{k}/\sqrt{3}\}$, representing those σ with $C = \frac{\pi}{2} + \frac{|\sigma|}{2}$, each of them is the half of a spherical rectangle with area $2|\sigma|$ and side-lengths of at least $\pi/3$, while the corner point $(\hat{k}, \hat{\delta})$ is exactly the $\widetilde{\sigma}_\theta$ with $|\widetilde{\sigma}_\theta| = |\sigma|$. (cf. Figure 8)

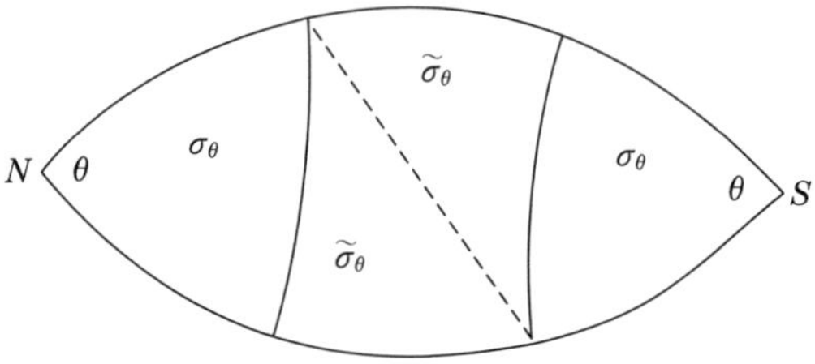

Figure 8

As indicated in Figure 8, $\theta \geq \alpha_0 = \arccos\frac{1}{3}$, a θ-lune is the assemblage of a pair of σ_θ and $\widetilde{\sigma}_\theta$.

Using the simple formulas for both the δ-deformations and the Lexall's deformations, it is not difficult to show the δ-monotonicity as well as the k-monotonicity of $\rho(\tau(\sigma, 2))$ with respect to area-preserving deformations, thus proving the following optimal upper bound area-wise estimate of $\rho(\tau(\sigma, 2))$, namely

Lemma I ([Hs-2]): Let σ be spherical triangles with edge-lengths of at least $\pi/3$ and containing their own circumcenters. In the cases that

$$2\arctan\frac{\sqrt{2}}{5} \leq |\sigma| \leq 2\arcsin\frac{1}{3}$$
$$(\textit{resp.}\ \ 2\arcsin\frac{1}{3} < |\sigma| \leq 0.97)$$

set σ_θ (resp. $\widetilde{\sigma}_\theta$) to be the triangles as indicated in Figure 8 and with $|\sigma|$ as its area. Then

$$\rho(\tau(\sigma, 2)) \leq \rho(\tau(\sigma_\theta, 2)) \ \ (\textit{resp.}\ \rho(\tau(\widetilde{\sigma}_\theta, 2))) \tag{22}$$

and equality holds when and only when σ is, itself, σ_θ (resp. $\widetilde{\sigma}_\theta$).

The combine graph of $\rho(\tau(\sigma_\theta, 2))$ (resp. $\rho(\tau(\widetilde{\sigma}_\theta, 2))$) as functions of $|\sigma_\theta|$ (resp. $|\widetilde{\sigma}_\theta|$) is as indicated in Figure 9:

The second stage optimal estimate: Collective area-wise upper bound estimates on $\widetilde{\rho}(St(\cdot))$ *for* $5\triangle$*-stars:*

Geometrically, the great majority of Type I configurations are of the combinatorial type of icosahedra (i.e. the twelve stars are all of $5\triangle$-type).

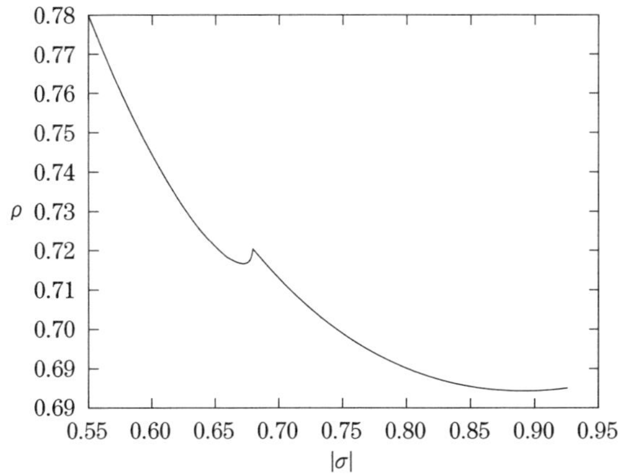

Figure 9

Let $S'(\Sigma)$ be such a Type I icosahedra and $\{St(A_i)\}$ be its twelve star-sub-configurations. Then

$$\sum_{A_i \in \Sigma} |St(A_i)| = 12\pi \quad (= 3|S_0|)$$

simply because each $|\sigma_j|$, $\sigma_j \in S'(\Sigma)$, occurs thrice in the above sum.

Set

$$\widetilde{\rho}(St(A_i)) := \sum_{\sigma_j \in St(A_i)} w^j \rho(\tau(\sigma_j, 2))/\widetilde{w}_i; \quad \widetilde{w}_i := \sum_{\sigma_j \in St(A_i)} w^j. \tag{23}$$

It is easy to check that

$$\sum_{i=1}^{12} \widetilde{w}_i \widetilde{\rho}(St(A_i)) / \sum_{i=1}^{12} \widetilde{w}_i = \sum_{j=1}^{20} w^j \rho(\tau(\sigma_j, 2)) / \sum_{j=1}^{20} w^j \tag{24}$$

because the numerator (resp. denominator) of the LHS are exactly thrice of that of the RHS. The above equations, intuitively, suggest that an *area-wise upper bound estimate* on $\widetilde{\rho}(St(\cdot))$ for such $5\triangle$-stars may provide a powerful tool and hence an advantageous way of achieving the optimal estimate of the RHS of (24) for Type-I icosahedra.

This naturally leads to the discovery of the following key-lemma of area-wise collective estimate of $\widetilde{\rho}(St(\cdot))$, namely

Lemma II : ([Hs-2]): Let $St(\cdot)$ be a $5\triangle$-star with edge-lengths of at least $\pi/3$, containing their circumcentres and $|St(\cdot)| \leq \pi$ (resp. $> \pi$ and at most equal to $\pi + 0.21672$.) Set $St_5^\circ(\theta)$ (resp. $St_5(\alpha_0, \theta)$) to be the one as indicated in Figure 10-(ii) (resp (i)) with the same area of $St(\cdot)$. Then

$$\widetilde{\rho}(St(\cdot)) \leq \widetilde{\rho}(St_5^\circ(\theta)) \ (resp.\ \ \widetilde{\rho}(St_5(\alpha_0, \theta))) \tag{25}$$

and equality holds when and only when $St(\cdot)$ consists of the same collection of quintuple triangles as that of $St_5^\circ(\theta)$ (resp. $St_5(\alpha_0, \theta)$).

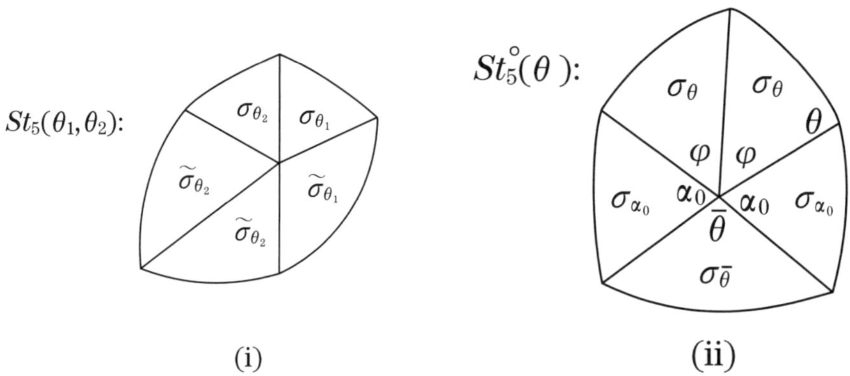

Figure 10

The graph of the above optimal area-wise collective estimate of $\widetilde{\rho}(St(\cdot))$, as a function of $|St(\cdot)|$ is as indicated in Figure 11.

We refer to [Hs-2] for the proofs of the above two lemmas as well as major role they play in the proof of Theorem I.

We mention here the proof of Theorem I in the most critical geometric situation, just as an example as well as an indication of the powerfulness of Lemma II as follows:

Note that the f.c.c and the h.c.p. are the only two local packings that $\overline{\rho}(s_0, L)$ actually achieve the maximality of $\pi/\sqrt{18}$. Geometrically, the h.c.p. is, in fact, an isolated point in the moduli space of congruence classes of Type-I configurations, while the f.c.c. is a cusp point with *1-dimensional tangent cone* in the 21-dimensional moduli space. First of all, both of them have triangulations of icosahedra type with all of their twelve stars of uniform area of π, and moreover, the f.c.c has a rather large neighborhood in the 21-dimensional moduli space consisting of Type-I icosahedra whose stars all have their areas quite close to π, say lying between $\pi - 0.07$ and

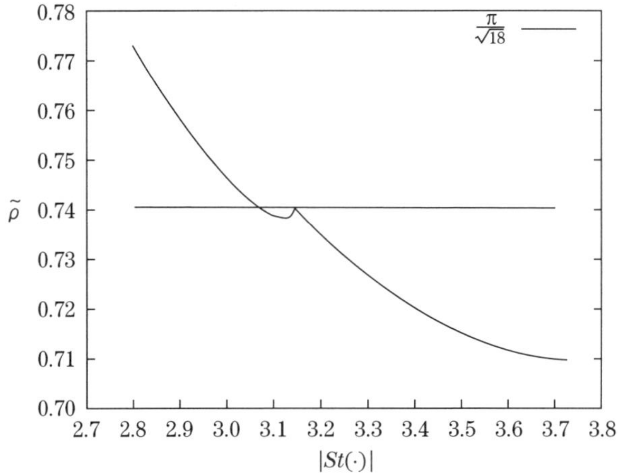

Figure 11

$\pi + 0.07$. Therefore, as a direct consequence of Lemma II, one has

$$\widetilde{\rho}(St(A_i)) \leq \frac{\pi}{\sqrt{18}}$$

for each star of such a Type-I icosahedra and equality holds when and only when $St(A_i)$ consisting of a pair of σ_{α_0} and a triple of $\widetilde{\sigma}_{\alpha_0}$. Hence, the f.c.c. is the only one in that large neighborhood achieving $\overline{\rho}(s_0, \alpha) = \pi/\sqrt{18}$.

6. Concluding Remarks

The brief discussion on the geometric invariant theory of the space in the article should be regarded as just a beginning of such an approach to solid geometry. However, the success in providing such elementary, cleancut solutions to the outstanding problems on various kinds of optimal densities on sphere packing [Hs-2] clearly indicates that it is indeed a good beginning towards vast prospects of such an approach.

References

Bar. W. Barlow, Probable Nature of the Internal Symmetry of Gystals, Nature 29(1883) 186-188.

Hs-1. Wu-yi Hsiang, Least Action Principle of Crystal Formation of Dense Packing Type and Kepler's Conjecture, Nankai Tracts in Math. Vol 3. World Scientific (2001).

Hs-2. Wu-yi Hsiang, Sphere Packing Problems Revisited, A New Local Invariant and Simple Proofs of Kepler's Conjecture and Least Action Principle of Crystal Formation (to appear).

Kep. Johannes Kepler, The Six-cornered Snowflake–A New Year's Gift (1611) [Paul Dry Books ISBN 978-1-58988-053-5 (2010)].

Pto. Ptolemy, Almagest

S-W. K. Schütte and B. L. van der Waerden, Das Problem der Dreizehn Kugeln, Math. Ann. 125(1953) 325-334.

Optimal Convergence Rate of the Binomial Tree Scheme for American Options and Their Free Boundaries

Lishang Jiang and Jin Liang*

Department of Mathematics, Tongji University, Shanghai 200092
People's Republic of China
jiang_lishang@yahoo.com.cn; liang_jin@mail.tongji.edu.cn

In this article, we review the recent developments of the researches on the optimal convergence rate of the Binomial Tree Method (BTM) scheme for an American put option which can be modeled as a variational inequality problem. The scheme is equivalent to Explicit Finite Difference Scheme (EFDS) under a condition. The optimal rate is obtained. Moreover, by using a modified EFDS scheme, the convergence rate of approximated free boundary to the actual free boundary is also obtained.

Estimate of a convergence rate for a numerical scheme always is interesting for taking numerical calculation using this scheme. With it, one can design accuracy region, estimate calculation time, check process, compare methods and so on. It is very significant to find out a convergence rate for a scheme. Financial calculations are not exceptions.

It is well known that an American option has an optimal exercise boundary, which is called a free boundary in PDE theory. Therefore pricing an American option is a nonlinear variational inequality problem. When the process has jumps, the pricing problem turns to a free boundary problem with an integral term in its partial differential equation. Finding the price and locating explicitly this free boundary are always significant in applications (e.g. Ref. 12). However, an American option usually fails to lead to analytical solutions even under simpler models without jumps, so that the numerical method becomes very important for its pricing and analysis.

Among the financial calculation schemes, binomial tree method (BTM) is very popular one. It is a discrete scheme in stochastic process framework

*This work is supported by National Natural Science Foundation of China (No.11271287).

and it is widely used in industry. It is not only suitable for European and American options, but also for Asian and other exotic options and interest derivatives. Under a probability framework, Amin *et al.* (Ref. 1) had consider the convergence, and obtained some result under a weak probability space. While in industry, many techniques used for this method are quite empirical without theoretical supports.

BTM method is proved to be equivalent to the explicit finite difference scheme (EFDS), however what condition can insure the convergence for using EFDS on solving an variational inequality problem is an open problem for a long time.[8] Therefore, find out the convergence rate for BTM theoretically on an American option is a concerned topic, which has both theoretical and practical significance.

Lamberton[18] obtained BTM convergence rate between $O(h^{2/3})$ and $O(h^{3/4})$ for an American Put Option at fixed time 0 (which is T comparing to our paper) or $O\big(h(\sqrt{|\log h|})^{4/5}\big)$ on additional conditions.

American option pricing in Merton's jump-diffusion model was developed by Zhang[26] as a variational inequality problem with a free boundary. Zhang[26] first applied an implicit-explicit mixed scheme on a jump diffusion model; the scheme was explicit on the integral term and implicit on the partial derivative terms in the equation of the problem. An error estimate was obtained by using energy estimate. In Refs. 9, 4 and 3, implicit/mixed schemes were used to calculate jump-diffusion model, where convergence was discussed. However, the convergence rate of these schemes remains an open interesting question.

Estimate the error on explicit scheme, usual variational scheme is not applicable. It has to use maximum principle. However, when we work on maximum-norm error estimates of EFDS for the variational inequality problem, there are three intrinsic difficulties to this discrete scheme: First, the optimal regularity of the solution $V(S,\tau)$ of the variational inequality (2) is only of $W^{2,1}_{\infty}$, which has not enough smoothness for the process. In fact, V_{SS} is bounded but not continuous at the free boundary, i.e. its truncation error near the free boundary is very difficult to estimate. Second, the initial datum $\widetilde{g}$ is only in W^1_{∞}, the derivative $\widetilde{g}'$ is bounded but not continuous. Third, for the problem with jump, it has a non-local jump-diffusion term, which can be reduced to an integro-partial differential problem. The method to overcome the first difficulty is partially motivated by the work,[16] which deals with Bellman's equations. We introduce a regularized solution V^{ε} to a penalized problem with a penalty $\beta_{\varepsilon}(\cdot)$. The error $|V - V_h|$ between the original solution and the finite difference solution can be estimated by the sum of two parts: $|V - V^\varepsilon|$ and $|V_h - V^\varepsilon|$; then we find a balance between ε and h to obtain the optimal convergence rate. For the second difficulty, we eliminate the singularity of the initial datum $\widetilde{g}$ by subtracting the corresponding European option price $U(S, \tau)$. Combining these two techniques, we obtain the error estimate $O(\Delta x)$ for $|V - V_h|$. The general method dealing with the integro-differential equation is used to solve the third difficulty. We believe that the method is also suitable for deriving error estimates of the BTM for other American style options.

In the passed 10 years, many people worked hard on this topic. Recently it has been well solved, and the optimal convergent rate has been found. The development can be described as following steps:

(1) Equivalence of the BTM and EFDS under a natural condition for an European/American option by Jiang and Dai;[11,13,14]
(2) Jiang and Dai also proved the uniformly convergence of EFDS with the condition $\sigma^2\Delta t/\Delta x^2 \leq 1$ for a vanilla American option under viscosity solution framework, where $\Delta x = \ln \delta S$, $\Delta t = t_{n+1} - t_n$, $\Delta x = x_{i+1} - x_i$, $\delta S = S_{i+1}/S_i$, n, i are difference node indexes. When the equality of the condition holds, i.e.

$$\sigma^2\Delta t/\Delta x^2 = 1, \tag{1}$$

it is the exactly critical case for BTM;[11,13,14]
(3) The equivalence of the BTM and EFDS along the characteristic line for an Asian option is proved by Jiang and Dai as well;[15]
(4) Qian *et al.* proved that the above results are also available to an American option with jumps;[24]
(5) First convergence rate $O((\Delta x)^{2/3} + (\Delta t)^{1/3})$ is obtained by Liang *et al.*. for an vanilla American option;[20,21]
(6) Hu et al. improved the convergence rate to the optimal as $O(\Delta x + \sqrt{\Delta t})$ for an vanilla American option;[10]
(7) Extension of the above result to an American option with jump-diffusion is made by Liang *et al.*;[19]
(8) Also Liang *et al.* proved the convergence rate is $O(\sqrt[4]{\Delta t})$ for the free boundary of an American option with a special definition of the approximated free boundary;[19]
(9) Chen *et al.* proved the same result with natural definition of the approximated free boundary in a modified EFDS scheme.[5]

In this paper, we review these works and the process.

1. Modeling

1.1. *American options*

An American put option is a contract which gives a holder a right to sell the underlying asset at any time before expiration date for a certain price. In the Black–Scholes framework, the valuation of an American put option is a free boundary problem.

Assume that the underlying asset price $\{S_t\}$ is a random process, which can be modeled by the following stochastic differential equation in a given probability space $(\Omega, \mathcal{F}, \mathcal{P})$ (see also in Ref. 17 by Kou):

$$\frac{dS_\tau}{S_\tau} = \mu d\tau + \sigma dW_\tau + d\Big(\sum_{j=1}^{\hat{N}_\tau} U_j\Big), \tag{1}$$

where μ, σ are positive constants; $\{W_\tau\}_{\tau>0}$ is the standard Brownian motion with

$$\mathrm{E}(dW_\tau) = 0, \qquad Var(dW_\tau) = d\tau;$$

$\{\hat{N}_\tau\}_{\tau>0}$ is a Poisson process with positive constant intensity λ, and the sequence $\{U_j\}_{j>1}$ are square integrable, independent, identically distributed random variable taking values in $(-1,\infty)$. So that $E[U_j] = E[U_1]$ for any integer $j > 0$. If $\lambda = 0$, it is a vanilla case.

If the corresponding American put option pricing at time τ is given by

$$V_\tau = V(S_\tau, \tau),$$

then the function $V(S,\tau)$ satisfies the following variational inequality (see e.g. Refs. 23, 24 and 26)

$$\min\big(r^{-1}\tilde{L}[V], V(S,\tau) - \tilde{g}(S)\big) = 0, \quad (S,\tau) \in \mathbb{R}^+ \times [0,T], \tag{2}$$

$$V(S,T) = \tilde{g}(S), \qquad S \in \mathbb{R}^1, \tag{3}$$

where

$$\begin{aligned}\tilde{L}[V] &= \frac{\partial V}{\partial \tau} + \frac{\sigma^2 S^2}{2}\frac{\partial^2 V}{\partial S^2} + (r - \lambda\omega)S\frac{\partial V}{\partial S} - (r+\lambda)V \\ &\quad + \lambda \int_{-1}^{\infty} V(S(1+z), t)\, d\tilde{N}(z), \\ \tilde{g}(S) &= \max\big(K - S, 0\big),\end{aligned}$$

$r, K, \sigma > 0$ being the interest rate, striking price and volatility of the option respectively. Also $\mu = r - \lambda\omega$, $\omega = \mathrm{E}U_1$. For the simplicity, we suppose that $K = 1$.

It is well known that no closed-form solutions are available even for the American vanilla option. In 1979, Cox and Rubinstein (see Ref. 6) were the first to propose the Binomial Tree Method (BTM), which is a discrete time model to value a vanilla options. Since then, it has been a very popular choice for computing the pricing of many options including American options. The method is widely used.

Let $\hat{\lambda}$ be the probability that a Poisson jump happens in small time interval h, i.e. approximately, $\hat{\lambda} = h\lambda$. Then the American put option with jump diffusion price $V_i^n = V(S_i^n, \tau_n)$, $i \in \mathbb{Z}$ can be defined by the following inverse induction process ([2]): $V_i^N = g(S_i^N)$, If V_i^{n+1} is already known, then V_i^n is defined by $V_i^n = \max\Big(\frac{1}{\rho}\Big((1-\hat{\lambda})\big(qV_{i+1}^{n+1} + (1-q)V_{i-1}^{n+1}\big) + \hat{\lambda}\sum_{l\in\mathbb{Z}} V_{l+1}^{n+1}\hat{p}_l\Big), g(S_i^n)\Big)$, where $0 \le n \le N-1$, $\rho = e^{rh}$, $u = e^{\sigma\sqrt{h}}$ and $q = \frac{(\rho-\hat{\lambda}(\omega+1))/(1-\hat{\lambda})-u^{-1}}{u-u^{-1}}$, $\hat{p}_l = \text{Prob}\Big(\ln(1+U_1) \in \Big[(l-\frac{1}{2})\sigma\sqrt{h}, (l+\frac{1}{2})\sigma\sqrt{h}\Big)\Big) = \tilde{N}\big(e^{(l+\frac{1}{2})\sigma\sqrt{h}} - 1\big) - \tilde{N}\big(e^{(l-\frac{1}{2})\sigma\sqrt{h}} - 1\big)$, where, $\tilde{N}(x)$ is the distribution of jump amplitude U_1, and $\tilde{N}(x) = N(e^x - 1)$ takes the form of norm distribution (for some μ, $\hat{\sigma} > 0$), i.e. $N(x) = \frac{1}{\hat{\sigma}\sqrt{2\pi}}\int_{-\infty}^{x} e^{-\frac{(\eta-\mu)^2}{2\hat{\sigma}^2}}d\eta$.

For the simplicity, in the rest of the paper, if it is not necessary, we will describe the results and proof ideas only on the problem without jumps.

1.2. *Variational inequality problem*

The pricing of an American option can be reduced to a variational problem (2)–(3). By changing variables

$$x = \log S, \qquad t = T - \tau, \qquad u(x,t) \equiv V(S,\tau), \qquad g(x) \equiv \tilde{g}(S), \quad (4)$$

the problem is reduced to

$$F[u] := \min\big(r^{-1}L[u], u - g(x)\big) = 0, \quad (x,t) \in \mathbb{R}^1 \times (0,T), \quad (5)$$

$$u(x,0) = g(x), \quad x \in \mathbb{R}^1, \quad (6)$$

where

$$L[u] = u_t - \frac{\sigma^2}{2}u_{xx} - \big(r - \frac{\sigma^2}{2}\big)u_x + ru, \quad (7)$$

$$g(x) = (1-e^x)^+, \qquad x \in \mathbb{R}^1. \quad (8)$$

For this variational problem, it is well known that there exists a unique solution in $C(\mathbb{R}^1 \times [0,T]) \cap W_{p,loc}^{2,1}(\mathbb{R}^1 \times (0,T])$ (e.g. see Ref. 7 §1.6), where $W_{p,loc}^{2,1}(\mathbb{R}^1 \times (0,T])$ consists of functions $u(x,t)$ which belong to $W_p^{2,1}((x_0 -$

$1, x_0 + 1) \times [\tau, T])$ for any $x_0 \in \mathbb{R}^1$ and any $0 < \tau < T$, Here we discuss the problem in spaces $W_p^{2,1}(Q)$ and $C^{m+\alpha,(m+\alpha)/2}(Q)$, for $Q = \Omega \times [0,T]$, $\Omega \subset \mathbb{R}^1$, $p \geq 1$, $0 \leq \alpha, \beta < 1$ and m is any non negative integer.

We introduce the regularized solution u^ε to the penalized problem

$$L[u^\varepsilon] = \beta_\varepsilon(g - u^\varepsilon), \quad (x,t) \in \mathbb{R}^1 \times (0,T), \tag{9}$$

$$u^\varepsilon(x,0) = g(x) = (1 - e^x)^+, \quad x \in \mathbb{R}^1, \tag{10}$$

where $\beta_\varepsilon(z) = \beta(z/\varepsilon)$, and β satisfies $\beta(z) = 0$ for $z < 0$, $\beta(z) = z - 1$ for $z > 2$, $\beta \in C^\infty(\mathbb{R})$, $\beta'(z) \geq 0$, $\beta''(z) \geq 0$, $\beta'(z) > 0$, $\beta''(z) > 0$, for $0 < z < 2$. By using well known theory, the existence of a classical solution $u^\varepsilon \in C^\infty$ to (9)–(10) for $\forall \varepsilon > 0$, and its convergence to the solution of (5), (6) can be derived in the same way as in [7, Chap. 1]. American option price (with jumps) and its free boundary has the following results. (Yang, Jiang and Bian [2006][25]):

Lemma 1.1. *The American put option price u is the solution of the variational inequality (4), (5), which can be decomposed into two parts: the first is a European option price U and the second is so called early exercise premium $(u - U)$. The optimal exercise boundary is a free boundary $x = s(t)$ which is continuous and strictly decreasing. The free boundary $x = s(t)$ divides the domain into two regions: the continuation region $x > s(t)$ and stopping region $x < s(t)$. In the stopping region, $u \equiv g$ and in the continuous region $u > g$ and $L[u] = 0$. Moreover, the solution u belongs to $W_\infty^{2,1}(\mathbb{R}^1 \times [\delta, T])$ for any $\delta > 0$. For $t > 0$, u, u_x are continuous across the free boundary.*

1.3. *Explicit difference scheme and approximated free boundary*

The corresponding explicit finite difference scheme (EFDS) for variational problem (2)–(3) is defined by

$$F_h[u_h] := \min\left\{r^{-1}L_h[u_h], u_h - g(x)\right\} = 0 \quad \text{for } (x,t) \in \mathbb{R}^1 \times [0,T], \tag{11}$$

$$u_h(x,t) = g(x) \quad \text{for } -h \leq t \leq 0,\ x \in \mathbb{R}^1, \tag{12}$$

where for $k, h > 0$,

$$
\begin{aligned}
L_h[u] &= \frac{u(x,t) - u(x,t-h)}{h} \\
&\quad - \frac{\sigma^2}{2} \frac{u(x+\sigma k, t-h) + u(x-\sigma k, t-h) - 2u(x,t-h)}{(\sigma k)^2} \\
&\quad + (\frac{\sigma^2}{2} - r) \frac{u(x+\sigma k, t-h) - u(x-\sigma k, t-h)}{2\sigma k} + ru(x,t) \\
&= \frac{u(x,t)}{h} + ru(x,t) + (\frac{1}{k^2} - \frac{1}{h}) u(x,t-h) \\
&\quad - \Big\{ \Big[\frac{1}{2k^2} - \frac{1}{2\sigma k} (\frac{\sigma^2}{2} - r) \Big] u(x+\sigma k, t-h) \\
&\qquad + \Big[\frac{1}{2k^2} + \frac{1}{2\sigma k} (\frac{\sigma^2}{2} - r) \Big] u(x-\sigma k, t-h) \Big\}.
\end{aligned}
$$

We assume that the following inequalities hold, where the left one is Courant–Friedrichs–Lax condition and the right one is easy to be satisfied when k is small enough:

$$
0 < \sqrt{h} \le k, \quad \Big| \frac{\sigma^2}{2} - r \Big| k \le \sigma. \tag{13}
$$

To simplify notations and without loss of generality $k = \sqrt{h}$ is used. The following result is proved by Qian, Xu, Jiang and Bian:[24]

Lemma 1.2. *Under the conditions (13), the EFDS (11) and (12) is uniformly convergent with respect to* $(x,t) \in \mathbb{R}^1 \times [0,T]$. *Moreover, when equality holds in (13), the scheme reduces to the Binomial Tree Method for the American option.*

For the proof of the approximation free boundary, we need another modified EFDS:

$$
\begin{cases}
F^h[u^h] := \min \big\{ L^h[u^h], u^h - g(x) \big\} = 0, & \text{in } \mathbb{R}^1 \times (0,T], \\
u_h(x,t) = g^+(x), & \text{in } \mathbb{R}^1 \times (-h, 0],
\end{cases} \tag{14}
$$

where L^h is the finite difference operator:

$$
\begin{aligned}
L^h[\varphi] &= \frac{\varphi(x,t+h) - \varphi(x,t)}{h} \\
&\quad - \frac{1}{k} \Big[e^{\frac{\mu+1}{2}k} \frac{e^{-k}\varphi(x+k,t) - \varphi(x,t)}{k} - e^{-\frac{\mu+1}{2}k} \frac{\varphi(x,t) - e^{k}\varphi(x-k,t)}{k} \Big] \\
&= \frac{1}{h} \Big\{ \varphi(x,t+h) - B[\varphi(\cdot,t)](x) \Big\},
\end{aligned} \tag{15}
$$

where

$$B[\psi](x) = a_0\psi(x) + a_+\psi(x+k) + a_-\psi(x-k), \tag{16}$$

$$a_0 = 1 - \frac{2h}{k^2}\cosh\frac{(\mu+1)k}{2}, \quad a_\pm = \frac{h}{k^2}e^{\pm\frac{(\mu-1)k}{2}}, \tag{17}$$

which requires that

$$0 < h \leq k^2/[2\cosh(\mu+1)k/2], \quad k > 0. \tag{18}$$

Lemma 1.3. *The difference of the solutions of u_h and u^h for two schemes of (11) and (14) has estimate:*

$$|u_h - u^h| \leq Ch.$$

Definition 1.1. For any $h > 0$, let u^h be the solution of the Problem (14). Define

$$s^h(t) = \begin{cases} 0, & \text{if } t \leq 0, \\ \max\{x | x > 0 \text{ and } u^h(x,t) = g(x)\}, & \text{otherwise.} \end{cases} \tag{19}$$

Definition 1.2. For any $h > 0$, let u_h be the solution of the Problem (11). Define $s_{C_0h}(t) = \min\{0, s^*_{C_0h}(t)\}$, where $s^*_h(t)$ is a solution to the equation

$$u_h(s^*_h(t), t) - g(s^*_h(t)) = C_0h^{1/2}, \tag{20}$$

with C_0 is some positive constant.

2. Optimal Convergence Rate of the BTM

2.1. *Comparison lemmas*

The operator has the following comparison lemmas (proof can be found in Ref. 10):

Lemma 2.1 (Comparison principle 1). *If $u_j \in C(\mathbb{R}^1 \times [0,T]) \cap W^{2,1}_{p,loc}(\mathbb{R}^1 \times (0,T])$ $(j = 1, 2,\ p \geq 2)$ are functions such that $u_1 - u_2$ is bounded from below by a constant, $\int_{\mathbb{R}^1} u_j(x+y,t)dN(y)$ $(j = 1, 2)$ are finite, and*

$$L[u_1] + c(x,t)u_1 \geq L[u_2] + c(x,t)u_2, \quad \textit{for } c(x,t) \geq 0,$$
$$(\textit{or } L[u_1] - \beta_\varepsilon(g-u_1) \geq L[u_2] - \beta_\varepsilon(g-u_2),) \quad \textit{for } (x,t) \in \mathbb{R}^1 \times (0,T),$$
$$u_1(x,0) \geq u_2(x,0) \quad \textit{for } x \in \mathbb{R}^1,$$

then

$$u_1(x,t) \geq u_2(x,t) \quad \textit{for } (x,t) \in \mathbb{R}^1 \times [0,T]. \tag{21}$$

Lemma 2.2 (Comparison principle 2). *If* $u_j \in C(\mathbb{R}^1 \times [0,T]) \cap W^{2,1}_{p,loc}(\mathbb{R}^1 \times (0,T])$ $(j = 1,2,\ p \geq 2)$ *are functions such that* $u_1 - u_2$ *is bounded from below by a constant,* ($\int_{\mathbb{R}^1} u_j(x+y,t)dN(y)$ $(j = 1,2)$ *are finite for jump case), and*

$$F[u_1] \geq F[u_2] \quad \textit{for} \ \ (x,t) \in \mathbb{R}^1 \times (0,T),$$
$$u_1(x,0) \geq u_2(x,0) \quad \textit{for} \ \ x \in \mathbb{R}^1,$$

then

$$u_1(x,t) \geq u_2(x,t) \quad \textit{for} \ \ (x,t) \in \mathbb{R}^1 \times [0,T]. \tag{22}$$

Lemma 2.3 (Comparison principle 3). *Let* F_h *and* F^h *be defined by (11) satisfying (13) and (14) satisfying (16) respectively. If* u *and* v *are piecewise continuous and satisfy*

$$F_h[u] \geq F_h[v] \quad \textit{or} \quad F^h[u] \geq F^h[v], \qquad \textit{for} \ \ (x,t) \in \mathbb{R}^1 \times (0,T], \tag{23}$$
$$u(x,t) \geq v(x,t) \quad \textit{for} \ \ -h < t \leq 0, \quad x \in \mathbb{R}^1, \tag{24}$$

then

$$u(x,t) \geq v(x,t) \quad \textit{for} \ \ (x,t) \in \mathbb{R}^1 \times (-h,T]. \tag{25}$$

2.2. *Decomposition*

Since our initial datum contains a cusp, it is convenient to use a European option price U as a comparison function, which is a solution to the problem

$$L[U] = 0 \quad \text{for} \ \ 0 < t < T, \ x \in \mathbb{R}^1, \tag{26}$$
$$U(x,0) = g(x) \quad \text{for} \ \ x \in \mathbb{R}^1. \tag{27}$$

The properties of U are shown in the following lemma:[19]

Lemma 2.4. *The European option price with jump diffusion* U *is the solution of the problem (26), (27), which satisfies:*

$$\|U\|_{L^\infty(\mathbb{R}^1\times[0,T])} + [U]_{C_t^{1/2}(\mathbb{R}^1\times[0,T])} + \|U_x\|_{L^\infty(\mathbb{R}^1\times[0,T])} \leq C, \tag{28}$$
$$\|[-U_t]^+\|_{L^\infty(\mathbb{R}^1\times[0,T])} + \|[-U_{xx}]^+\|_{L^\infty(\mathbb{R}^1\times[0,T])} \leq C, \tag{29}$$
$$U_{xx}(x,t), \ U_t(x,t) \leq \frac{C}{\sqrt{t}} \exp\Big(-\frac{x^2}{Ct}\Big), \tag{30}$$
$$\|U\|_{C^{m+\alpha,(m+\alpha)/2}(\mathbb{R}^1\times[\tau,T])} \leq C\tau^{-(m-1+\alpha)/2}, \ m \geq 1, \ \alpha \in [0,1). \tag{31}$$

We decompose u^ε into two terms:

$$u^\varepsilon = U + v^\varepsilon. \tag{32}$$

Then v^ε does not have any singularities at $t = 0$, and

$$L[v^\varepsilon] = \beta_\varepsilon(1 - e^x - U - v^\varepsilon), \quad \text{for } (x,t) \in \mathbb{R}^1 \times (0,T), \tag{33}$$

$$v^\varepsilon\Big|_{t=0} \equiv 0. \tag{34}$$

It is not difficult to get the estimates

$$0 \geq \frac{\partial u^\varepsilon(x,t)}{\partial x} \geq -1, \quad \text{for } (x,t) \in \mathbb{R}^1 \times (0,T), \tag{35}$$

$$\beta_\varepsilon(g - u^\varepsilon) \leq r, \quad g - u^\varepsilon \leq \max(r+1, 2)\varepsilon. \tag{36}$$

From these estimate, one comes ([22, Theorem 4.30, p. 79]).

Lemma 2.5. *For $(x,t) \in \mathbb{R}^1 \times (0,T)$, there hold*

$$\|v^\varepsilon\|_{C^{1+\alpha,(1+\alpha)/2}(\mathbb{R}^1 \times [0,T])} \leq C, \quad v_t^\varepsilon(x,t) \leq C, \quad v_{xx}^\varepsilon(x,t) \leq C, \tag{37}$$

where the constant C is independent of ε (but may depend on T).

2.3. *Estimates of Solutions and Truncation Error for Penalized Problems*

We shall derive some basic estimates for the penalized problem (9), (10) here. Proofs can be found in Ref. 19.

By using the bounds on β_ε, the decomposition, derivatives of the equation, cut-off function, Lemma 2.5, comparison lemmas and interpolation theory, it can have

Lemma 2.6. *The solution u^ε of the problem (9), (10) satisfies, for $0 < \alpha < 1$,*

$$\|u^\varepsilon\|_{C^{2+\alpha,(2+\alpha)/2}(\mathbb{R}^1 \times [\tau,T])} \leq C\tau^{-(1+\alpha)/2} + C(\varepsilon\sqrt{\tau}\,)^{-\alpha/2}, \tag{38}$$

where $C = C_\alpha$ is independent of ε and τ.

In order to estimate the truncation error between differential operator L and finite difference operator L_h for regularized solutions of penalized problem, we need the following main estimate:

Lemma 2.7 (main estimates). *For any $1/3 < \eta < 1/2$, there holds*

$$|u_{tt}^\varepsilon| + |u_{xxxx}^\varepsilon| + |u_{xxt}^\varepsilon| \leq Ct^{-3/2} + C\varepsilon^{-1}t^{-1+\eta}, \tag{39}$$

where $C = C_\eta$ is independent of ε.

To avoid possible difficulties near $t = 0$, we let $w^\varepsilon(x,t) = u^\varepsilon(x, t+3h)$. For $w^\varepsilon(x,t)$, we have the estimate of truncation error between differential operator L and finite difference operator L_h as follows

Lemma 2.8 (truncation error estimate).

$$\left| L_h[w^\varepsilon] - L[w^\varepsilon] \right|_{t\geq\tau} \leq C\big((\tau+2h)^{-3/2} + \varepsilon^{-1}(\tau+2h)^{-1+\eta}\big)h. \tag{40}$$

2.4. *Optimal Convergence Rate*

Our main theorem is the following:

Theorem 2.9 (Main Theorem 1). *Under the assumption (13) (or (16)), the solution of the problem (11), (12) (or (14)) is convergent to the solution of the problem (5), (6) with error rate estimates of $O(h^{1/2})$.*

In order to establish this theorem, we shall apply the discrete comparison lemma presented to the following auxiliary functions:

$$\Phi_h^\pm(x,t) = w^\varepsilon \pm \left[C\varepsilon^{-1}h\int_{-h}^{t} \frac{d\xi}{(\xi+2h)^{1-\eta}} + Ch\int_{-h}^{t}\frac{d\xi}{(\xi+2h)^{3/2}} + \hat{C}\varepsilon + \hat{C}\sqrt{h}\right],$$

which satisfies that

$$\Phi_h^-(x,t) \leq u_h(x,t) \leq \Phi_h^+(x,t), \qquad \mathbb{R}^1 \times [0,T] \tag{41}$$

where the constant C and constant $\hat{C}$ are some positive constants. By it, we have

Lemma 2.10 (main lemma). *Let u^ε be the solution of the problem (9), (10) and u_h be the solution of the problem (11), (12). Then*

$$|u_h - u^\varepsilon| \leq Ch\varepsilon^{-1} + C\varepsilon + C\sqrt{h}, \quad (x,t) \in \mathbb{R}^1 \times [0,T],$$

where C is independent of h and ε.

Proof of Theorem 2.9. Combining all the estimates, we get

$$|u-u_h| \leq |u-u^\varepsilon| + |u^\varepsilon - u_h| \leq C\varepsilon + \left\{Ch\varepsilon^{-1} + C\varepsilon + C\sqrt{h}\right\},$$

where C is independent of ε and h. With Lemma 1.3, upon taking $\varepsilon = \sqrt{h}$, we have

$$|u-u_h| < C_0 h^{1/2}, \quad |u-u^h| < C_0h^{1/2}. \tag{42}$$

This completes the proof of this theorem. □

Remark 2.1. The European option admits an explicit solution. We can use the finite difference scheme on this explicit solution to calculate the convergence rate. It is not difficult to get

$$U_h(0,h) - U(0,h) = (\frac{1}{2} - \frac{1}{\sqrt{2\pi}})\sigma\sqrt{h} + o(\sqrt{h}).$$

This means that the finite difference scheme on a European option cannot have a better convergence rate than $O(\sqrt{h})$, unless $\sigma \equiv 0$. We conclude that the convergence rate of the finite difference scheme on a European option cannot be better than $O(\sqrt{h})$.

Remark 2.1 shows that the $O(h^{1/2})$ rate is the optimal convergence rate, i.e. the error rate of the approximation process by the BTM cannot be better than $O(h^{1/2})$.

3. Convergence Rate of the Free Boundary

The optimal convergence rate for the solution is used to study the convergence rate for the free boundary. For the approximated free boundary defined in Definition 1.2, non-degeneracy of continuous operator can be used directly to obtain the convergence rate (as proof of the following Lemma 3.3). That is, for the original EFDS including BTM, if C_0 is the one defined in (42), we have

Theorem 3.1. *Let $s_{C_0h}(t)$ be defined in Definition 1.2 and $s(t)$ be the free boundary of the Problem (5), (6), then*

$$|s_{C_0h}(t) - s(t)| \leq Ch^{1/4}, \tag{43}$$

where C is some positive constant. If the problem is with jump, we need additional condition $r - \lambda\int_0^\infty (e^y - 1)\,dN(y) > 0$.

Since C_0 is a positive constant but not clear, Definition 1.2 is not easy to use in the calculation. we should come back to consider original approximated free boundary as defined in Definition 1.1. To overcome a technical difficulty, a modified EFDS defined in (14) with the condition (18) is applied, a new definition of an approximated of a *slightly shifted* free boundary for a more general model with jump is given in Ref. 19. It is also proved that this slightly shifted approximated free boundary is actually within an error of $\Delta x^{1/2}$ to the actual free boundary.

The key to obtain the convergent rate for the free boundary is to get some non-degeneracy estimates of the continuous and finite difference operators near the free boundary.[5] Especially it is not easy to get the estimate of the finite difference operator (Lemma 3.3), which is proved with the help of an auxiliary function $\varphi_k(y) = \frac{e^{\mu k}-1}{e^{\mu k}-e^{-k}} e^y + \frac{1-e^{-k}}{e^{\mu k}-e^{-k}} e^{-\mu y} - 1$. Therefore the following Lemmas play important roles, where $\mu = 2r/\sigma^2$. The details of the proofs can be found in Ref. 5:

Lemma 3.2 (non-degeneracy lemma 1). *If $\{u(x,t), s(t)\}$ is the solution and the free boundary of the problem (5), (6), then for each $x \geq s(t)$, $t > 0$, there holds*

$$u(x,t) - g(x) \geq \frac{\mu}{2(1+\mu)}[x - s(t)]^2.$$

Lemma 3.3 (non-degeneracy lemma 2). *If $\{u^h(x,t), s^h(t)\}$ is the solution and the free boundary of the problem (14), then for each $x \geq s^h(t)$, $t > 0$, there holds*

$$u^h(x,t) - g(x) \geq \varphi_k(x - s^h(t) - k) \geq \frac{\mu}{2(\mu+1)}[x - s^h(t) - k]^2.$$

By using the above lemmas, we are ready to obtain the convergence rate for the approximated free boundary. The proof follows the positions of the approximated free boundary $s^h(t)$ and the exact free boundary $s(t)$. For any $0 < t < T$, if $s(t) > s^h(t)$, then

$$\begin{aligned} C_0 h^{1/2} &\geq C_0 h^{1/2} + u^h(s^h(t), t) - g(s^h(t)) \\ &\geq u(s^h(t), t) - g(s^h(t)) \ = \ v(s^h(t), t) \geq \frac{\mu}{2(1+\mu)}(s^h(t) - s(t))^2, \end{aligned}$$

where C_0 is the one in (42) of Theorem 2.9, otherwise

$$\begin{aligned} C_0 h^{1/2} &\geq C_0 h^{1/2} + u(s(t), t) - g(s(t)) \\ &\geq u^h(s(t), t) - g(s(t)) \geq \frac{\mu}{2(\mu+1)}[s(t) - s^h(t) - k]^2. \end{aligned}$$

In the both cases, by choosing proper constant C_1, we have

Theorem 3.4 (Main Theorem 2). *The difference of the approximated free boundary $s_h(t)$ and the real one $s(t)$ has estimate:*

$$|s^h(t) - s(t)| \leq C_1 h^{1/4}. \tag{44}$$

Remark 3.1. We see that the condition (18) is not cover (1). That means, the new scheme cannot lead to BTM. So, the convergent rate of the free boundary in BTM is still open, though we guess it would have the same convergence rate as ours.

References

1. Amin, K., Khanna, A.: Convergence ofAmerican option values from disete to continuous-time financial models. Math. Financ. 4, 289C304 (1994).
2. Amin, K., Jump Diffusion Option valuation in Discrete Time, *J. Finan.* **48** (1993) 1833-1863.
3. Almendral, A. & Oosterlee, C. O., Numerical Valuation of Options with Jumps in the Underlying, *Applied Numerical Mathematics*, **53** (2005) 1-18.
4. Briani, M., La Chioma, C. & Natalini, R., Convergence of Numerical Schemes for Viscosity Solutions to Integro-Differential Degenerate Parabolic Problems Arising in Financial Theory, *Numerische Mathematik*, **98** (2004) 607-646.
5. Chen, X. F., Hu B. & Liang, J. Convergence Rate of Approximated Free Boundaries of American Options by Explicit Finite Method, working paper 2012.
6. Cox, J. & Rubinstein, M., Option Pricing: A Simplified Approach, *J. Finan. Econ.* **7** (1979) 229-263.
7. Friedman, A., "Variational Principles and Free Boundary Problems", John Wiley & Sons, New York 1982.
8. Glowinski, R., Lions, J. L. & Trémolières, R., "Numerical Analysis of Variational Inequalities", North-Holland Amsterdam, 1981
9. d'Halluin, Y., Forsyth, P. A. & Labahn, G. A Penalty Method for American Options with Jump Diffusion Processes *Numerische Mathematik*, **97** (2004) 321-352
10. Hu, B., Jiang, L. & Liang, J., Optimal Convergence Rate of the Explicit Finite Difference Scheme for American Options, *J. Comp. Appl. Math.*, **230** (2009) 583-599.
11. Jiang, L.,"Mathematical Modeling and Methods for Option Pricing", Higher Education Press, Beijing 2003. World Scientific, 2005.
12. Jiang, L., Existence and Differentiability of the Solution of a two-phase Stefan Problem for Quasi-linear Parabolic Equations, *Chinese Math. Acta* **7** (1965) 481-496.
13. Jiang, L. & Dai, M, Convergence of binomial tree method for American options, Proceedings of PDE and their Applications, World Scientific (1999) 106-119.
14. Jiang, L. & Dai, M, Convergence of the Explicit Difference Scheme and the Binomial Tree Method for American Options, *J. Comp. Math.* **22** (2004) 371-380.
15. Jiang, L. & Dai, M, Convergence of Binomial Tree Methods for European/American Options Path-Depedent Options, *SIAM J. Numer. Anal.* **42** (2004) 1094-1109.

16. Krylov, N.V., On the Rate of Convergence of Finite-Difference Approximations for Bellman's Equations with Variable Coefficients, *Probab. Theory Relat. Fields* **117** (2000) 1-16.
17. Kou, S. G., A Jump-Diffusion Model for Option Pricing, *Management Science*, **48** (2002) 1086-1101.
18. Lamberton, D., Error Estimates for the Binomial Approximation of American Put Options, *The Ann. Appl. Prob.* **8** (1998) 206-233.
19. Liang, J., Hu, B. & Jiang L., Optimal Convergence Rate of the Binomial Tree Scheme for American Options with Jump Diffusion and Their Free Boundaries, *SIAM Financial Mathematics*, **1** (2010) 30-65.
20. Liang, J., Hu, B., Jiang, L. & Bian, B., On the Rate of Convergence of the Binomial Tree Scheme for American Options, Numer. Math. **107** (2007), 333-352.
21. Liang, J., On the Convergence Rate of the Binomial Tree Scheme for an American Option with Jump-Diffusion, *Numerical Mathematics - A Journal of Chinese University* **30** (2008) 76-96.
22. Lieberman, G.M., Second Order Parabolic Differential Equations, World Scientific, 1996.
23. Pham, H., Optimal Stopping, Free Boundary and American Option in a Jump-Diffusion Model, *Appl. Math. Opt.* **35** (1995) 145-164.
24. Qian, X., Xu, C., Jiang, L. & Bian, B., Convergence of the Binomial Tree Method for American Options in Jump-Diffusion Model, *SIAM J. Numer. Anal.*, **42** (2005) 1899-1913.
25. Yang, C., Jiang, L. & Bian, B., Free Boundary and American Options in a Jump-Diffusion Model, European J. of Applied Mathematics, **17** (2006) 95-127.
26. Zhang, X., Numerical Analysis of American Option Pricing in a Jump-Diffusion Model, *Mathematics of Operations Research,* **22** (1997) 668-690.

Rademacher Φ Function, Jacobi Symbols, Quantum and Classical Invariants of Lens Spaces*

Bang-He Li

Key Laboratory of Mathematics Mechanization, National Center for Mathematics and Interdisciplinary Sciences, Academy of Mathematics and Systems Science, Chinese Academy of Sciences, Beijing 100190, China
Libh@amss.ac.cn

Tian-Jun Li

Math. Dept., University of Minnesota, Minneapolis, MN 55455, USA
tjli@math.umn.edu

We describe several applications of the explicit formulas for $SU(2)$ and $SO(3)$ quantum invariants τ_r and ξ_r of lens spaces in [LL]. By using formulas for τ_2 and ξ_3 we derive a formula of the Rademacher Φ function mod 24, which appears in the transformation of Dedekind η function under the modular group $PSL(2, Z)$. Via τ_3, Brown invariants of lens spaces, as well as μ invariants of lens spaces mod 8, are presented in terms of Jacobi symbols. Formulas for $\mu-$invariants of lens spaces involving the Φ function are also given.

1. Introduction

Let $\eta(z)$ be the Dedekind eta function defined for z in the upper half plane. Then, for

$$A = \begin{pmatrix} a & b \\ c & d \end{pmatrix} \in PSL(2, \mathbb{Z}) \quad \text{with } c \neq 0,$$

$$\eta(Az) = e_{24}^{\Phi(A)} (\frac{cz + d}{i \ \text{sign}(c)})^{\frac{1}{2}} \eta(z)$$

*1. The first author supported by a NKBRPC (2011CB302400) and the National Center for Mathematics and Interdisciplinary Sciences, CAS. 2. The second author supported by NSF grant numbers DMS-1021956, DMS-1103976.

where e_r stands for $\exp(\frac{2\pi i}{r})$ and $\Phi : PSL(2,\mathbb{Z}) \longrightarrow \mathbb{Z}$ is the Rademacher Φ function defined by

$$\Phi(A) = \begin{cases} \frac{a+d}{c} - 12s(a,c), \text{ if } c \neq 0 \\ \\ \frac{b}{d}, \qquad\qquad\qquad\quad \text{if } c = 0 \end{cases}$$

where $s(a,c)$ is the Dedekind sum (cf. [KM1])

To evaluate $\eta(Az)$, we only need to compute $\Phi(A)$ mod 24. In this paper we offer such a formula without involving the Dedekind sum (see Theorem 2).

We achieve this by first evaluating Φ_3 and Φ_8, where $\Phi_k(A) = \Phi(A) \pmod{k}$.

The case $c = 0$ is trivial. The following is the simple formula for Φ_3 when $c \neq 0$:

$$\Phi_3(A) = \begin{cases} c(a+d), & \text{if } c \equiv \pm 1 \pmod 3 \\ \\ ab, & \text{if } c \equiv 0 \pmod 3 \end{cases}. \tag{1}$$

We will see in section 3 that it is often easier to derive properties of Φ_3 from (1).

The novelty is that we drive the formulas for Φ_3 and Φ_8 via the quantum invariants τ_r and ξ_r. Quantum invariants of 3−manifolds, first introduced by Witten in 1988 using path integrals and Chern-Simons gauge theory, depend on a choice of gauge group and a root of unity. The existence of such invariants for the gauge group $SU(2)$ was established by Reshetikhin and Turaev in terms of quantum group and Kirby calculus in [RT]. The Reshetikhin–Turaev invariants $\tau_r(M)$ with $r \geq 3$ in [RT] for an oriented, connected, closed 3-manifold M were subsequently refined and extended to $r = 2$ by Kirby and Melvin in [KM2]. Moreover, for small values of r, $\tau_r(M)$ is expressed in terms of classical invariants in [KM1], [KM2], [KMZ]. Using the Kauffman bracket, Blanchet, Habegger, Masbaum and Vogel introduced the invariant in [BHMV] for x an r−th primitive root of unit, which is a very useful variation of $\tau_r(M)$. The invariant $\xi_r(M,x)$ in [L] is a modification of $\theta_r(M,x)$: $\xi_r(M,x) = 2^\nu\theta_r(M,-x)$ for odd $r > 1$ with ν the first Betti number of M.

We will apply the elementary formulas for τ_r and ξ_r of lens spaces in [LL], which involves the Φ function. In this case we write $A = \begin{pmatrix} q & -p^* \\ p & q^* \end{pmatrix} \in SL(2,\mathbb{Z})$ with $p > 0$, and take M to be the corresponding lens space $L(p,q)$. Specifically, to evaluate $\Phi_3(A)$ and $\Phi_8(A)$, we will rely respectively on the facts that $\tau_2(M) = 1$ and $\xi_3(M) = 1$. The first is stated in [KM2] (p.474), and the second follows from the fact $\theta_3(M,x) = 1$ stated in [BHMV].

We further apply the formulas for Φ_{24} and τ_3 to evaluate the Brown invariant. The Brown invariant $\beta(M) \in \mathbb{Z}_8$, defined in [KM2] for any 3-manifold M whose μ invariants are all congruent mod 4. In this case,

$$\beta(M) = \sigma(A) - \lambda(A) \pmod 8,$$

where $\lambda(A)$ is the $\mathbb{Z}_8$ Arf invariant of A introduced by Brown in [B], viewed as a $\mathbb{Z}_4-$valued quadratic form on the $\mathbb{Z}_2-$inner product space given by A. It is noted in [KM2] that β is well defined and β is an oriented homotopy invariant.

It is also noted in [KM2] that for lens spaces, β is defined for $L(p,q)$ with $p \not\equiv 2 \pmod 4$. For odd p introduce $\delta_4(p) = \pm 1$ with $\delta_4(p) \equiv p \pmod 4$. Let $(\frac{q}{p})$ denote the Jacobi symbol.

Theorem 1. *Suppose $p > 0$. The Brown invariant of $L(p,q)$ is given by*

$$\beta(L(p,q)) = \begin{cases} 2\delta_4(p)(\frac{q}{p}) - \delta_4(p) - 1, & \text{if } p \text{ is odd} \\ \\ -2\delta_4(q)(\frac{p}{q}) + \delta_4(q), & \text{if } p \text{ is divisible by } 4. \end{cases}$$

Finally, we deal with the μ invariants, which are $\mathbb{Z}_{16}-$valued. For each spin structure θ on M, the μ invariant is defined to be the signature mod 16 of any compact spin 4$-$manifold W bounded by (M,θ). This is well defined by Rohlin's theorem.

For odd $p > 0$, $L(p,q)$ is a $\mathbb{Z}_2$ homology sphere and only one μ-invariant, which is also called the Rokhlin invariant.

For even $p > 0$, $L(p,q)$ has two spin structures and hence two μ-invariants. Kirby and Melvin [KM1] actually define integral lifts of the μ invariants and denote them by $\mu_0(L(p,q))$ and $\mu_1(L(p,q))$. By abuse of notations we will use the μ_0 and μ_1 but regard them as integers mod 16. The set of unordered pair $\{\mu_0(L(p,q)), \mu_1(L(p,q))\}$ mod 16 is called the μ-invariant of $L(p,q)$ for even p.

It is shown in [KM1], [KM2] that the μ invariant is closely related to the *Brown* invariant. We apply Theorem 1 to obtain mod 8 formulas of the μ invariant in terms of Jacobi symbols in Theorem 3. In Proposition 2 we deduce a number of oriented homotopy invariance properties of μ mod 8 from Theorem 3.

In the last part of the paper we provide explicit formulas of $\{\mu_0(L(p,q)), \mu_1(L(p,q))\}$ mod 16 via τ_4 in Theorem 5.

2. Quantum Invariants of Lens Spaces

2.1. τ_r *and* ξ_r

We briefly review the quantum invariants τ_r and we adopt the notation of [KM2]. If an oriented manifold M is obtained by surgery on a framed link L in S^3, $\tau_r(M)$ is defined in terms of the values of the Jones polynomial of L at e_r. τ_r has a number of nice properties properties. For instance, $\tau_r(-M) = \overline{\tau_r(M)}$ and $\tau_r(S^3) = 1$.

Now let M be the lens space $L(p,q)$ with $p > 0$ and $0 < |q| < p$. M can be described by surgery on any simple chain link

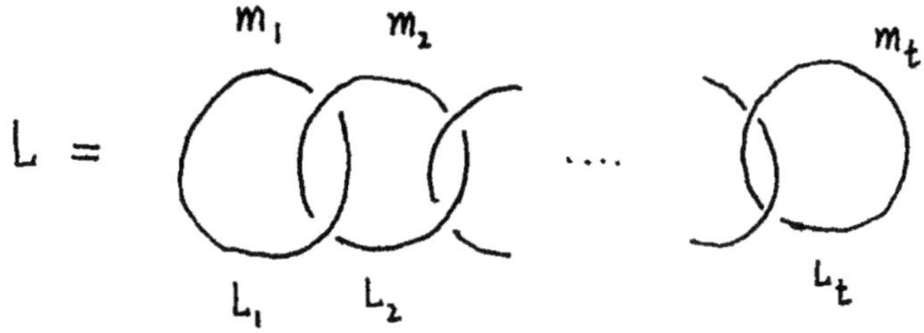

with the framings m_i obtained from a continued fraction expansion

$$p/q = [m_1, m_2, \cdots, m_t] = m_t - \cfrac{1}{m_{t-1} - \ddots - \cfrac{1}{m_2 - \cfrac{1}{m_1}}}. \tag{2}$$

If $q > 0$ we can assume that $m_i \geq 2$.

Direct formulas for $\tau_r(L(p,q))$ with $r \geq 2$ and $\xi_r(L(p,q)) = \xi_r(L(p,q), e_r)$ with odd $r \geq 3$ were obtained in [LL]. For τ_r, it was derived from Jeffrey's formula [J] via explicit evaluation of generalized Gauss sum $\sum_{j=1}^{a} e_a^{bj^2+cj}$ in Theorems 2.1 and 2.2 in [LL].

Notice the formulas in [LL] are presented for $p, q \geq 0$, and involve Jacobi symbol $(\frac{a}{b})$ and Dedekind sum. For $L(p,q)$ with $p > 0, q \leq 0$, $\tau_r(L(p,q)) = \overline{\tau_r(L(p,-q))}, \xi_r(L(p,q)) = \overline{\xi_r(L(p,-q))}$. With these understood, the formulas in [LL] are also true for $L(p,q)$ with $p > 0$ and $q \leq 0$ (If $q = 0$, then $p = 1$, $L(1,0) = S^3$, the formulas work also).

2.2. *Jacobi symbols, homotopy equivalence and Dedekind sum*

We review Jacobi symbols and discuss how they are related to the homotopy classification of Lens spaces.

The Jacobi symbol (cf. [S], pp. 442-443) is the product of the Legendre symbols corresponding to prime factors of b. Recall that if l is an odd prime number and a is not divisible by l, the Legendre symbol $(\frac{a}{l})$ is 1 if a is a quadratic residue mod l, and is -1 otherwise. For odd $b < 0$, if $a \geq 0$, let $(\frac{a}{b}) = (\frac{a}{-b})$ and $(\frac{0}{\pm 1}) = 1$. Like the Legendre symbol, it also has the reciprocity law: If a and b are odd and coprime, then

$$(\frac{a}{b}) = (-1)^{\frac{a-1}{2}\frac{b-1}{2}}(\frac{b}{a}). \tag{3}$$

For an integer m let $\delta_3(m) = 0$ or ± 1 such that $m \equiv \delta_3(m) \pmod 3$.

Lemma 1.

Let a be an odd integer not divisible by 3. (1) $(\frac{a}{3}) = \delta_3(a) = \dfrac{e_3^{2a} - e_3^{-2a}}{e_3^2 - e_3^{-2}} = \dfrac{e_{12}^{2a} - e_{12}^{-2a}}{e_{12}^2 - e_{12}^{-2}}$,

(2)$(\frac{3}{a}) = \delta_4(a)\delta_3(a)$,

Proof. 1 is straightforward. 2 follows from the reciprocity law.

$$(\frac{3}{a}) = (-1)^{\frac{a-1}{2}}(\frac{a}{3}) = \delta_4(a)\delta_3(a).$$

□

It is well known that two lens spaces $L(p,q_1)$ and $L(p,q_2)$ are oriented homotopy equivalent if and only if $q_1 q_2$ is a quadratic residue mod p.

Prop 1. Suppose $L(p,q_1)$ and $L(p,q_2)$ are oriented homotopy equivalent.

- If p is odd, then $(\frac{q_1}{p}) = (\frac{q_2}{p})$,

- If p is divisible by 4, then $(\frac{p}{q_1}) = (\frac{p}{q_2})$ and $\delta_4(q_1) = \delta_4(q_2)$.

To prove this proposition we will need two basic facts about quadratic residues:

1. if a is a residue modulo n, then a is a residue modulo p^k for every prime power dividing n.

2. If a is an odd number and $m = 8$, or some higher power of 2, then a is a residue modulo m if and only if $a \equiv 1 \pmod 8$.

Proof. The p odd case is clear.

From now on we suppose p is divisible by 4. Write $p = 2^l\bar{p}$ with $\bar{p}$ odd. By the first fact above, q_1q_2 is a residue mod $\bar{p}$ and a residue mod 2^l.

Since $l \geq 2$, we have $q_1q_2 \equiv 1 \pmod 4$, which is the same as $\delta_4(q_1) = \delta_4(q_2)$.

By the definition of Jacobi symbol,

$$(\frac{p}{q_i}) = (\frac{\bar{p}}{q_i})(\frac{2^l}{q_i}). \tag{4}$$

Since q_1q_2 is a residue mod $\bar{p}$, we have

$$(\frac{q_1}{\bar{p}}) = (\frac{q_2}{\bar{p}}).$$

Since $\delta_4(q_1) = \delta_4(q_2)$, by the reciprocity law (3), we further have

$$(\frac{\bar{p}}{q_1}) = (\frac{\bar{p}}{q_2}). \tag{5}$$

If l is even, then $(\frac{2^l}{q_1}) = (\frac{2^l}{q_2}) = 1$. If l is odd, then

$$(\frac{2^l}{q_i}) = (\frac{2}{q_i}) = (-1)^{\frac{q_i^2-1}{2}}.$$

Since now l is at least 3 and q_1q_2 is a residue mod 2^l, by the second fact about quadratic residue, we have $q_1q_2 \equiv 1 \pmod 8$. Thus $q_1 \equiv q_2 \pmod 8$. Therefore for any $l \geq 2$, we have $(\frac{2^l}{q_1}) = (\frac{2^l}{q_2})$, and together with (4) and (5) we have

$$(\frac{p}{q_1}) = (\frac{p}{q_2}).$$

□

Remark 1. Notice that if $p \equiv 2 \pmod 4$, then $(\frac{p}{q})$ is not an oriented homotopy invariant. $L(10,7)$ and $L(10,3)$ are oriented homotopy invariant, but $(\frac{10}{7}) = (\frac{3}{7}) = -1$ while $(\frac{10}{3}) = (\frac{1}{3}) = 1$.

We finally remark that it is more convenient for us to use the following normalization of the Dedekind sum: $S(q,p) = 12s(q,p)$. It can be defined via ((2):

$$S(q,p) = \sum_{1}^{n} m_i - 3n + (q+q^*)/p.$$

3. Φ Function Mod 24

In this section, we calculate Φ_{24}. Let $A = \begin{pmatrix} q & -p^* \\ p & q^* \end{pmatrix} \in SL(2,\mathbb{Z})$ with $p > 0$. Since Φ is defined on $PSL(2,Z)$, we can assume $p > 0$.

Theorem 2. *Φ_{24} is calculated by $\Phi_{24}(A) = 16\Phi_3(A) + 9\Phi_8(A)$, where $\Phi_3(A)$ is given by (1), and*

$$\Phi_8(A) = \begin{cases} p(q+q^*) - 2\delta_4(p)(\frac{2q}{p}) - \delta_4(p) + 3, & \text{if } p \equiv 1 \pmod 2 \\[2ex] \frac{p}{2}\frac{q+q^*-2}{2} - 2(\frac{q}{p/2}) + 2\delta_4(p/2) + 1, & \text{if } p \equiv 2 \pmod 4 \\[2ex] \frac{p}{4}\frac{q+q^*+2\delta_4(q)}{4} - 2\delta_4(q)(\frac{2q}{p/4}) & \\ \quad -\delta_4(p/4) + 1, & \text{if } p \equiv 4 \pmod 8 \\[2ex] -qp^* + 2\delta_4(q)(\frac{p/2}{q}) + \delta_4(q), & \text{if } p \equiv 0 \pmod 8. \end{cases}$$

We will compute Φ_3 and Φ_8 in the next two subsections.

For later applications, we give here the explicit formula of Φ_{24} in the case of odd p. If p is odd and $p \not\equiv 0 \pmod 3$ then

$$\Phi_{24} = p(q+q^*) + 6\delta_4(p)(\frac{2q}{p}) - 9\delta_4(p) + 3, \tag{6}$$

and if p is odd and $p \equiv 0 \pmod 3$ then

$$\Phi_{24} = -16qp^* + 9p(q+q^*) + 6\delta_4(p)(\frac{2q}{p}) - 9\delta_4(p) + 3. \tag{7}$$

3.1. Φ_3 *and* ξ_3

Proof of Theorem 2–Φ_3 part. Consider the lens space $L(p,q)$. We divide into two cases.

Case 1. $p \equiv \pm 1 \pmod 3$. By the first case of Theorem 4.1 in [LL], we have

$$1 = \xi_3(L(p,q)) = (\frac{p}{3}) e_3^{S(q,p)} e_p^{-3'(q+q^*)} \frac{e_3^{2p'} - e_3^{-2p'}}{e_3^2 - e_3^{-2}},$$

where $p^*p + q^*q = 1$ and $p'p + 3'3 = 1$. Take $p' = p$ and $3' = \frac{(1-p^2)}{3}$. By Lemma 1 (1),

$$\begin{aligned} 1 &= e_3^{S(p,q)} e_p^{-3'(q+q^*)} = e_3^{S(p,q)} e_p^{(p^2-1)(q+q^*)/3} = e_3^{S(p,q)-(q+q^*)/p} e_3^{p(q+q^*)} \\ &= e_3^{-\Phi + p(q+q^*)}. \end{aligned}$$

Thus $\Phi_3 \equiv p(q+q^*) \pmod 3$.

Case 2. $p \equiv 0 \pmod 3$. By the second case of Theorem 4.1 in [LL] again, we have

$$\begin{aligned} 1 = \xi_3(L(p,q)) &= -(\frac{q}{3}) e_3^{S(p,q)} e_{3p}^{-(q+q^*+\delta_3(q)p^*p)} \frac{i\delta_3(q)\sqrt{3}}{e_3^2 - e_3^{-2}} \\ &= \delta_3(q)(\frac{q}{3}) e_3^{-\Phi - \delta_3(q)p^*}. \end{aligned}$$

By Lemma 1 (1), we have $\Phi_3 = -qp^* \pmod 3$. □

Let

$$\begin{pmatrix} a'' & b'' \\ c'' & d'' \end{pmatrix} = \begin{pmatrix} a & b \\ c & d \end{pmatrix} \begin{pmatrix} a' & b' \\ c' & d' \end{pmatrix}$$

then (cf. [RG], p. 51)

$$\Phi \begin{pmatrix} a'' & b'' \\ c'' & d'' \end{pmatrix} = \Phi \begin{pmatrix} a & b \\ c & d \end{pmatrix} + \Phi \begin{pmatrix} a' & b' \\ c' & d' \end{pmatrix} - 3\text{sign}(cc'c'').$$

From this one sees that

$$\Phi_3 : PSL(2,\mathbb{Z}) \longrightarrow \mathbb{Z}_3$$

is a homomorphism. And from (1), it is clearly that if $A \equiv B \pmod 3$, then $\Phi_3(A) = \Phi_3(B)$, which is not easily seen from the definition of Φ.

Also, it is obvious now that if $\Phi_3(A) = 0$, then

$$A \pmod 3 = \begin{pmatrix} 1 & 0 \\ 0 & 1 \end{pmatrix}, \begin{pmatrix} 0 & -1 \\ 1 & 0 \end{pmatrix}, \begin{pmatrix} 1 & 1 \\ 1 & -1 \end{pmatrix} \text{ or } \begin{pmatrix} -1 & 1 \\ 1 & 1 \end{pmatrix}.$$

As pointed in [RG], p. 54, the original proof is long as it requires enumerating all possible matrices (mod 3).

3.2. Φ_8 *and* τ_2

Proof of Theorem 2–Φ_8 part. We divide into 4 cases.

Case 1. $p \equiv 1 \pmod 2$. By Theorem 3.1, I.1 in [LL], we have

$$1 = \tau_2(L(p,q)) = (\frac{2q}{p})\varepsilon(p)e_8^{S(q,p)}e_p^{(-(q+q^*)8')}\frac{e_8^{2p'} - e_8^{-2p'}}{e_8^2 - e_8^{-2}},$$

where $8'8 + p'p = 1$ and $\varepsilon(p) = i^{(1-\delta_4(p))/2}$. Take $p' = p$ and $8' = \frac{(1-p^2)}{8}$. Then

$$1 = e_8^{-\Phi+p(q+q^*)}\delta_4(p)(\frac{2q}{p})\varepsilon(p) = e_8^{-\Phi+p(q+q^*)+2(1-\delta_4(p)(\frac{2q}{p})) + (1-\delta_4(p))}.$$

Case 2. $p \equiv 2 \pmod 4$. Let $\bar{p} = p/2$. Notice that the formula for $p \equiv 2 \pmod 4$ and $c = 2$ in Theorem 3.1, III. 2) in [LL] is

$$\tau_r(L(p,q)) = (\frac{qr/2}{\bar{p}})\epsilon(\bar{p})e_{4r}^{S(q,p)}e_p^{-q^*(r/2)^*4^*(q^2+1)/2}\sqrt{2}\frac{e_{4r}^b - e_{4r}^{-b}}{e_{4r}^2 - e_{4r}^{-2}},$$

where $q^*q \equiv 1 \pmod p$, $(r/2)(r/2)^* + (\bar{p})^*\bar{p} = 1, 4^* = (2+p)^2/16$, and $b = 4^*4(\bar{p})^* - 2 - \bar{p}$. But for our purpose, we need to take $4^* = (2 - \delta_4(\bar{p})p)^2/16$ then the formula for b will be $b = 4^*4(\bar{p})^* + 2\delta_4(\bar{p}) - \bar{p}$.

Now for $r = 2$, we can take $(r/2)^* = 1$ and $\bar{p}^* = 0$. Then

$$1 = \tau_2(L(p,q)) = (\frac{q}{\bar{p}})\epsilon(\bar{p})e_8^{S(q,p)}e_{8p}^{-q^*4^*4(q^2+1)}\frac{e_8^{2\delta_4(\bar{p})-\bar{p}} - e_8^{-2\delta_4(\bar{p})+\bar{p}}}{\sqrt{2}i}.$$

Since $e_8^{2\delta_4(\bar{p})-\bar{p}} - e_8^{-2\delta_4(\bar{p})+\bar{p}} = e_8^{2\delta_4(\bar{p})-\bar{p}}\sqrt{2}e_8^{\delta_4(\bar{p})}, 4^*4 = 1 + 2\bar{p}^2 - \delta_4(\bar{p})p$, $4^* \equiv 0 \pmod 4$, we have

$$\begin{aligned} e_{8p}^{-q^*4^*4(q^2+1)} &= e_{8p}^{-(q+q^*)4^*4+4^*4pp^*} = e_{8p}^{-(q+q^*)}e_8^{\frac{-(q+q^*)}{2}(\bar{p}-2\delta_4(\bar{p}))} \\ &= e_{8p}^{-(q+q^*)}e_8^{\frac{(q+q^*)}{2}\bar{p}}. \end{aligned}$$

Therefore

$$1 = e_8^{2(1-(\frac{q}{\bar{p}})) + 1 - \delta_4(\bar{p}) - \Phi - \bar{p}\frac{q+q^*}{2} + 3\delta_4(\bar{p}) - \bar{p} - 2},$$

and $\Phi_8 = 1 - \bar{p} + 2\delta_4(\bar{p}) - 2(\frac{q}{\bar{p}}) - \bar{p}\frac{q+q^*}{2} \pmod 8$.

Case 3. $p \equiv 4 \pmod 8$. By the formula in Theorem 3.1, III. 4) in [LL], we have

$$1 = \tau_2(L(p,q)) = (\frac{2q}{p/4})\varepsilon(p/4)e_8^{S(q,p)}e_{4p}^{\delta_4(q)}e_p^{-q^*8^*(q+\delta_4(q))^2}(-i)\delta_4(q),$$

where $8^*8 + (p/4)^2 = 1$. Now

$$e_{4p}^{\delta_4(q)}e_p^{-q^*8^*(q+\delta_4(q))^2} = e_{4p}^{\delta_4(q)}e_{8p}^{-8^*8(q(1-pp^*)+q^*+2\delta_4(q)(1-pp^*))}$$

$$= e_{8p}^{(1-8^*8)(q+q^*+2\delta_4(q))+2\delta_4(q)} = e_{8p}^{-8^*8(q+q^*+2\delta_4(q))-(q+q^*)}$$
$$= e_8^{\frac{(q+q^*+2\delta_4(q))}{4}\frac{p}{4} - \frac{q+q^*}{p}},$$

and by Lemma 7.1 in [LL], $q + q^* + 2\delta_4(q) \equiv 0 \pmod 4$. Thus

$$\Phi_8 = 1 - \delta_4(p/4) - 2(\frac{2q}{p/4})\delta_4(q) + \frac{p}{4}\frac{q+q^*+2\delta_4(q)}{4} \pmod 8.$$

Case 4. $p \equiv 0 \pmod 8$. According to the formula in Theorem 3.1, III, 3) in [LL], we have

$$\tau_2(L(p,q)) = (\frac{p/2}{q})\overline{\epsilon(q)}e_8^{S(q,p)}e_{4p}^{-\delta_4(q)}(1+i)e_{8p}^{-q^*(q-\delta_4(q))^2}\frac{-\sqrt{2}\delta_4(q)}{2i}$$

(a typo in [LL] has been corrected here). Since

$$e_{4p}^{-\delta_4(q)}e_{8p}^{-q^*(q-\delta_4(q))^2} = e_{8p}^{-(1-p^*p)q-q^*+2\delta_4(q)(q^*q-1)}$$

$$= e_{8p}^{-q-q^*+p^*pq-2\delta_4(q)p^*p} = e_8^{-\frac{q+q^*}{p}+p^*(q-2\delta_4(q))}$$

and $q - 2\delta_4(q) \equiv -q \pmod 8$, we have

$$\Phi_8 = \delta_4(q)(1 + 2(\frac{p/2}{q})) - p^*q \pmod 8.$$

□

4. τ_3 and the Brown Invariant

In this section, we prove Theorem 1 for the Brown invariant. The starting point is: when $\tau_3(M) \neq 0$,

$$\tau_3(M) = \sqrt{2}^{b(M)} e_8^{-\beta(M)},$$

where $b(M) = rkH^1(M, \mathbb{Z}_2)$. By Remark 6.2 in [KM1], τ_3 is an oriented homotopy invariant. It is determined by the first Betti number and the linking pairing on $\mathrm{Tor}H_1$. Thus β is also an oriented homotopy invariant.

Proof of Theorem 1. Case 1. p odd $\not\equiv 0 \pmod 3$. In this case we will show that

$$\beta(L(p,q)) = -1 - \delta_4(p) + 2\delta_4(p)(\frac{q}{p}). \tag{8}$$

By Theorem 3.1, I 1) in [LL],

$$\tau_3 = (\frac{3q}{p})\epsilon(p) e_{12}^S e_p^{-(q+q^*)(12)'} \frac{e_{12}^{2p'} - e_{12}^{-2p'}}{e_{12}^2 - e_{12}^{-2}},$$

where $(12)'12 + p'p = 1, p \pmod{12} \equiv 1, 5, 7, 11$ implies that $p^2 \equiv 1 \pmod{12}$, thus we may take $p' = p$. Apply (6) we get

$$e_{12}^S e_p^{-(q+q^*)(12)'} = e_{12}^{-\Phi + p(q+q^*)} = e_{12}^{-6\delta_4(p)(\frac{2q}{p}) + 9\delta_4(p) - 3} = e_8^{6 - 2\delta_4(p) - 4\delta_4(p)(\frac{2q}{p})}.$$

Hence, using $\epsilon(p) = i^{(1-\delta_4(p)/2}$ and Lemma 1 (1),

$$-\beta(L(p,q)) = 2 - 2\delta_3(p)(\frac{3q}{p}) + 1 - \delta_4(p) + 6 - 2\delta_4(p) - 4\delta_4(p)(\frac{2q}{p}).$$

By Lemma 1 (2) we have (8).

Case 2. p odd $\equiv 0 \pmod 3$. By Theorem 3.1, I 3) in [LL],

$$\tau_3 = (\frac{q}{p/3})\epsilon(p/3) e_{12}^S e_{12p}^{-q^*4^*4(q-\delta_3(q))^2 - 2\delta_3(q)} \frac{-\sqrt{3}\delta_3(q)}{e_{12}^2 - e_{12}^{-2}},$$

where $4^*4 \equiv 1 \pmod{p/3}$. Let $4^*4 = (p+1)^2$ and

$$X = e_{12}^S e_{12p}^{-q^*4^*4(q-\delta_3(q))^2 - 2\delta_3(q)} = e_{12}^{-\Phi + (q+q^* - 2\delta_3(q))(1-4^*4)/p + 4^*4p^*(q - 2\delta_3(q))}.$$

Apply (7), and notice that $(1 - 4^*4)/p = -(p+2)$, $16q - 4^*4(q - 2\delta_3(q))$ and $-(p + 2 - 9p)(q + q^* - 2\delta_3(q))$ are divisible by 12, we get

$$X = e_4^{3 - \delta_4(p)(1 + 2(\frac{2q}{p})) + 2}.$$

By Lemma 1 (1) we have

$$\tau_3 = i(\frac{q}{p})\varepsilon(p/3)X = e_4^{1+1-(\frac{q}{p})+\frac{1}{2}(1+\delta_4(p))+1-\delta_4(p)+2}$$

and

$$\beta(L(p,q)) = 2(\frac{q}{p}) - 3 + \delta_4(p) = -1 - \delta_4(p) + 2\delta_4(p)(\frac{q}{p}) \pmod 8.$$

Case 3. When p is divisible by 4, we apply the reciprocity law (4.17) in [KM1]:

$$\beta(L(p,q)) = -1 - \beta(L(q,p)) = \delta_4(q) - 2\delta_4(q)(\frac{p}{q}).$$

□

We can easily read off properties of β from Theorem 1.

(i) $\beta(L(p,-q)) = -\beta(L(p,q))$. Notice that $(\frac{-1}{p}) = \delta_4(p)$. If $\delta_4(p) = 1$ then $\beta = 0, 4 \pmod 8$. So $\beta = -\beta$. If $\delta_4(p) = -1$, then $\beta(L(p,-q)) = 2\delta_4(p)(\frac{-q}{p}) = -\beta(L(p,q))$.

(ii) β is an oriented homotopy invariant. This follows directly from Proposition 1.

(iii) $\beta(L(p,q))$ distinguishes oriented homotopy type when p is prime.

Example 1. β also distinguishes oriented homotopy type when $p = 12$.

$\beta(L(12,5)) = 1-2(\frac{12}{5}) = 1-(-2) = 3$, $\beta(L(12,7)) = -3$, $\beta(L(12,1)) = 1 - 2 = -1$ since $(\frac{12}{1}) = 1$, $\beta(L(12,11)) = 1$.

5. μ Invariants

5.1. μ *invariants mod 8 and Jacobi symbols*

For lens spaces, μ-invariant mod 8 are closely related to the Brown invariant β ([KM1], [KM2]). Together with Theorem 1 we obtain the following formulas of μ mod 8 in terms of Jacobi symbols.

Theorem 3. *When p is odd,*

$$\mu(L(p,q)) = -1 - \delta_4(p) + 2\delta_4(p)(\frac{q}{p}) + \frac{p^2-1}{2} \pmod 8.$$

When p is even and $q > 0$,

$$\mu_1(L(p,q))) = \delta_4(q) - 2\delta_4(q)(\frac{p}{q}) + \frac{q^2-1}{2} \pmod 8$$

$$\mu_0(L(p.q)) \ = \delta_4(q) - 2\delta_4(q)(\frac{p}{q}) + \frac{q^2-1}{2} + pq \pmod 8$$

Proof. It is shown in [KM2] (see also [KM1], (4.16)) that if p is odd, then

$$\beta(L(p,q)) = \mu(L(p,q)) + (p^2-1)/2 \pmod 8. \tag{9}$$

The formula for μ mod 8 follows immediately from this and (8).

If p is even, then q is odd. From the reciprocity law for μ_1 in [KM2],

$$\mu_1(L(p,q)) = -\text{sign}(q) - \mu(L(q,p)).$$

Since q is odd, the formula for μ_1 follows from (9)

The formula for μ_0 follows from Assertion 4.9 (ii) in [KM1]:

$$\mu_0(L(p,q)) - \mu_1(L(p,q)) \equiv pq \pmod 8. \tag{10}$$

□

Corollary 4. *The μ invariants mod 4 are given by*

$$\mu(L(p,q)) \pmod 4 = \begin{cases} 0 & \textit{if } p \equiv 1 \pmod 4, \\ 2, & \textit{if } p \equiv -1 \pmod 4, \end{cases}$$

$$\{\mu_0(L(p,q)), \mu_1(L(p,q))\} \pmod 4 = \begin{cases} \{1,-1\}, & \textit{if} p \equiv 2 \pmod 4, \\ \{-\delta_4(q), -\delta_4(q)\}, & \textit{if} p \equiv 0 \pmod 4. \end{cases}$$

Now we discuss oriented homotopy invariance of μ mod 8. Recall that if $L(p,q_1)$ and $L(p,q_2)$ are oriented homotopy equivalent, then q_1q_2 is a quadratic residue of p.

Example 2. It is known that for odd p, μ is not an oriented homotopy invariant: $L(7,1)$ and $L(7,2)$ are oriented homotopy equivalent, but $\mu(L(7,1)) = 6$ and $\mu(L(7,2)) = 14$.

Prop 2. The following are oriented homotopy invariants of lens spaces.

(i) $\mu(L(p,q)) \pmod 8$ for p odd.

(ii) $\mu_0(L(p,q)) + \mu_1(L(p,q)) \pmod 8$ for even p.

(iii) The unordered pair $\{\mu_0(L(p,q)), \mu_1(L(p,q))\} \pmod 8$ for p divisible by 4.

Proof. (i) It follows from (9) and the fact that β is an oriented homotopy invariant.

(ii) First we observe that $\mu_0(L(p,q))+\mu_1(L(p,q)) \equiv pq-2\delta_4(q) \pmod 8$ since q^2-1 is divisible by 8 and $4\delta_4(q)(\frac{p}{q}) = 4\delta_4(q) \pmod 8$.

Suppose $L(p,q_1)$ and $L(p,q_2)$ are oriented homotopy equivalent.

If p is divisible by 4, $\delta_4(q_1)=\delta_4(q_2)$ by Proposition 1.

Suppose $p=2 \pmod 4$. If $p=2 \pmod 8$ then $pq-2\delta_4(q)=2(q-\delta_4(q))=0 \pmod 8$. If $p=-2 \pmod 8$ then $pq-2\delta_4(q)=-2(q+\delta_4(q))=4 \pmod 8$.

(iii) When p is divisible by 4, the Brown invariant is defined and an oriented homotopy invariant. So it suffices to show that the set

$$T(L(p,q)) = \{(q^2-1)/2,\ (q^2-1)/2+pq\}$$

is an oriented homotopy invariant.

If p is divisible by 8, q_1q_2 is a quadratic residue of 8. Then $q_1q_2 \equiv 1 \pmod 8$. So $q_1 \equiv q_2 \pmod 8$ and $T(L(p,q_1)) = T(L(p,q_2))$. Now assume that p is divisible by 4 but not divisible by 8. Then $pq=4$ and $q_1 \equiv q_2 \pmod 4$. Hence the set $T=(0,4)$.

□

For $p \equiv 2 \pmod 4$, the unordered pair $\{\mu_0(L(p,q)), \mu_1(L(p,q))\}$ mod 8 is not an oriented homotopy invariant: $L(38,7)$ and $L(38,5)$ are oriented homotopy equivalent, but have different μ invariants mod 8.

Example 3. Since $(\frac{38}{7}) = -1$, $\mu_1(L(38,7)) = -1+2(-1) = -3 = 13 \pmod 8$, $\mu_0(L(38,7)) = -3+2 = -1 \pmod 8$.

Similarly, since $(\frac{38}{5}) = -1$, $\mu_1(L(38,5)) = 1-2(-1)+4 = 3 \pmod 8$, $\mu_0(L(38,5)) = 3+(-2) = 1 \pmod 8$.

5.2. *μ invariants and τ_4*

In the previous subsection we obtain rather simple formulas for μ invariants mod 8. For the original $\mathbb{Z}_{16}$–valued μ invariants, there is a recursive algorithm [NR]. There is also a formula in terms of continued fraction expansion of $-p/q$ with all m_i even in [FG] (p. 96, (2.14), (2.15)). We present here a new formula in terms of the Rademacher Φ function and Jacobi symbols.

Theorem 5. *The μ invariants can be expressed in terms of the Rademacher Φ function and Jacobi symbols. Precisely,*

(1) If $p \equiv 8 \pmod{16}$ *and* $q \equiv \pm 1 \pmod 8$, *or* $p \equiv 0 \pmod{16}$ *and* $q \equiv \pm 3$ $\pmod 8$,

$$\{\mu_0, \mu_1\} = \{\delta_4(q) - 2\delta_4(q)(\frac{p}{q}) + \frac{q^2-1}{2},\quad \delta_4(q) - 2\delta_4(q)(\frac{p}{q}) + \frac{q^2-1}{2} + 8\}. \tag{11}$$

(2) If $p \equiv 0 \pmod{16}$ *and* $q \equiv \pm 1 \pmod 8$,

$$\mu_0 = \mu_1 \equiv -5\Phi + 5(q - 2\delta_4(q))p^* - 4(\frac{p}{q}) - 2\delta_4(q) + 4. \tag{12}$$

(3) If $p \equiv 8 \pmod{16}$ *and* $q \equiv \pm 3 \pmod 8$, *then with* $\bar{p} = p/8$,

$$\mu_0 = \mu_1 = -5\Phi + 5(2\delta_4(\bar{p}) - \bar{p})\frac{q + q^* - 2\delta_4(q)}{8} - 4(\frac{2q}{\bar{p}}) + 4\delta_4(q) - 2\delta_4(\bar{p}) + 6. \tag{13}$$

(4) If $p \equiv 4 \pmod 8$, *then with* $\bar{p} = p/4$,

$$\{\mu_0, \mu_1\} = \{-5\Phi + 5(2\delta_4(\bar{p}) - \bar{p})\frac{q + q^* - 2\delta_4(q)}{4} - 4(\frac{q}{\bar{p}}) + 4\delta_4(q) + 6\delta_4(\bar{p}) - 2 \pm 6\}. \tag{14}$$

(5) If $p \equiv 2 \pmod 4$, *then with* $\bar{p} = p/2$,

$$\{\mu_0, \mu_1\} = \{-5\Phi + 5\bar{p}\frac{q + q^* + 2}{2} - 4\delta_4(q)(\frac{2q}{\bar{p}}) + (1 - \bar{p}^2)p^*q - 2\delta_4(\bar{p}) + 2 + \frac{pq + q^2 - 1}{2} \pm \frac{pq + q^2 - 1}{2}\}. \tag{15}$$

(6) If p *is odd then*

$$\mu = -5p^2\Phi + 5p(q + q^*). \tag{16}$$

Remark 2. The formula (16) in the p odd case is known. Theorem 5 can be viewed generalizations to all even p.

Example 4. $L(38, 7)$

$\Phi = -3$, $q^* = 11$, $p^* = -2$, $(\frac{14}{19}) = -1$.

So $\mu_0 = -3$, $\mu_1 = \mu_0 + pq + 0 = -3 + 10 = 7$.

Example 5.

$L(100, 9)$

$\Phi = -1$, $q^* = 89$, $\bar{p} = 25$.

$\mu_0 = 5 + 5(2 - 25)8 + 6 - 2 + 6 = 17 + 6 = 7$, $\mu_1 = 17 - 6 = 11$.

Example 6.

$L(64,25)$

$\Phi = 3$, $p^* = -16$.

$\mu_0 = \mu_1 = -1$.

Example 7.

$L(8,5)$

$\Phi = 2$, $q^* = 5$.

$\mu_0 = \mu_1 = -10 + 5 - 4 + 6 + 4 - 2 = -1$.

Proof of Theorem 5. The starting point is the following formula of Kirby and Melvin [KM2]:

$$\tau_4(L(p,q)) = \begin{cases} e_{16}^{-3\mu(q/p)}, & \text{for odd } p \\ \\ e_{16}^{-3\mu_0(q/p)} + e_{16}^{-3\mu_1(q/p)}, & \text{for even } p \end{cases} \tag{17}$$

1. In this case, $\tau_4 = 0$ by [Y]. Hence $\mu_0 = \mu_1 + 8$ and (11) follows from Theorem 3.

2. In this case, $|\tau_4(L(p,q))| = 2$ by [Y]. Hence $\mu_0 = \mu_1$, and $e_{16}^{-3\mu_0} = e_{16}^{-3\mu_1} = \frac{1}{2}\tau_4$.

Let $\varepsilon = \pm 1$ be such that $q + \varepsilon \equiv 0 \pmod 8$, then $\varepsilon = -\delta_4(q)$, and by Theorem 3.1, III 3) in [LL], we have

$$\frac{1}{2}\tau_4 = (\frac{p/4}{q})\overline{\varepsilon(q)}e_{16}^{S}e_{8p}^{\varepsilon}e_{16p}^{-q^*(q+\varepsilon)^2}e_8(-i)\varepsilon.$$

Since $e_{16p}^{-q^*(q+\varepsilon)^2} = e_{16p}^{-(q+q^*-qp^*p+2\varepsilon-2\varepsilon p^*p)}$ we have

$$\frac{1}{2}\tau_4 = (\frac{p/4}{q})\overline{\varepsilon(q)}e_{16}^{-\Phi+(q+2\varepsilon)p^*}e_8(-i)\varepsilon.$$

Hence

$$-3\mu_0 = -3\mu_1 \equiv 4(1-(\frac{p}{q}))-2(1-\delta_4(q))+2-4+4(1+\delta_4(q))-\Phi+(q-2\delta_4(q))p^*$$

and (12) follows.

3. We have $|\tau_4| = 2$ by [Y], hence $\mu_0 = \mu_1$.

Let $\bar{p} = p/8$. Notice that $\bar{p}$ is odd. By Theorem 3.1, III 4) in [LL],

$$\frac{1}{2}\tau_4 = (\frac{2q}{\bar{p}})\varepsilon(\bar{p})e_{16}^{S}e_{8p}^{-\delta_4(q)}e_{16p}^{-q^*8^*8(q-\delta_4(q))^2}\delta_4(q)i,$$

where $8^*8 + \bar{p}^2 = 1$. Hence

$$e_{16p}^{-q^*8^*8(q-\delta_4(q))^2} = e_{16p}^{-8^*8(q+q^*-2\delta_4(q))+8^*8p^*p(q-2\delta_4(q))} = e_{16p}^{(\bar{p}^2-1)(q+q^*-2\delta_4(q))+8^*8p^*p}.$$

From $qq^*+pp^* = 1$, we deduce that $\delta_4(q) = \delta_4(q^*)$ and $8p^* \equiv q+q^*-2\delta_4(q) \pmod{16}$. Thus

$$e_{16p}^{\bar{p}^2(q+q^*-2\delta_4(q))+8^*8p^*p} = e_{16}^{(q+q^*-2\delta_4(q))(1-\bar{p}^2+\bar{p})/8}.$$

It is easy to see that $1-\bar{p}^2+\bar{p} \equiv 2\delta_4(\bar{p}) - \bar{p} \pmod{16}$ we get then

$$-3\mu_0 = -\Phi + (2\delta_4(\bar{p}) - \bar{p})\frac{q+q^*-2\delta_4(q)}{8} - 4(\frac{2q}{\bar{p}}) + 6 + 4\delta_4(q) - 2\delta_4(\bar{p}),$$

and (13) follows.

4. $|\tau_4| = \sqrt{2} = |e_{16}^{-3\mu_0}(1 + e_{16}^{3(\mu_0-\mu_1)})|$ by [Y]. Thus $\mu_0 - \mu_1 \equiv \pm 4 \pmod{16}$.

Let $\bar{p} = p/4$. Notice that $\bar{p}$ is odd. Now by Theorem 3.1, III 5) in [LL], we have

$$\tau_4/\sqrt{2} = i\delta_4(q)(\frac{q}{\bar{p}})\varepsilon(\bar{p})Z$$

with $Z = e_{16}^{S}e_{16p}^{-2\delta_4(q)-q^*4^*4(q-\delta_4(q))^2}$, where $4^*4 \equiv 1 \pmod{\bar{p}}$. Let $4^*4 = (1-\delta_4(\bar{p})\bar{p})^2$, then $4^*4 = 0 \pmod{16}$ and

$$Z = e_{16}^{S}e_{16p}^{-2\delta_4(q)-4^*4(q+q^*-2\delta_4(q)-p^*p(q-2\delta_4(q)))} = e_{16}^{-\Phi+(2\delta_4(\bar{p})-\bar{p})(q+q^*-2\delta_4(q))/4}.$$

From this, we see that

$$\tau_4 = (e_{16}^{-2} + e_{16}^{2})e_{16}^{4\delta_4(q)-4(\frac{q}{\bar{p}})-2\delta_4(\bar{p})+(2\delta_4(\bar{p})-\bar{p})(q+q^*-2\delta_4(q))/4+6-\Phi}$$

and (14) follows.

5. Let $\bar{p} = p/2$. By Theorem 3.1, II.2) in [LL],

$$\tau_4(L(p,q)) = (\frac{2q}{\bar{p}})\varepsilon(\bar{p})e_{16}^{S(q,p)}e_p^{-q^*8^*\frac{q^2+1}{2}}\sqrt{2}\frac{e_{16}^{\bar{p}} - e_{16}^{-\bar{p}}}{e_8 - \overline{e_8}},$$

where $8^*8 + \bar{p}^2 = 1$. On the other hand,

$$\tau_4(L(p,q)) = e_{16}^{-3\mu_0(q/p)}(1 - e_{16}^{3(\mu_0-\mu_1)+8}).$$

Since $|1-e_{16}^a| = |1-e_{16}^b|$ is equivalent to $a \equiv \pm b \pmod{16}$, and

$$\mu_0 - \mu_1 \equiv pq + 8k \pmod{16}$$

for $k = 0$ or 1 by (10), we have $3pq + (k+1)8 \equiv hp \pmod{16}$ for $h = 1$ or -1. It is easier to see that $h = -\delta_4(q)$ and $k = (q^2-1)/8 \pmod 2$. Thus

$$\mu_0 - \mu_1 \equiv pq + q^2 - 1 \pmod{16}. \tag{18}$$

Since $e_p^{-q^*8^*(q^2+1)/2} = e_{16p}^{-8^*8(q(1-p^*p)+q^*)} = e_{16p}^{(\bar{p}^2-1)(q+q^*)+8^*8p^*pq}$, we have

$$\tau_4(L(p,q)) = (\frac{2q}{\bar{p}})\varepsilon(\bar{p})(-i)e_{16}^{-\Phi}e_{16}^{p(q+q^*)/4+8^*8p^*q}(e_{16}^{\bar{p}} - e_{16}^{-\bar{p}}).$$

Thus $e_{16}^{-3\mu_0} = (\frac{2q}{\bar{p}})\varepsilon(\bar{p})(-i)e_{16}^{-\Phi+p(q+q^*)/4+8^*8p^*q+\bar{p}}$ and

$$-3\mu_0 = (-\Phi+p(q+q^*)/4+8^*8p^*q+\bar{p})-4+2(1-\delta_4(\bar{p}))+4(1-\delta_4(q)(\frac{2q}{\bar{p}})).$$

Since q is odd, (15) follows from the formula above and (18).

6. (16) follows directly from the well known formula of μ:

$$3\mu(L(p,q)) = p^2S(q,p) \pmod{16}. \tag{19}$$

In fact, the p odd case of (17) could be used with the formula for τ_4 in [LL] to give a new proof of (19). □

References

[BHMV] C. Blanchet, N. Habegger, G. Masbaum, P. Vogel: Three- manifold invarinats derived from the Kauffman bracket, Topology, 31 (1992), 685-699.

[B] E. Brown, Generalization of the Kervaire invariant. Ann. Math. **95** (1972), 368-383.

[FG] D. S. Freed, R. E. Gompf : Computer calculation of Witten's 3-manifold invariants, Comm. Math. Phys. **141** (1991), 79-117.

[J] L. C. Jeffrey, Chern-Simons-Witten invariants of lens spaces and torus bundles, and the semi-classical approximation, Comm. Math. Phys. **147** (1992), No.3, 563-604.

[KM1] R. Kirby, P. Melvin : Dedekind sums, mu-invariants and signature cocycle Math. Ann. **299** (1994) 231-267.

[KM2] R. Kirby, P. Melvin : The 3-manifold invariants of Witten and Reshetikhin-Turaev for $sl(2,C)$, Invent. Math. **105**. (1991), 473-545.

[KMZ] R. Kirby, P. Melvin, X. R. Zhang : Quantum invariants at the sixth root of unity,Comm. Math. Phys. **151** (1993) 607-617.

[L] B. H. Li, Relations among Chern-Simons-Witten-Jones invariants, Science in China, series A, **38** (1995), 129-146.

[LL] B. H. Li, T. J. Li : Generalized Gauss Sums, Chern-Simons-Witten-Jones invariants of lens spaces J. Knot theory and its Ramification, **5** (1996), 183-224.

[NR] W. D. Neumann, F. Raymond, Seifert manifolds, plumbing, μ-invariant and orientation reversing maps, Lecture Notes in Math. **664** (1978), 162-195.

[RG] H. Rademacher, E. Grosswald : Dedekind Sums, Carus Math. Mono. No. 16, Math. Assoc. Amer. (1972).

[RT] N. Yu. Reshetikhin, V. G. Turaev : Invariants of 3-manifolds via link polynomials and quantum groups, Invent. Math. **103** (1991), 547-597.

[S] G. Shimura, On modular forms of half integral weight, Ann. Math. **97** (1973), 440-481.

[Y] S. Yamada, The absolute value of Witten invariants of Lens space, J. Knot Theory Ramifications, **4** (1995), 319-328.

Historical Review on the Roles of Mathematics in the Study of Aerodynamics

Jiachun Li

Institute of Mechanics, CAS, Beijing, China, 100190

jcli05@imech.ac.cn

Mathematics and mechanics are twins and were concurrently developing in history. In the times of classic mechanics, mathematicians always were great masters of mechanics. The progresses of aerodynamics with many great mathematicians and applied mathematicians involved further demonstrated the close link between two disciplines in the 20th century. The present article is primarily focused on the advances of aerodynamics in the period of aeronautical engineering from low to hypersonic speeds. Correspondingly, singular perturbation theories, hodograph method, mixed type and hyperbolic PDE, shock capture scheme in CFD etc. were developing. The persuasive facts became additional paradigm of excellent combination of mathematics and mechanics. Finally, we foresee potential significant directions in future compressible flow study and expect further collaboration of scientists in mathematics and mechanics communities.

Keywords: Perturbation method, Hodograph, hyperbolic and mixed type equation, Shock capture scheme.

The success of the first mankind's powered flight by Brother Wright in 1903 was an epoch-making event, marking the start of aeronautics and aerospace era. However, engineers were immediately confronting a challenging issue of how to scientifically design an aircraft. Since the famous D'Alemberg paradox implies that the drag of a vehicle based on ideal fluid assumption vanishes and people also had little knowledge about the lift of a wing. The difficulty then was that the Navier-Stokes equation for viscous fluids usually lacks analytical solution even for a simple airfoil and was also unable to be numerically solved prior to the advent of advanced computers (Li 1995).

1. Boundary Layer and Wing Theories

Prandtl (1904) at the Department of Applied Mechanics, Gottingen University of Germany experimentally discovered in a water flume that the effects of viscosity are merely restricted to a very thin layer adjacent to the solid wall for large Re flows and then proposed in time the boundary layer theory for drag estimation. Considering great contribution in enhancing understanding of viscous flows and remarkably promoting the progress of aeronautical engineering, mechanics community unanimously regarded Prandtl's BL theory as a milestone of modern mechanics. As for lift, Rayleigh was the earliest to explain Magnus effect by additional circulation around a body, which can be determined by Kutta (1902)–Joukowski (1907) condition, in an incoming flow with $L = \rho U_\infty \Gamma$. Although Lanchester initiated the study of lift for a wing of finite span, Prandtl (1918) was the first to present its mathematical theory. The vortex system consisting of bound vortex at the wing surface and free vortex trailing from wing tip and extending downstream was assumed responsible for downwash and lift generation. Prandtl furthermore found that the wing of elliptic eddy distribution has the minimal induced drag.

As a matter of fact, both BL and wing theories belong to the category of singular perturbation (Van Dyke 1964, Li & Zhou 1998). The boundary layer approach dealing with a DE with a small parameter in the highest derivative term was broadly applied and extended known as Matched Asymptotic Expansion Method, which enormously enriched the contents of applied mathematics.

Since then people witnessed the rising of aviation industry manufacturing varieties of commercial and military aircrafts at the flight speed of a few hundred km/h. Under these circumstances, the research of compressible flows was put on agenda.

We may usually divide compressible flows into the following regimes: (1) subsonic flows mean that the density variation can no longer be neglected; (2) transonic and supersonic flow regime is characteristic of the appearance of shock waves and aerothermodynamic effects in the flow field; (3) hypersonic flow regime should consider aerothermochemistry effects due to internal freedom excitation (Von Karman 1963).

2. Subsonic and Transonic Flows

As you know, the relative variation of density in an isentropic flow is proportional to M^2, where M denotes Mach number. People considered compressibility effects firstly with the help of perturbation theory when a thin or slender vehicle flying at low Mach number is concerned with. Then, the drag and lift of a wing and slender body at small angle of attack can be estimated theoretically.

However, as the speed of an aircraft is growing to 700 $\sim$ 800 km/h when the disturbance is no longer negligibly small, the air flow will obey a nonlinear potential equation like:

$$\left(1-\frac{\phi_x^2}{a^2}\right)\phi_{xx}-2\frac{\phi_x\phi_y}{a^2}\phi_{xy}+\left(1-\frac{\phi_y^2}{a^2}\right)\phi_{yy}=0,$$

which turns out sufficiently difficult to solve. The most effective approach in two dimension cases is the hodograph method by virtue of exchanging the positions of dependent and independent variables. For gas jet problem, Chaplygin (1904) derived a hodograph equation in term of stream function ψ with the module q and argument θ of velocity vector taken as new independent variables:

$$q^2\psi_{qq}+q(1+M^2)\psi_q+(1-M^2)\psi_{\theta\theta}=0,$$

which is evidently linear and then the superposition principle can be applied again. Of course, the difficulty now is how to find a counterpart in the physical plan corresponding to the solution in the velocity plan. Anyway, we can list some of solved problems by this approach (Kuo 1954):

1) Ringleb solution representing a compressible flow turning 180° around a flat plate;

2) Subsonic plane jets;

3) Subsonic flows around an elliptic airfoil with or without rotation;

4) Karman-Tsien formula relating pressure coefficients for compressible and incompressible flows around an airfoil as an effective design tool in aeronautical engineering at that time (Von Karman 1941).

When looking at the solutions above, people found that the continuous mixed subsonic and supersonic flow can coexist with local maximum Mach number $M_{\max}=2.5$ for Ringleb solution and $M_{\max}=1.25$ and 1.22 for subsonic flows $M_\infty=0.6$ and 0.7 around an elliptic airfoil of thickness 0.6 (Kuo 1953). Such kind of mixed subsonic and supersonic flows can be maintained until shock waves appear. Therefore, the study of typical mixed type PDE, namely, the Tricomi equation

$$\phi_{\eta\eta}-\eta\phi_{\xi\xi}=0,$$

was used to solve transonic flows around a wedge or in the nozzle.

In reality, some kind of discontinuity may appear in unsteady compressible flows as well as in the mixed supersonic and subsonic flow field. The physical mechanism for a shock wave to emerge is attributed to the accumulation of compressible disturbance. On the other hand, the mathematical cause of discontinuity is the occurrence of limiting lines when the Jacobi determinate of hodograph transformation vanishes. As a result, Tsien and Guo defined the Mach number for sonic region first to appear as the lower critical Mach number, whereas they called the Mach number for shock wave to occur as the upper critical Mach number. Actual shock waves tend to occur at the Mach number between lower and upper critical Mach numbers due to flow instability when a flow transit from supersonic to subsonic speed.

When there is a shock wave in the flow field, potential assumption wouldn't be justified any longer because isentropic and irrotational conditions break down. People turned to examine quasi-linear hyperbolic equation and its solution. Actually, Riemann problem describing the evolution of an initial step discontinuity in one dimensional air can be regarded as the earliest study on compressible flows. Courant & Friedrichs (1948) made a comprehensive summary on the research of two categories of flows with discontinuity, namely one dimensional unsteady flow and two dimensional transonic or supersonic flows. Their classic book dealt with propagation of rarefied and compressible waves, the formation of shock waves, and reflection of gas dynamic waves from a free surface or solid wall (including regular and Mach reflection), wave-wave interaction etc.

Gu *et al.* (1961, 1962, and 1963) mathematically examined initial or boundary value problems for hyperbolic equation system of 1+1 or 2 dimensions with 3 dependent functions. By reducing to an integral equation, their uniformly convergence and thus the existence of local solution were proved. The results were applied to one dimensional, cylindrically or spherically symmetric gas motions driven by a piston, flood evolution in a river and two dimensional supersonic flows. The proposed successive approximate method may as well serve as an effective tool to find solutions. Friedrich's positive symmetric theory (1958) for mixed type DE was further extended to DE of higher dimension by Gu.

3. Compressible Viscous Flows

As far as skin friction and heat transfer are concerned, we have to give up the assumption of ideal gas and employ compressible boundary layer theory for viscous gases, the complexities of which came from apparent increment in internal energy accompanied by density and viscosity variations in thickened BL (Tsien 1938).

In order to simultaneously solve both momentum and energy equations effectively based on the transformation method, people at first handled compressible flows around a flat plate under the assumption:

$$\mu \propto T, \text{ and } \rho\mu = C$$

for perfect gas throughout the boundary layer. If the Dorodnitsyn–Howarth transformation is applied, we are able to derive an ODE system corresponding to them with similarity solutions as below:

$$f''' + ff'' = 0$$

$$g'' + \mathbf{Pr} fg' + \frac{(\gamma-1)M_\infty^2}{4}\mathbf{Pr} f'^2 = 0.$$

Obviously, f satisfies the Blasius equation and g as the solution of an inhomogeneous linear equation can be explicitly expressed in terms of f. In this way, the skin friction and wall temperature of compressible boundary layer over an insulated flat plate can be given as:

$$C_f = \frac{0.664}{\sqrt{Re_x}}\sqrt{C}, C = (1 + 0.36(\gamma-1)M_\infty^2\sqrt{Pr})^{-(1-\omega)}, \omega = 0.7 \sim 0.9$$

$$T_w = T_\infty(1 + \frac{\gamma-1}{2}M_\infty^2\sqrt{Pr}).$$

That is, the friction coefficient is a bit smaller than the value of incompressible one and the recovery temperature at the wall is smaller than stagnation one at moderate Mach number for air, where Pr number represents the ratio of viscous and heat diffusions. In the same way, we are able to solve the problem at an isothermal plate. It is easy to understand that the recovery temperature for insulated plate or peak temperature for isothermal plate is higher for media with larger Pr number. Since the momentum equation and the energy equation take the similar form, analogy theory may help us to deduce some very useful arguments such as the Reynolds analogy relating expressions of velocity and temperature profiles and coefficients of heat transfer and skin friction: $St = 0.5C_f Pr^{2/3}$ (Schlichting 1950)

In the days of breaking through "sonic barrier", wind tunnel tests revealed that the chief criminal of transonic flight failure was attributed to the occurrence of shock at the surface of an airfoil. And then, scientists paid attention to the critical influences of a shock on its aerodynamic performances to answer how aerodynamic stall took place. Although there were solutions about regular and Mach reflections of an incident shock, they weren't consistent with experiments. Consequently, the interaction between shock wave and boundary layer became the frontier in aerodynamics at that time.

Further including viscous and heat transfer effects based on previous studies, Kuo (1953) assumed a potential outflow with pressure disturbance and a viscous inner layer with proper velocity profile. The momentum integration method gave overall tendency in pressure variation and separation, while the approach of differential equation provided the details of flow pattern in the boundary layer. An approximate solution in theory for shock-laminar boundary layer interaction was qualitatively satisfactory with following conclusions:

1) The interaction may induce apparent variation in flow pattern: A series of oblique shock appear prior to the main shock; there will be a bump nearby the incident point;

2) Pressure disturbance decays exponentially upstream in the distance of tens of momentum thicknesses. However, overpressure takes place right behind the shock and then gradually drops to the value of regular reflection for inviscid fluids;

3) When shock strength is strong enough, there will be separation ahead of incident point with backward flow adjacent to the wall. How the flow separates also depends on M and Re to a certain extent. Sometimes, reattachment can occur.

4. Shock Wave Capturing Schemes in CFD

As you have seen in the previous paragraphs, the most salient feature of unsteady, transonic and supersonic flows is the appearance of shock waves, the capture of which obviously is a formidably hard task. Since the continuous flow field is separated by an unknown shock, prior to and behind which the Rankine-Hugoniot relations between physical quantities should be satisfied, tedious and time-consuming fitting procedures to locate the shock by try and error were commonly followed in 1950s. As a matter of fact, a realistic shock wave has a thickness of length scale about molecular free path when gas viscosity is taken into account. Based on this concept, Von Neumann and Richtmyer (1957) proposed to introduce a term of artificial viscosity:

$$q = \begin{cases} \rho b^2 \triangle x^2 (\frac{\partial u}{\partial x})^2, & \frac{\partial u}{\partial x} < 0 \\ 0, & \frac{\partial u}{\partial x} \geq 0, \end{cases}$$

where b is an adjustable constant when solving the inviscid Euler equation so that the shock wave can be identified as the position of large gradient and thus automatically captured. The target of this technique is (1) the width of shock should be restricted in one grid scale; (2) shock wave relations should be satisfied; (3) the computation in the continuous region will not be affected.

Since 1960s, people found that this kind of artificial viscosity can be introduced numerically by constructing some kinds of FD scheme with numerical dissipation. As a result, the most challenging task for CFD during recent decades was to work out high performance schemes with adequate numerical dissipation to accurately capture shock without virtual oscillation. The major effective approaches were:

(1) Upwind scheme and its high order counterparts;

(2) Flux Vector Splitting Scheme(FVS) for multidimensional problems accounting for disturbance propagation along characteristic directions;

(3) Godunov type scheme (1959) based on the exact solution of the Riemann problem capable of accurately representing wave evolution;

(4) Total Variation Diminish Scheme (TVD) $TV(v^{n+1}) \leq TV(v^n)$ by Harten (1983), a new concept to eliminate nonphysical oscillation;

(5) Essentially No-Oscillation (ENO) (Harten 1989) and Weighted Essentially No-Oscillation Scheme (WENO) (Liu 1994, Jiang & Shu 1996), $TV(v^{n+1}) \leq TV(v^n) + (h^r)$, an idea for reconstruction of high order numerical flux based on adaptive or weighted stencil selection to avoid loss of accuracy.

Pirrozoli (2011) systematically reviewed foregoing advances so that higher accuracy was achieved by upwind scheme along with filtering and physical conservation in smooth region and virtual oscillation was diminished by hybrid scheme and nonlinear filter with the help of varieties of shock sensors in the vicinity of discontinuities. Shock capture in unsteady compressible laminar and turbulent flows with complicated geometry remains to be most attractive topic in this regard.

5. Concluding Remark

In the 21th century, China has achieved great success in aerospace engineering in manned flight, walking out of capsule and rendezvous/docking. The future plans are the establishment of space station, deep space exploration and new vehicles for transportation between space and continents. New issues for compressible flows such as: unsteady complicated flows with separation, vortices, turbulence and their control; thermal environment and protection of near space vehicles and combustion/chemical flows in the scramjet engine etc. are most challenging. We believe that the collaboration between mathematics and mechanics communities will continue to play indispensable roles in the process ahead.

References

1. Chaplygin S.A. (1904), On gas jets, Annals 21, Phys. Math. Div., Imper. Univ. Moscow.
2. Courant R. & Friedrichs K.O. (1948), Supersonic flow and shock waves, Interscience Pub. Inc., New York.
3. Friedrichs K.O. (1958), Symmetric positive differential equations, Comm. Pure Appl. Math., 11:333-418.
4. Godunov S.K. (1959) A difference scheme for numerical solution of discontinuous solution of hyperbolic hydrodynamic equations, Math. Sbornik, 47:271-306.
5. Gu C.H., Li D.Q. & Hou Z.Y. (1961) Discontinuous initial value problem for a quasi-linear hyperbolic equation systems, Acta Mathematica Sinica (1)(2), 11.
6. Gu C.H., Li D.Q. & Hou Z.Y. (1962) Discontinuous initial value problem for a quasi-linear hyperbolic equation systems (3), Acta Mathematica Sinica, 12:132-143.
7. Gu C.H. (1963) A boundary value problem of a hyperbolic equation system and its applications, Acta Mathematica Sinica, 13:32-48.
8. Harten A. (1983) high resolution scheme for hyperbolic conservation law, J. Comput. Phys. 49:357-393.
9. Harten A. (1989) ENO schemes with subcell resolution, J. Comput. Phys.,83:148-184.
10. Joukowski N.E. (1907) On adjunctive vortices, Communications, 112:12-25
11. Jiang G.S. & Shu C.W. (1996) Efficient implement weighted ENO schemes, J. Comput. Phys. 126:202-228.
12. Kuo Y.H. (1953) Reflection of a weak shock wave from boundary layer over a flat plate, NACA TN2868, 2869.
13. Kuo Y.H. & Sears W.R. (1954) Plane subsonic and transonic potential flows, in General Theory of High Speed Aerodynamics, Sec.F: 490-582.
14. Kutta M.W. (1902) Lift in fluid flows, Repot on aeronaut. Tech. 6:133-135. Li J.C.(1995) Retrospect and prospects of fluid mechanics, Adv. Mech. 25:442-450.
15. Li J.C. & Zhou X.C. (1998) Asymptotic methods in mathematical physics, Science Press.
16. Liu X.D., Osher S. & Chan T. (1994) Weighted essentially non-oscillatory schemes, J. Comput. Phys. 115:200-212.
17. Pirrozoli S. (2011) Numerical methods for high speed flows, Ann, Rev. Fluid Mech. 43:161-194.

18. Prandtl L. (1904) Boundary layer in fluids with small viscosity, Proc. of the 3rd Int. Math. Con., Heidberg: 484-491.
19. Prandtl L. (1918) Lift theory, Gottingen Bulletin, Math. Phys. Class: 451-477.
20. Richtmyer R.D. (1957) Difference method for initial problems, Interscience Pub. Inc., New York.
21. Schlichting H. (1950) Boundary layer theory, 281-316.
22. Tsien H.S. (1938) Boundary layer in compressible fluids, J. Aero. Sci., 5:227-232.
23. Van Dyke M. (1964) Perturbation theory in fluid mechanics, Academic Press.
24. Von Karman T. (1941) Compressibility effects in aerodynamics, J. Aero. Sci., 8:337-356.
25. Von Karman T. (1963) From low speed aerodynamics to astronautics, Pergmon press.

Postscript: The review article on mathematical roles in the study of aerodynamics is especially dedicated to the professor C.H. Gu for his remarkable contribution in promoting mathematical research and applications in this area. Expecting China's needs in aerospace engineering, he had no hesitation to switchs his majority from differential geometry to PDE when he was studying in Russia by the end of 1950s. A mechanics class in the Department of Mathematics, Fudan University was formed in 1958 as soon as he came back in China. Except for systematically planning fundamental curriculums and editing mechanics textbooks, he himself gave two most important courses on "High Speed Aerodynamics" and "PDE of Mixed Type" and supervised a few seminars on wing theory, etc., which I as a student earnestly attended. Hundreds of professional students were trained during the 50 years to meet the needs of different industrial sectors for this new specialty. Therefore, Professor Gu no doubt was the founder of Fudan's mechanics. On the other hand, his research in hyperbolic and mixed type differential equations was pioneering and outstanding and consists of a significant part of his scientific achievements in PDE and mathematical physics.

Toward Chern–Simons Theory of Complexes on Calabi–Yau Threefolds

Jun Li

Shanghai Center of Mathematical Science
and Department of Mathematics, Stanford University
jli@math.stanford.edu

Professor Gu was an internationally renowned mathematician. I first learned of his name when I was a junior in high school. In the next several years, I studied at Fudan University, attended his lectures and his courses. During these years, he had paid special attention to our undergraduate and graduate studies, and made sure that we received the best mathematical education. He had been constantly supporting my research career throughout the years.

My mathematical research is also influenced by his work. Back when I was at high school, I already learned that Prof. Gu and several of his colleagues had been working on gauge theory, since its introduction in mathematical physics. Hardly a coincidence, a significant portion of my research works are related to the gauge theory and its later development. During my visit at Fudan in late 90's, Prof. Gu and I had a lengthy discussion of his work on gauge theory and integrable system. Prof. Gu had always been at the forefront of mathematical research.

In this article, I will report on some of the recent progress in the study of gauge theory on Calabi–Yau threefolds.

1. Donaldson–Thomas Invariants and Their Categorifications

Let (Y, ω) be a Calabi–Yau threefold. In Ref. 23, R. Thomas first introduced the holomorphic Chern–Simons functional

$$CS : \mathcal{A} \longrightarrow \mathbb{C}$$

on the space $\mathcal{A}$ of semiconnections on a fixed smooth complex vector bundle E and showed that the critical locus $dCS = 0$ is the space of integrable semiconnections on E. (This will be recalled in the next section.) Let $\mathcal{G}$ be the gauge group of smooth automorphisms of E. Passing to the gauge equivalence classes and assuming

$$\mathfrak{X} \subset (dCS = 0) \subset \mathcal{B} := \mathcal{A}/\mathcal{G}$$

is an open and compact subset, then the Donaldson–Thomas (DT) invariant of $\mathfrak{X}$ is the localized Euler class

$$e(T^\vee \mathcal{B}, dCS)_{\mathfrak{X}},$$

using the Fredholm theory of the section dCS.

In general, let $(Y, \mathcal{O}(1), \omega)$ be a projective Calabi–Yau threefold, and let $\mathfrak{X}$ be the moduli of stable sheaves of given rank and Chern classes c, which we fix throughout this paper. In the same paper, Thomas proved that the moduli space $\mathfrak{X}$ admits a perfect obstruction theory, thus applying Refs. 2 and 17, he obtained the virtual cycle $[\mathfrak{X}]^{\text{virt}} \in A_0\mathfrak{X}$, which is of dimension zero because of Y is Calabi-Yau. He then defined the DT invariant

$$DT(c) = \deg[\mathfrak{X}]^{\text{virt}},$$

when $\mathfrak{X}$ is projective. The DT invariants are invariant under deformations of Y.

Later, Behrend[1] discovered that the DT invariant $DT(\mathfrak{X})$ is motivic in that there is a constructible function $\nu_{\mathfrak{X}} : \mathfrak{X} \to \mathbb{Z}$, called the Behrend function, such that $DT(\mathfrak{X})$ is the Euler number of $\mathfrak{X}$ weighted by $\nu_{\mathfrak{X}}$:

$$DT(\mathfrak{X}) = \sum_m m \cdot e_c(\nu_{\mathfrak{X}}^{-1}(m)).$$

The Behrend function satisfies the following property: If an (analytic) open neighborhood U of $x \in \mathfrak{X}$ is isomorphic as complex spaces (complex analytic schemes) to the critical locus

$$\mathfrak{X}_f := (df = 0) \subset V$$

of a holomorphic function $f : V \to \mathbb{C}$ on a complex manifold V, then

$$\nu_{\mathfrak{X}}(x) = (-1)^{\dim V - 1}(1 - e(F)),$$

where F is the Milnor fiber of f. For example, if $f(x_1, \cdots, x_n) = x_1^2 + \cdots + x_n^2$ on $\mathbb{C}^n$, the Milnor fiber at 0 is homotopically equivalent to S^{n-1}, and $\nu_{\mathfrak{X}}(0) = 1$.

In Ref. 10, Joyce and Song generalized DT invariants to the cases when the moduli space has semistable sheaves. The generalized DT invariants are defined as the Euler number of $\mathfrak{X}$ weighted by the Behrend function and some combinatorial coefficients. The generalized invariants are deformation invariant and satisfy wall crossing formula.

In their study of generalized DT invariants, Joyce and Song proved that for every $x \in \mathfrak{X}$, there is a finite dimensional complex manifold V and a holomorphic function $f : V \to \mathbb{C}$ such that the critical locus $df = 0$ is analytically equivalent to an open neighborhood of x in $\mathfrak{X}$. In fact, V is chosen as a complex submanifold of the space $\mathcal{A}$ of semiconnections on a hermitian vector bundle E and f is the restriction of the holomorphic Chern-Simons (CS for short) functional from Ref. 23. We call such (V, F) with an open embedding $(df = 0) \to \mathfrak{X}$ a CS-chart of $\mathfrak{X}$.

In Ref. 14 , Kontsevich and Soibelman introduced another generalization of DT theory which contains the Joyce–Song theory modulo some conjectures. Their invariants take values in a generalized form of the Grothendieck ring of varieties. The key feature is that the Behrend function should be replaced by the motivic Milnor fiber and one should take motives instead of Euler numbers. They also establish wall crossing formulas for the motivic DT invariants for change of stability conditions in the given Calabi–Yau category of dimension three.

The DT invariants were originally defined as integers and for projective moduli spaces of stable sheaves. Later they were generalized to objects in the derived category and to motivic invariants. The progress in this way nowadays is called *categorification*.

"There is a natural sequence of classes of mathematical objects, of increasing complexity: integers, vector spaces, categories, 2-categories, ... Moving leftwards in this sequence is easy, and loses information: to a vector space we associate its dimension, and so on. Categorification roughly means moving rightwards in this sequence in some problem, adding information, turning integer invariants into vector spaces, etc.".[19]

Given a holomorphic function f on a complex manifold V, one has the perverse sheaf $\phi_f(\mathbb{Q}[\dim V - 1])$ of vanishing cycles supported on the critical locus of f and the Euler number of this perverse sheaf at a point x equals $\nu_{\mathfrak{X}_f}(x)$. This motivated Joyce and Song to raise the question: "*Let $\mathfrak{X}$ be the moduli space of simple coherent sheaves on Y. Does there exist a natural perverse sheaf $P^\bullet$ on the underlying analytic variety $X = \mathfrak{X}_{red}$ which is locally isomorphic to the sheaf $\phi_f(\mathbb{Q}[\dim V - 1])$ of vanishing cycles for (f, V) CS-charts of $\mathfrak{X}$?*" ([10, Question 5.7]).

In Ref. 12, an affirmative answer is obtained.[a]

[a] In Ref. 12, the authors sometimes need to pass to a finite étale cover to bypass a technical requirement for the existence of such perverse sheaf. Via an argument of Okunkov, such technical condition is always satisfied for quasi-projective moduli space.

Theorem 1. *Let $\mathfrak{X}$ be a quasi-projective moduli space of simple sheaves on a smooth projective Calabi–Yau 3-fold Y with universal family $\mathcal{E}$ and let $X = \mathfrak{X}_{red}$ be the reduced scheme of $\mathfrak{X}$. Then there exists a perverse sheaf $P^\bullet$ on X which is locally isomorphic to the perverse sheaf $\phi_f(\mathbb{Q}[\dim V - 1])$ of vanishing cycles for the CS-charts of $\mathfrak{X}$.*

Indeed, the same for mixed Hodge modules is proved; namely, there is a polarized mixed Hodge module $M^\bullet$ on X whose underlying perverse sheaf is $rat(M^\bullet) = P^\bullet$.

As an application of Theorem 1, the hypercohomology $\mathbb{H}^i(X, P^\bullet)$ of $P^\bullet$ gives us the *DT (Laurent) polynomial*

$$DT_t^Y(X) = \sum_i t^i \dim \mathbb{H}^i(X, P^\bullet)$$

such that $DT_{-1}^Y(X)$ is the ordinary DT invariant by Ref. 1. Following the work of Hosono–Saito–Takahashi,[8] the authors also introduced a perverse sheaf analogue of Gopakumar–Vafa invariants[7] in Ref. 12.

Our proof of the categorification theorem is via gauge theory. Recently, Joyce *et al.* have proved the existence of perverse and motivic perverse sheaves for a more general class of spaces; for precise statements see Refs. 4, 5 and 9. It will be interesting to generalized the gauge theory construction to cover the case of moduli of derived objects. For this, the first task is to generalize the notion of Chern–Simons functional to the space of semiconnections on complexes of vector bundles and construct CS-charts.

In 2008, Behrend lectured on his joint work with Getzler in finding the Chern–Simons functional on the space of "semiconnections" of a complex of vector bundles, and constructing CS-charts for the moduli space of complexes on a Calabi–Yau threefold.[3]

In this note, we will first review the Chern–Simons functional of complexes of Behrend–Getzler. We will also construct explicitly CS-charts for the complexes that are resolutions of sheaves. As our construction shows, it is extendable to cover a large class of gluable complexes. Sections 2–4 are based on the note of Ref. 3. The construction in Sections 4–5 is our own work. Though the result proved in this paper is covered by what was announced in Ref. 3, as the details of the construction of CS-charts in Ref. 3 is not available at present, this article will provide a useful reference for further development of this subject.

The author thanks Y-H. Kiem for helping preparing some background materials and for useful discussion. This work was partially supported by the grant NSF-1104553 and DMS-1159156.

2. Chern–Simons Functional and CS Charts

In this subsection, we recall the Chern–Simons (CS) functional on the space of connections over a Calabi–Yau (CY) threefold.[23] Its generalization to the space of complexes will be discussed in the next section.

Let $(Y, \mathcal{O}(1), \omega)$ be a smooth projective Calabi–Yau 3-fold over $\mathbb{C}$, where ω is a nowhere vanishing holomorphic $(3,0)$-form on Y. Let E be a smooth complex vector bundle on Y. In this paper, a smooth semiconnection on E is a differential operator $\bar\partial : \Omega^0(E) \to \Omega^{0,1}(E)$ satisfying the $\overline{\partial}$-Leibniz rule. Here $\Omega^{0,k}(E)$ is the space of smooth $(0,k)$-forms on Y taking values in E. Following the notation in gauge theory, we denote $ad\, E = E^\vee \otimes E$. Thus fixing a smooth semiconnection $\overline{\partial}_0$, the space $\mathcal{A}$ of smooth semiconnections on E is an affine space, which is isomorphic to $\Omega^{0,1}(adE)$ after a choice of smooth $\overline{\partial}_0 \in \mathcal{A}$, via

$$\overline{\partial}_0 + \cdot : \Omega^{0,1}(ad\, E) \longrightarrow \mathcal{A}, \quad \mathfrak{a} \mapsto \overline{\partial}_0 + \mathfrak{a}. \tag{2.1}$$

We let $\mathcal{G}$ be the gauge group consisting of smooth sections of $\mathrm{Aut}(E)$ modulo the subgroup $\mathbb{C}^* = \{c \cdot \mathrm{id} \mid c \in \mathbb{C}^*\}$. It acts on $\mathcal{A}$ via the gauge-transformation

$$g \cdot (\overline{\partial}_0 + \mathfrak{a}) = (g^{-1})^*(\overline{\partial}_0 + \mathfrak{a}) = g \circ (\overline{\partial}_0 + A) \circ g^{-1}. \tag{2.2}$$

The quotient

$$\mathcal{B} = \mathcal{A}/\mathcal{G}$$

is the space of gauge-equivalent classes of semiconnections.

An element $\bar\partial \in \mathcal{A}$ is integrable if its curvature $F(\bar\partial) := (\bar\partial)^2$ vanishes. Note that if $\bar\partial$ is integrable and $\mathfrak{a} \in \Omega^{0,1}(ad\, E)$, then $F(\bar\partial + \mathfrak{a}) = \bar\partial\mathfrak{a} + \mathfrak{a} \wedge \mathfrak{a}$. An integrable smooth semiconnection $\overline{\partial}$ defines a holomorphic vector bundle $(E, \overline{\partial})$.

We recall the CS functional on the CY threefold Y. Picking a (reference) integrable $\bar\partial \in \mathcal{A}$, the CS functional is

$$CS : \mathcal{A} \to \mathbb{C}, \quad CS(\bar\partial + \mathfrak{a}) = \frac{1}{4\pi^2} \int_Y \mathrm{tr}\left(\frac{1}{2}(\bar\partial\mathfrak{a}) \wedge \mathfrak{a} + \frac{1}{3}\mathfrak{a} \wedge \mathfrak{a} \wedge \mathfrak{a}\right) \wedge \omega.$$

This is a cubic polynomial in $\mathfrak{a}$ whose quadratic part is

$$CS_2(\bar\partial + \mathfrak{a}) = \frac{1}{4\pi^2} \int_Y \mathrm{tr}\big(\frac{1}{2}(\bar\partial\mathfrak{a}) \wedge \mathfrak{a}\big) \wedge \omega.$$

Since the directional derivative of CS at $\overline{\partial} + \mathfrak{a}$ in the direction of $\mathfrak{b}$ is

$$d\,CS(\overline{\partial} + \mathfrak{a})(\mathfrak{b}) = \frac{1}{4\pi^2} \int_Y \mathrm{tr}(\mathfrak{b} \wedge F_{\overline{\partial}+\mathfrak{a}}) \wedge \omega,$$

$d\,CS(\overline{\partial}+\mathfrak{a})=0$ if and only if $F(\overline{\partial}_{\mathfrak{a}})=0$, which is when $\overline{\partial}+\mathfrak{a}$ is integrable. Thus the subspace $\mathcal{A}^{int}$ of integrable smooth semiconnections is the critical locus of CS:

$$\mathcal{A}^{int}=(d\,CS=0)\subset\mathcal{A}.$$

Let $\mathcal{B}^{int}=\mathcal{A}^{int}/\mathcal{G}\subset\mathcal{B}$. Since CS is $\mathcal{G}$-equivariant, it descends to

$$cs:\mathcal{B}\to\mathbb{C},$$

whose critical locus is $\mathcal{B}^{int}$.

To argue that the critical loci of the CS functional defines the complex analytic structure of the space of gauge-equivalence classes of integral semiconnections, we can work with the completions of $\mathcal{A}$ and $\mathcal{G}$ under appropriate Sobolev norms. Following Ref. 10 (See also Ref. 12), we fix a pair of integers $s\geq 4$ and $\ell>6$, and form the completion $\mathcal{A}_s$ (resp. $\mathcal{G}_s$) of $\mathcal{A}$ (resp. $\mathcal{G}$) under the Sobolev norm L_s^ℓ. (L_s^ℓ is the sum of L^ℓ-norms of up to s-th partial derivatives.) Both $\mathcal{A}_s$ and $\mathcal{G}_{s+1}$ are smooth complex Bamach manifolds, and $\mathcal{G}_{s+1}$ acts on $\mathcal{A}_s$, via the gauge-transformations (2.2), holomorphically. By Sobolev embedding theorem, the functional CS extends to a $\mathcal{G}_{s+1}$-invariant homomorphic CS-functional

$$CS:\mathcal{A}_s\longrightarrow\mathbb{C}.$$

As Y is projective, one can define the space of equivalence classes of complex structures $(E,\overline{\partial})$ algebraically. For S any (finite type) scheme over $\mathbb{C}$, we define $\mathcal{M}_E(S)$ be the collection of all holomorphic vector bundles $\mathcal{E}$ over $Y\times S$ such that for a smooth line bundle L on S, $\mathcal{E}\cong_{C^\infty}\mathrm{pr}_Y^*E\otimes\mathrm{pr}_S^*L$ as C^∞-vector bundles. For a morphism $S\to S'$, the pullback defines a transformation $\mathcal{M}_E(S')\to\mathcal{M}_E(S)$. For $\mathcal{E}$ and $\mathcal{E}'\in\mathcal{M}_E(S)$, we define a arrow $\sigma:\mathcal{E}\to\mathcal{E}'$ be an isomorphism of vector bundles. This way, $\mathcal{M}_E$ is an Artin stack, called the moduli stack of complex structures on E.

Alternatively, by replacing S with complex analytic spaces, we obtain a similarly defined analytic Artin stack, which we denote by $\mathcal{M}_E^{an}$. By GAGA, $\mathcal{M}_E^{an}$ is the underlying complex space (analytic scheme) of $\mathcal{M}_E$.

Proposition 2.[10] Let $\xi\in\mathcal{A}^{int}$ be any point. Let $\xi\in\mathcal{V}\subset\mathcal{A}_s$ be a complex Banach submanifold so that $T_\xi\mathcal{V}\oplus T_\xi(\mathcal{G}_{s+1}\cdot\xi)=T_\xi\mathcal{A}_s$ and $T_\xi\mathcal{V}\cap T_\xi(\mathcal{G}_{s+1}\cdot\xi)$ is finite dimensional. Then there is an open neighborhood $\xi\in\mathcal{V}'\subset\mathcal{V}$ such that $\mathcal{U}:=\mathcal{V}'\cap(d(CS|_{\mathcal{V}'})=0)$ is a finite dimensional complex analytic space, and the family of semiconnections on $\mathcal{A}_s$ induces a smooth morphism

$$\mathcal{U}\longrightarrow\mathcal{M}_E^{an}.$$

Further, all such $\mathcal{U} \to \mathcal{M}_E^{an}$ form an atlas of $\mathcal{M}_E^{an}$.

To utilize the tool of vanishing cycle operator of a holomorphic function on a finite dimensional complex manifold, it is desirable to replace $\mathcal{V}$ by a finite dimensional complex submanifold. In the following, we adopt the convention that for $f : V \to \mathbb{C}$ a holomorphic function, we denote by $\mathfrak{X}_f = (df = 0) \subset V$ the complex analytic space defined by the vanishing of df. Given a morphism $\mathfrak{X}_f \to \mathcal{M}_E^{an}$, we say it is a smooth chart if the morphism is smooth.

Definition 3. Let $\xi \in \mathcal{M}_E^{an}$ be a closed point. A *CS chart* for $\xi \in \mathcal{M}_E^{an}$ is a pair $(V, f = cs \circ \imath)$, where $V \subset \mathcal{A}_s$ is a (finite) dimensional complex submanifold and $\imath : V \to \mathcal{A}_s$ is the inclusion, such that the canonical morphism $\mathfrak{X}_f \to \mathcal{M}_E^{an}$ is smooth and its image contains ξ.

We have the following existence theorem.

Theorem 4.[10] *The stack $\mathcal{M}_E^{an}$ has a smooth atlas by finite dimensional CS charts.*

Corollary 5. *Let $\mathcal{X} \subset \mathcal{M}_E^{an}$ be an open substack so that every $\xi \in \mathcal{X}$ has* $\mathrm{Aut}(\xi) = \mathbb{C}^*$. *Let X be the coarse moduli scheme of $\mathcal{X}$. Then we can find an atlas of X consisting of CS charts.*

3. Gauge theory of Complexes of Vector Bundles

In this section, we introduce the parallel theory that will lead us to the Chern–Simons functional on the space of semiconnections on a complex of vector bundles on Y.

We begin with a fixed complex of locally free sheaves of $\mathcal{O}_Y$-modules

$$\mathcal{E}^\bullet = [\mathcal{E}^{k_0} \to \mathcal{E}^{k_0+1} \to \cdots \to \mathcal{E}^0]. \tag{3.1}$$

As usual, we will not distinguish a locally free sheaf from its associated holomorphic vector bundle. We denote by E^k the underlying smooth (C^∞) vector bundle of $\mathcal{E}^k$; denote by $\bar\partial_k$ the semiconnection on E^k that gives the holomorphic structure of $\mathcal{E}^k$, and denote by $b_k : \mathcal{E}^k \to \mathcal{E}^{k+1}$ the holomorphic map in the complex $\mathcal{E}^\bullet$, which is a smooth section of $\mathcal{E}^{k\vee} \otimes \mathcal{E}^{k+1}$.

We introduce

$$\Omega^{0,j}(E^\bullet) = \bigoplus_{k=k_0}^{0} \Omega^{0,j}(E^k) \quad \text{and} \quad \Omega^{0,\bullet}(E^\bullet) = \bigoplus_{j=0}^{3} \Omega^{0,j}(E^\bullet).$$

Here $\Omega^{0,j}(E^k)$ is the space of smooth $(0,j)$-forms taking values in the vector bundle E^k. We introduce

$$\mathcal{S}^{j;k,l}(E^\bullet)=\Omega^{0,j}(E^k\otimes E^{l\vee}) \quad \text{and} \quad \mathcal{S}^i(E^\bullet)=\bigoplus_{j+k-l=i}\mathcal{S}^{j;k,l}(E^\bullet).$$

By assigning elements $\varphi\in\mathcal{S}^i$ with grading $[\varphi]=i$,

$$\mathcal{S}^\bullet(E^\bullet)=\bigoplus_i \mathcal{S}^i(E^\bullet)$$

becomes a graded vector space.

We introduce a graded operation $\odot$:

$$(\alpha\otimes s)\odot(\beta\otimes t)=(-1)^{[s][\beta]}(\alpha\wedge\beta)\otimes(s\circ t)$$

for $\alpha\otimes s$ and $\beta\otimes t\in\mathcal{S}^\bullet(E^\bullet)$ so that both s and β have pure gradings. (We adopt the convention that φ will have pure grading whenever $[\varphi]$ appears in the formula.) In case $\alpha\otimes s\in\mathcal{S}^\bullet(E^\bullet)$ while $\beta\otimes t\in\Omega^{0,\bullet}(E^\bullet)$, then the same formula holds with $s\circ t$ replaced by $s\cdot t$, with the understanding that the pairing

$$\Omega^0(E^k\otimes E^{l\vee})\times\Omega^0(E^m)\longrightarrow\Omega^0(E^k)$$

is the application of homomorphisms when $l=m$, and is trivial otherwise. We define the bracket

$$[x,y]=x\odot y-(-1)^{[x][y]}y\odot x, \quad x,y\in\mathcal{S}^\bullet(E^\bullet).$$

We next introduce the space of semiconnections of $E^\bullet$. First, the semiconnections $\bar\partial_k$ extends to $\mathbb{C}$-linear maps

$$\bar\partial:\Omega^{0,i}(E^\bullet)\longrightarrow\Omega^{0,i+1}(E^\bullet)$$

satisfying the $\bar\partial$-Leibnize rule. (We adopt the convention that for (z_i) a local holomorphic coordinates of Y, $\bar\partial v=d\bar z_i\wedge\frac{\partial}{\partial\bar z_i}v$.) Also, for $x\in\mathcal{S}^1(E^\bullet)$, we agree $(\bar\partial+x)s=\bar\partial s+x\odot s$.

Definition 6. A semiconnection of $E^\bullet$ is an operation of the form

$$\bar\partial_x:=\bar\partial+x:\Omega^0(E^\bullet)\longrightarrow\Omega^{0,1}(E^\bullet), \quad x\in\mathcal{S}^1(E^\bullet).$$

The curvature of the semiconnection $\bar\partial_x$ is the composite

$$(\bar\partial_x)^2=\bar\partial_x\circ\bar\partial_x:\Omega^0(E^\bullet)\longrightarrow\Omega^{0,2}(E^\bullet).$$

It is direct to check that semiconnections satisfy the $\overline{\partial}$-Leibnize rule:

$$\overline{\partial}_x(f\varphi) = f\overline{\partial}_x\varphi + \overline{\partial}f \wedge \varphi.$$

Indeed, it holds for $\overline{\partial}_x = \overline{\partial}$ (i.e. $x = 0$); the general case follows immediately.

Proposition 7 (Curvature forms and the Bianchi identity). *For a semiconnection $\overline{\partial}_x$ of $E^\bullet$, there is a $F(\overline{\partial}_x) \in \mathcal{S}^2(E^\bullet)$, called the curvature form of $\overline{\partial}_x$, so that $(\overline{\partial}_x)^2 = F(\overline{\partial}_x)$, and satisfies the Bianchi identity $\overline{\partial}_x F(\overline{\partial}_x) = 0$.*

Proof. First, we comment that for a semiconnection $\overline{\partial}_x$, the relations

$$\overline{\partial}_x(\varphi \cdot v) = \overline{\partial}_x(\varphi) \odot v + (-1)^{[\varphi]}\varphi \odot \overline{\partial}_x(v), \quad \varphi \in \mathcal{S}^\bullet(E^\bullet),\ v \in \Omega^{0,\bullet}(E^\bullet)$$

define a graded (degree one) operation

$$\overline{\partial}_x : \mathcal{S}^k(E^\bullet) \longrightarrow \mathcal{S}^{k+1}(E^\bullet)$$

satisfying the $\overline{\partial}$-Leibnize rule. More explicitly,

$$\overline{\partial}_x\varphi = \overline{\partial}\varphi + x \odot \varphi - (-1)^{[\varphi]}\varphi \odot x = \overline{\partial}\varphi + [x, \varphi]. \tag{3.2}$$

Thus the Bianchi identity takes the form

$$\overline{\partial}_x F(\overline{\partial}_x) = \overline{\partial}F(\overline{\partial}_x) + [x, F(\overline{\partial}_x)] = 0. \tag{3.3}$$

We now calculate the curvature of a semiconnection $\overline{\partial}_x$:

$$(\overline{\partial}_x)^2(s) = \overline{\partial}_x(\overline{\partial}s + x\odot s) = (\overline{\partial}x)\odot s - x\odot\overline{\partial}s + x\odot\overline{\partial}s + x\odot x\odot s = (\overline{\partial}x + x\odot x)\odot s.$$

This proves that for

$$F(\overline{\partial}_x) = \overline{\partial}x + x \odot x \in \mathcal{S}^2(E^\bullet),$$

$(\overline{\partial}_x)^2 s = F(\overline{\partial}_x) \odot s$ for all $s \in \Omega^0(E^\bullet)$.

We now prove the Bianchi identity. For $s \in \Omega^0(E^\bullet)$, we have

$$F(\overline{\partial}_x) \odot \overline{\partial}_x s = (\overline{\partial}_x)^3 s = \overline{\partial}_x\big(F(\overline{\partial}_x) \odot s\big) = \overline{\partial}_x F(\overline{\partial}_x) \odot s + F(\overline{\partial}_x) \odot \overline{\partial}_x s.$$

Thus the Bianchi identity follows. □

We fix an (integrable) semiconnection $\overline{\partial} = \oplus\overline{\partial}_k$ throughout the remainder of this paper.

We denote by $\mathcal{A}$ the space of semiconnections on $E^\bullet$:

$$\mathcal{A} = \{\overline{\partial}_x = \overline{\partial} + x \mid x \in \mathcal{S}^1(E^\bullet)\}.$$

This space is independent of the choice of the reference $\overline{\partial}$, and becomes an infinite dimensional complex Banach manifold, after a completion under a choice of Sobolev norm.

4. Chern–Simons Functional on the Space of Semiconnections

We define the (super)-trace

$$\mathrm{tr}_s : \mathcal{S}^\bullet(E^\bullet) \longrightarrow \Omega^{0,\bullet}, \quad \mathrm{tr}_s(\varphi) = \sum_k (-1)^k \,\mathrm{tr}\big(\varphi|_{E^k}\big).$$

It has the usual commuting property

$$\mathrm{tr}_s(\varphi \odot \psi) = (-1)^{[\varphi][\psi]}\,\mathrm{tr}_s(\psi \odot \varphi). \tag{4.1}$$

Following Behrend-Getzler,[3] we define the Chern–Simons functional $CS : \mathcal{A} \to \mathbb{C}$ via

$$CS(x) = \frac{1}{2}\int_Y \mathrm{tr}_s(\overline{\partial} x \odot x) \wedge \omega + \frac{1}{3}\int_Y \mathrm{tr}_s(x \odot x \odot x) \wedge \omega.$$

Clearly, it is a cubic polynomial on $\mathcal{A}$, thus holomorphic on $\mathcal{A}$.

We calculate its first order variation:

$$\delta\, CS(x)(\dot{x}) = \frac{1}{2}\int_Y \mathrm{tr}_s(\overline{\partial}\dot{x} \odot x + \overline{\partial} x \odot \dot{x}) \wedge \omega + \int_Y \mathrm{tr}_s(x \odot x \odot \dot{x}) \wedge \omega.$$

Here we have used the commuting property of tr_s to obtain the last term. Because $\overline{\partial}\dot{x} \odot x - \dot{x} \odot \overline{\partial} x = \overline{\partial}(\dot{x} \odot x)$, the first integral is

$$\frac{1}{2}\int_Y \mathrm{tr}_s(\overline{\partial}(\dot{x} \odot x) + \dot{x} \odot \overline{\partial} x + \overline{\partial} x \odot x) \wedge \omega = \int_Y \mathrm{tr}_s(\overline{\partial} x \odot \dot{x}) \wedge \omega.$$

Therefore,

$$\delta\, CS(x)(\dot{x}) = \int_Y \mathrm{tr}_s\big((\overline{\partial} x + x \odot x) \odot \dot{x}\big) \wedge \omega = \int_Y \mathrm{tr}_s(F(\overline{\partial}_x) \odot \dot{x}) \wedge \omega. \tag{4.2}$$

Proposition 8. The critical locus of the Chern–Simons functional CS is

$$\mathcal{A}^{int} := (dCS = 0) = (F(\overline{\partial}_x) = 0) \subset \mathcal{A}.$$

We have two observations on this space. First, the space $\mathcal{S}^j(E^\bullet)$ has a direct decomposition

$$\mathcal{S}^j(E^\bullet) = \bigoplus_{i=0}^{3} \mathcal{S}^j_i(E^\bullet), \quad \mathcal{S}^j_i(E^\bullet) = \bigoplus_k \Omega^{0,i}(E^{k-i+j} \otimes E^{k\vee}).$$

We let $\mathcal{S}^j_-(E^\bullet) = \oplus_{j\ge i}\mathcal{S}^j_i(E^\bullet)$ and $\mathcal{S}^j_+(E^\bullet) = \oplus_{j<i}\mathcal{S}^j_i(E^\bullet)$. For $x \in \mathcal{S}^1(E^\bullet)$, we often write $x_i = \pi_i(x) \in \mathcal{S}^{1;i}(E^\bullet)$ be the corresponding components of x in $\mathcal{S}^1_i(E^\bullet)$. Given $x = x_0 + x_1 \in (F(\overline{\partial}_x) = 0) \cap \mathcal{S}^1_-(E^\bullet)$, we calculate

$$F(\overline{\partial}_x) = (\overline{\partial}_{x_1})^2 + \overline{\partial}_{x_1} x_0 + x_0 \odot x_0.$$

Thus $F(\overline{\partial}_x) = 0$ implies that all three terms are zero. Let $x_{1,k}$ be the component of x_1 in $\Omega^0(E^k \otimes E^{k\vee})$; then $(\overline{\partial}_{x_1})^2 = 0$ implies that

$$\mathcal{E}^k_{x_{1,k}} = (E^k, \overline{\partial}_k + x_{1,k})$$

are holomorphic vector bundles. Let $x_{0,k}$ be the component of x_0 in $\Omega^0(E^{k+1} \otimes E^{k\vee})$; then $\overline{\partial}_{x_1} x_0 = 0$ implies that $x_{0,k} : \mathcal{E}^k_{x_{1,k}} \to \mathcal{E}^{k+1}_{x_{1,k+1}}$ are holomorphic. Finally, $x_0 \odot x_0 = 0$ implies that $x_{0,k+1} \circ x_{0,k} = 0$ for all k. Thus

$$\mathcal{E}^\bullet_x = [\mathcal{E}^{k_0}_{x_{1,k_0}} \xrightarrow{x_{0,k_0}} \cdots \xrightarrow{x_{0,-2}} \mathcal{E}^{-1}_{x_{1,-1}} \xrightarrow{x_{0,-1}} \mathcal{E}^0_{x_{1,0}}].$$

is a complex of locally free sheaves. We have the following result of Behrend-Getzler.

Proposition 9. The space of complexes of locally free sheaves $\mathcal{F}^\bullet = [\mathcal{F}^{k_0} \to \cdots \to \mathcal{F}^0]$ together with smooth isomorphisms $\eta^k : F^k \cong E^k$, $k_0 \le k \le 0$ is parameterized by the space

$$\mathcal{A}^{int}_- := (F(\overline{\partial}_x) = 0) \cap \mathcal{S}^1_- \subset \mathcal{A}.$$

5. Sheaves and their Complexes of Resolutions

In this section, we fix a choice of Chern classes $c = (\mathrm{rk}, c_1, c_2, c_3)$ and form the moduli of Gieseker stable sheaves of rank rk and Chern classes c_i. We denote by $\mathcal{M}(c)^{\mathrm{st}}$ such a moduli space. By Refs. 6 and 18, $\mathcal{M}(c)^{\mathrm{st}}$ is quasi-projective. For simplicity, we pick c so that sheaves in $\mathcal{M}(c)^{\mathrm{st}}$ have positive (pure) dimensions.

We pick a reference $[\mathcal{E}] \in \mathcal{M}(c)^{\mathrm{st}}$ and form its standard resolutions as follows. We pick a sufficiently large n_0 and form

$$[b_0 : \mathcal{E}^0 \to \mathcal{E}] := [H^0(\mathcal{E}(n_0)) \otimes \mathcal{O}_Y(-n_0) \to \mathcal{E}],$$

where the arrow b_0 is induced by the tautological evaluation, and b_0 is surjective since n_0 is large. (Here $\mathcal{E}(n) = \mathcal{E} \otimes_{\mathcal{O}_Y} \mathcal{O}_Y(n)$.) We then pick a sufficiently large $n_{-1} > n_0$ and form

$$[b_{-1} : \mathcal{E}^{-1} \to \mathcal{E}^0] := [H^0(\ker(b_0)(n_{-1})) \otimes \mathcal{O}_Y(-n_{-1}) \to \mathcal{E}^0],$$

where b_{-1} is the composition of the evaluation with the inclusion $\ker(b_0) \to \mathcal{E}^0$. We let $\mathcal{E}^{-2} = \ker(b_{-1})$ and let $b_{-2} : \mathcal{E}^{-2} \to \mathcal{E}^{-1}$ be the inclusion. This implies

$$\mathcal{E}^\bullet = [\mathcal{E}^{-2} \xrightarrow{b_{-2}} \mathcal{E}^{-1} \xrightarrow{b_{-1}} \mathcal{E}^0] \tag{5.1}$$

is a locally free resolution of $\mathcal{E}$ because $n_{-1} \gg n_0 \gg 1$, Y is a smooth threefold and $\mathcal{E}$ is a sheaf of pure positive dimension.

We let $\mathcal{S}^\bullet(E^\bullet)$, $\mathcal{A}$ and $\mathcal{A}_-^{int} \subset \mathcal{A}^{int}$ be the spaces as constructed in the previous section with $\mathcal{E}^\bullet$ in (3.1) the complex $\mathcal{E}^\bullet$ in (5.1). We let $\mathcal{A}^{st} \subset \mathcal{A}_-^{int}$ be the open subspace of all $[\mathcal{E}_x^\bullet] \in \mathcal{A}_-^{int}$ so that $h^0(\mathcal{E}_x^\bullet)$ are stable. Then the correspondence $[\mathcal{E}_x^\bullet] \mapsto h^0(\mathcal{E}_x^\bullet)$ defines a surjective holomorphic map

$$\mathcal{A}^{st} \longrightarrow \mathcal{M}(c)^{\mathrm{st}}. \tag{5.2}$$

Indeed, it is the quotient by the gauge group $\mathcal{G}_-$ which we define now. We let $\mathcal{G}_k = \mathrm{Aut}(E^k)$ be the group of smooth (fiberwise) automorphisms of E^k. For $g = (g_i) \in \mathcal{G} = \mathcal{G}_{-2} \times \mathcal{G}_{-1} \times \mathcal{G}_0$ and $x = x_0 + x_1 \in \mathcal{A}_-$, we define $x^g \in \mathcal{A}_-$ be such that

$$\begin{array}{ccccc}
(E^{-2}, \bar\partial_{x_{1,-2}}) & \xrightarrow{x_{0,-2}} & (E^{-1}, \bar\partial_{x_{1,-1}}) & \xrightarrow{x_{0,-1}} & (E^0, \bar\partial_{x_{1,0}}) \\
\Big\downarrow{\scriptstyle g_{-2}} & & \Big\downarrow{\scriptstyle g_{-1}} & & \Big\downarrow{\scriptstyle g_0} \\
(E^{-2}, \bar\partial_{(x^g)_{1,-2}}) & \xrightarrow{(x^g)_{0,-2}} & (E^{-1}, \bar\partial_{(x^g)^{-1}_{1,-1}}) & \xrightarrow{(x^g)_{0,-1}} & (E^0, \bar\partial_{(x^g)_{1,0}})
\end{array}$$

is a commutative diagram of smooth vector bundles and preserves the semi-connections. Explicitly,

$$\bar\partial_{x_{1,k}} + (x^g)_{1,k} = g_k \circ \bar\partial_{x_{1,k}} \circ g_k^{-1} \quad \text{and} \quad (x^g)_{0,k} = g_{k+1} \circ x_{0,k} \circ g_k^{-1}.$$

We let $\mathbb{C}^* \to \mathcal{G}$ be $c \mapsto (c \cdot \mathrm{id}_{E^k})$. The images of $\mathbb{C}^*$ act trivially on $\mathcal{A}_-$, thus $\mathcal{G}_- = \mathcal{G}/\mathbb{C}^*$ acts on $\mathcal{A}_-$. Repeating the argument in Ref. 20, $\mathcal{G}_-$ acts holomorphically on $\mathcal{A}_-$, (5.2) is $\mathcal{G}_-$-equivariant and is a quotient map.

To make precise of this statement, we introduce the notion of complex Banach manifolds. As we will fix $E^\bullet$ in the remainder of this paper, we will drop the reference $E^\bullet$ from $\mathcal{S}^\bullet(E^\bullet)$, and use $\mathcal{S}^{j;k,l}$ to denote $\mathcal{S}^{j;k,l}(E^\bullet)$, etc. We fix smooth hermitian metrics h_k on E^k; we fix a pair of integers $s \geq 4$ and $\ell > 6$, and form the completion $\mathcal{S}_s^{j;k,l}$ of $\mathcal{S}^{j;k,l}$ under the Sobolev norm L_s^ℓ, and define $\mathcal{S}_s^\bullet$ accordingly. (Here L_s^ℓ is the sum of L^ℓ-norms of up to s-th partial derivatives, and our convention is to use subscript s to denote the L_s^ℓ-completions of the corresponding spaces. For instance, $\mathcal{G}_s$ is the L_s^ℓ-completion of $\mathcal{G}$.) This way, $\mathcal{G}_{-,s+1} = \mathcal{G}_{-,s+1}/\mathbb{C}^*$ is a complex Banach group, $\mathcal{S}_{-,s}^1$ is a complex Banach manifold, and $\mathcal{G}_{-,s+1}$ acts holomorphically on $\mathcal{S}_{-,s}^1$.

Because the Chern–Simons functional $CS : \mathcal{A} \to \mathbb{C}$ is a cubic polynomial, it extends to $CS : \mathcal{A}_s \cong \mathcal{S}_s^1 \to \mathbb{C}$, and its restriction to $\mathcal{A}_{-,s} \cong \mathcal{S}_{-,s}^1$ is invariant under $\mathcal{G}_{-,s+1}$. Let $\mathcal{A}_{-,s}^{si} \subset \mathcal{A}_{-,s} \cong \mathcal{S}_{-,s}^1$ be the open subset of $x \in \mathcal{S}^1_{-,s}$ so that $\mathrm{stab}_{\mathcal{G}_{-,s+1}}(x) = \{1\}$. Thus $\mathcal{B}^{si}_{-,s} := \mathcal{A}^{si}_{-,s}/\mathcal{G}_{-,s+1}$ is a complex Banach manifold and CS descends to a holomorphic

$$cs_- : \mathcal{B}^{si}_{-,s} \longrightarrow \mathbb{C}.$$

Proposition 10. We define $\mathcal{B}^{st} = \mathcal{A}^{st}_s/\mathcal{G}_{-,s+1}$,[b] which is an open subspace of $(d\,cs_- = 0) \subset \mathcal{B}^{si}_s$. The map (5.2) induces an isomorphism of complex spaces $\mathcal{B}^{st}_s \cong \mathcal{M}(c)^{st}$.

Proof. The proof is a repetition of the argument in Ref. 20, combined with the fact that for $x \in \mathcal{A}^{st}_{-,s}$, $\mathrm{Hom}(\mathcal{E}^\bullet_x, \mathcal{E}^\bullet_x) = \mathrm{Hom}(h^0(\mathcal{E}^\bullet_x), h^0(\mathcal{E}^\bullet_x)) = \mathbb{C}$. □

Using this description, one gets a local chart of $\mathcal{M}(c)^{\mathrm{st}}$ near $[\mathcal{E}]$ immediately. For $b = (b_{-2}, b_{-1}, b_0) \in \mathcal{S}^1_{-,s}$ of the complex (5.1), it forms a complex

$$\mathcal{S}^0_{-,s+1} \xrightarrow{\overline{\partial}_b} \mathcal{S}^1_{-,s} \xrightarrow{\overline{\partial}_b} \mathcal{S}^2_{-,s-1}.$$

We form their formal adjoints, using the hermitian metrics h_k and the fixed Kahler metric on Y:

$$\mathcal{S}^2_{-,s+1} \xrightarrow{\overline{\partial}^{*-}_b} \mathcal{S}^1_{-,s} \xrightarrow{\overline{\partial}^{*-}_b} \mathcal{S}^0_{-,s-1}.$$

We pick an $\epsilon > 0$, and form

$$V^\epsilon_- := \{x \in \mathcal{S}^1_{-,s} \mid \overline{\partial}^{*-}_b x = \overline{\partial}^{*-}_b F(\overline{\partial}_x) = 0,\ \| x \|_s < \epsilon\}.$$

Proposition 11. For sufficiently small $\epsilon > 0$, V^ϵ_- is a smooth finite dimensional complex manifold, and the induced map

$$U^\epsilon := (F(\overline{\partial}_x) = 0) \cap V^\epsilon_- \longrightarrow \mathcal{M}(c)^{\mathrm{st}}$$

is an open embedding of complex spaces whose image contains $[\mathcal{E}] \in \mathcal{M}(c)^{\mathrm{st}}$.

Proof. The proof is parallel to that of Ref. 10, using that $\mathrm{Hom}(\mathcal{E}^\bullet, \mathcal{E}^\bullet) = \mathbb{C}$. □

As can be seen easily, $(F(\overline{\partial}_x) = 0) \cap V^\epsilon_-$ is different from the critical locus of $CS|_{V^\epsilon_-}$. To obtain a CS-chart of $[\mathcal{E}]$, we will use a balanced version of V^ϵ_-. We let $\overline{\partial}^*_b : \mathcal{S}^j_s \to \mathcal{S}^{j-1}_{s-1}$ be the formal adjoint of $\overline{\partial}_b : \mathcal{S}^{j-1}_{s+1} \to \mathcal{S}^j_s$.

Definition-Proposition 12. For sufficiently small $\epsilon > 0$,

$$V^\epsilon = \{x \in \mathcal{S}^1_s \mid \overline{\partial}^*_b x = \overline{\partial}^*_b F(\overline{\partial}_x) = 0,\ \| x \|_s < \epsilon\}$$

[b] We drop the subscript s from $\mathcal{B}^{st}$ since it is independent of the choice of s.

is a smooth finite dimensional complex manifold. We let $f^\epsilon = CS|_{V^\epsilon} : V^\epsilon \to \mathbb{C}$, and define

$$\mathfrak{X}_{f^\epsilon} := (df^\epsilon = 0) = (F(\overline{\partial}_x) = 0) \cap V^\epsilon.$$

In the next section, we will show that for sufficiently small $\epsilon \ll \epsilon'$, U^ϵ is an open subspace of $\mathfrak{X}_{f^{\epsilon'}}$. This way, U^ϵ admits a CS-chart.

We close this section by describing the tangent space of U^ϵ and $\mathfrak{X}_{f^\epsilon}$ at $0(=\overline{\partial}_b)$. First, because $\mathcal{E}^\bullet$ is a standard resolution of $\mathcal{E}$, it is easy to check that

$$T_{\overline{\partial}_b} U^\epsilon = \operatorname{Ext}^1_{K(Y)}(\mathcal{E}^\bullet, \mathcal{E}^\bullet). \tag{5.3}$$

By the construction of $\mathfrak{X}_{f^\epsilon}$, letting $\mathcal{H}^1(E^\bullet, \overline{\partial}_b) = \ker\{\Box_{\overline{\partial}_b} : \mathcal{S}^1_s \to \mathcal{S}^1_s\}$, we have

$$T_{\overline{\partial}_b} \mathfrak{X}_{f^\epsilon} = \mathcal{H}^1(E^\bullet, \overline{\partial}_b). \tag{5.4}$$

By Hodge theory, $\mathcal{H}^1(E^\bullet, \overline{\partial}_b)$ is canonically isomorphic to the H^1 of the complex

$$\cdots \longrightarrow \mathcal{S}^0 \xrightarrow{\overline{\partial}_b} \mathcal{S}^1 \xrightarrow{\overline{\partial}_b} \mathcal{S}^2 \longrightarrow \cdots,$$

which we denote by $H^1(\mathcal{S}^\bullet, \overline{\partial}_b)$. By Corollary 14, we have the canonical isomorphism $\operatorname{Ext}^1_{K(Y)}(\mathcal{E}^\bullet, \mathcal{E}^\bullet) \cong H^1(\mathcal{S}^\bullet, \overline{\partial}_b)$. Thus we have canonical isomorphisms

$$\mathcal{H}^1(E^\bullet, \overline{\partial}_b) \cong H^1(\mathcal{S}^\bullet, \overline{\partial}_b) \cong \operatorname{Ext}^1_{K(Y)}(\mathcal{E}^\bullet, \mathcal{E}^\bullet). \tag{5.5}$$

6. CS-charts via Resolution Complexes

We first prove a vanishing result.

Lemma 13. *Let $\mathcal{E}^\bullet$ be the complex* (5.1). *Then*
(1). $\operatorname{Ext}^{i<3}(\mathcal{E}^k, \mathcal{E}^{k-j}) = 0$, $j \geq 1$;
(2). $b_{-2} : \operatorname{Ext}^3(\mathcal{E}^0, \mathcal{E}^{-2}) \to \operatorname{Ext}^3(\mathcal{E}^0, \mathcal{E}^{-1})$ is injective.

Proof. Since Y is a Calabi–Yau threefold, $\mathcal{E}^k = \mathcal{O}_Y(-n_k)^{\oplus m_k}$ for $k = 0, -1$, and $n_{-1} \gg n_0 \gg 1$, we have $\operatorname{Ext}^2(\mathcal{E}^k, \mathcal{E}^{k-1}) = 0$ for $k - j \neq -2$. Thus to prove (1) we need to verify that $\operatorname{Ext}^{i<3}(\mathcal{E}^k, \mathcal{E}^{-2}) = 0$ for $k = 0, -1$.

We let $\mathcal{F} = \ker\{\mathcal{E}^0 \to \mathcal{E}\}$, where $\mathcal{E} = h^0(\mathcal{E}^\bullet)$. It fits into the exact sequences

$$0 \to \mathcal{E}^{-2} \to \mathcal{E}^{-1} \to \mathcal{F} \to 0 \quad \text{and} \quad 0 \to \mathcal{F} \to \mathcal{E}^0 \to \mathcal{E} \to 0. \tag{6.1}$$

Tensoring the first with $\mathcal{E}^{-1\vee}$, and taking cohomology groups, we obtain the long exact sequence

$$\to \mathrm{Ext}^{i-1}(\mathcal{E}^{-1},\mathcal{F}) \to \mathrm{Ext}^{i}(\mathcal{E}^{-1},\mathcal{E}^{-2}) \to \mathrm{Ext}^{i}(\mathcal{E}^{-1},\mathcal{E}^{-1}) \to \mathrm{Ext}^{i}(\mathcal{E}^{-1},\mathcal{F}) \to .$$

In the construction of the standard resolution of $\mathcal{E}$, we can choose n_{-1} large so that $\mathrm{Ext}^{i>0}(\mathcal{E}^{-1},\mathcal{F})=0$. Combined with $H^{i>0}(\mathcal{E}^{-1\vee}\otimes\mathcal{E}^{-1})=0$, and that $\mathrm{Hom}(\mathcal{E}^{-1},\mathcal{E}^{-1}) \to \mathrm{Hom}(\mathcal{E}^{-1},\mathcal{F})$ is an isomorphism, we obtain $\mathrm{Ext}^{i}(\mathcal{E}^{-1},\mathcal{E}^{-2})=0$. The proof of $\mathrm{Ext}^{i<3}(\mathcal{E}^{0},\mathcal{E}^{-2})=0$ is similar, and will be omitted. This proves (1).

We now prove (2). Using the exact sequences (6.1), we obtain the long exact sequences

$$0=\mathrm{Ext}^{2}(\mathcal{E}^{0},\mathcal{F}) \longrightarrow \mathrm{Ext}^{3}(\mathcal{E}^{0},\mathcal{E}^{-2}) \longrightarrow \mathrm{Ext}^{3}(\mathcal{E}^{0},\mathcal{E}^{-1}) \longrightarrow$$

and

$$0=\mathrm{Ext}^{1}(\mathcal{E}^{0},\mathcal{E}) \longrightarrow \mathrm{Ext}^{2}(\mathcal{E}^{0},\mathcal{F}) \longrightarrow \mathrm{Ext}^{2}(\mathcal{E}^{0},\mathcal{E}^{0})=0.$$

Here we used that $n_0 \gg 1$. This proves (2). □

We let $P_+ : \mathcal{S}^i \to \mathcal{S}^i_+$ be the projection via the direct sum decomposition $\mathcal{S}^i=\mathcal{S}^i_-\oplus\mathcal{S}^i_+$.

Corollary 14. *Let $\alpha\in\mathcal{S}^1$ such that $P_+(\overline{\partial}_b\alpha)=0$, then there is a $\beta\in\mathcal{S}^0_+$ so that $P_+(\alpha+\overline{\partial}_b\beta)=0$.*

Proof. Let $\alpha\in\mathcal{S}^1$ be such that $P_+(\overline{\partial}_b\alpha)=0$. Decomposing $\alpha=\alpha_0+\cdots+\alpha_3$, where $\alpha_i=\pi_i(\alpha)$, we have $\overline{\partial}\alpha_3=0$ and $b\odot\alpha_3+\overline{\partial}\alpha_2=0$. Adding Lemma 13, we have $[\alpha_3]=0\in\mathrm{Ext}^3(\mathcal{E}^0,\mathcal{E}^{-2})$. Thus there is an $\beta_{-2}\in\Omega^{0,2}(E^0\otimes E^{-2\vee})$ such that $(\alpha+\overline{\partial}_b\beta_2)_3=0\in\mathcal{S}^1_3$. Repeating the same argument, we see that (1) of Lemma 13 implies that there is a $\beta_1\in\mathcal{S}^0_1$ such that $P_+(\alpha+\overline{\partial}_b\beta_1+\overline{\partial}_b\beta_2)=0$. This proves the Corollary. □

In the following, for Z a complex space, we let $z\in Z$ be a general point and for a holomorphic $\xi: Z\to\mathcal{S}^1$, we use $\xi(z)$ to emphasize its dependence on z. We prove an integrability Lemma.

Lemma 15. *Let $0\in Z$ be a pointed complex space and let $\mathfrak{m}\subset\mathcal{O}_Z$ be the ideal sheaf of $0\in Z$. Let $\xi: Z\to\mathcal{S}^1$ be a holomorphic map so that $\xi(0)=0$ and $(\overline{\partial}_{b+\xi})^2=0$ on Z. Suppose $P_+(\xi)\equiv 0 \bmod \mathfrak{m}^k$ for some integer $k\ge 1$. Then there is a holomorphic $\delta: Z\to\mathfrak{m}^k\cdot\mathcal{S}^1$ such that $P_+(\xi+\delta)\equiv 0 \bmod \mathfrak{m}^{k+1}$ and $(\overline{\partial}_{b+\xi}+\delta)^2=0$ on Z.*

Proof. Because $P_+(\xi) \equiv 0 \mod \mathfrak{m}^k$, by Corollary 14, possibly after shrinking $0 \in Z$, we can find $\eta : Z \to \mathfrak{m}^k \cdot \mathcal{S}^1_+$ such that

$$P_+(\xi(z) + \overline{\partial}_{b+\xi}\eta) \equiv 0 \mod \mathfrak{m}^{k+1}.$$

We then let

$$\delta = \overline{\partial}_{b+\xi}\eta + \frac{1}{2}[\overline{\partial}_{b+\xi}\eta, \eta] + \eta \odot \overline{\partial}_{b+\xi}\eta \odot \eta.$$

Since $\delta \equiv \overline{\partial}_{b+\xi}\eta \mod \mathfrak{m}^{k+1}$, we have $P_+(\xi + \delta) \equiv 0 \mod \mathfrak{m}^{k+1}$.

Next, using $(\overline{\partial}_{b+\xi})^2 = 0$, after simplification, we obtain

$$(\overline{\partial}_{b+\xi} + \delta)^2 = -\frac{1}{4}\big(\overline{\partial}_{b+\xi}\eta \odot \eta \odot \overline{\partial}_{b+\xi}\eta \odot \eta + \overline{\partial}_{b+\xi}\eta \odot \eta \odot \eta \odot \overline{\partial}_{b+\xi}\eta$$
$$+\eta \odot \overline{\partial}_{b+\xi}\eta \odot \overline{\partial}_{b+\xi}\eta \odot \eta + \eta \odot \overline{\partial}_{b+\xi}\eta \odot \eta \odot \overline{\partial}_{b+\xi}\eta\big).$$

Here we have used that the images of η lie in the union of $\Omega^{0,j}(E^{k+1} \otimes E^{k\vee})$ while $\overline{\partial}_{b+\xi}\eta$ lie in the union of $\Omega^{0,j}(E^k \otimes E^{k\vee})$. Therefore, because $\mathcal{E}^k = 0$ for $k < -2$, any monomial of η and $\overline{\partial}_{b+\xi}\eta$ that contains at least three η must be trivial.

But the right hand side of the above identity is

$$\overline{\partial}_{b+\xi}\big(-\overline{\partial}_{b+\xi}\eta\odot\eta\odot\eta\odot\eta - \eta\odot\overline{\partial}_{b+\xi}\eta\odot\eta\odot\eta + \eta\odot\eta\odot\overline{\partial}_{b+\xi}\eta\odot\eta + \eta\odot\eta\odot\eta\odot\overline{\partial}_{b+\xi}\eta\big),$$

which is trivial by the comment above. This proves that $(\overline{\partial}_{b+\xi} + \delta)^2 = 0$ on Z. □

Lemma 16. *Let $0 \in Z$ be a complex space, and let $\xi : Z \to \mathcal{S}^1$ be a holomorphic map such that $\xi(0) = 0$ and $(\overline{\partial}_{b+\xi})^2 = 0$ on Z. Let $\mathcal{I} \subset \mathcal{O}_{Z\times\mathcal{G}_-\mathcal{S}^0_+}$ be the ideal sheaf of $Z \times \mathcal{G}_- \times 0 \subset Z \times \mathcal{G}_- \times \mathcal{S}^0_+$. Then there is a holomoprhic $\delta : Z \times \mathcal{G}_- \times \mathcal{S}^0_+ \to \mathcal{S}^1$ such that*

$$\delta(z, g, a) - \overline{\partial}_{(b+\xi(z))^g} a \in \mathcal{I}^2 \cdot \mathcal{S}^1 \quad \text{and} \quad (\overline{\partial}_{(b+\xi(z))^g} + \delta(z, g, a))^2 = 0.$$

Proof. We define

$$\delta(z, g, a) = \overline{\partial}_{(b+\xi(z))^g} a + \frac{1}{2}[\overline{\partial}_{(b+\xi(z))^g} a, a] + a \odot \overline{\partial}_{(b+\xi(z))^g} a \odot a.$$

Then $\delta(z, g, a) - \overline{\partial}_{(b+\xi(z))^g} a \in \mathcal{I}^2 \cdot \mathcal{S}^1$, and the proof of the previous Lemma shows that $(\overline{\partial}_{(b+\xi(z))^g} + \delta(z, g, a))^2 = 0$. This proves the Lemma. □

Proposition 17. Let U^ϵ and $\mathfrak{X}_{f^\epsilon}$ be as defined in Proposition 11 and 12. Then there is an $\epsilon' > 0$ so that there is holomorphic map of complex spaces $\rho : U^{\epsilon'} \to \mathfrak{X}_{f^\epsilon}$ such that $\rho(0) = 0$, such that $d\rho(0) = \text{id}$ under the isomorphisms (5.3), (5.4) and (5.5).

Proof. We apply Lemma 11 to $Z = U^\epsilon$, $0 \in U^\epsilon$ and $\xi = \mathrm{id} : U^\epsilon \to \mathcal{S}^1$ to obtain a homolorphism map $\delta : U^\epsilon \times \mathcal{G}_- \times \mathcal{S}^0_+ \to \mathcal{S}^1$. We let $\tilde{\delta}$ be defined by

$$\overline{\partial}_{(b+\xi(z))^g} + \delta(z, g, a) = \overline{\partial}_b + \tilde{\delta}(z, g, a).$$

Thus $(\overline{\partial}_b + \tilde{\delta}(z, g, a))^2 = 0$ on $U^\epsilon \times \mathcal{G}_- \times \mathcal{S}^0_+$.

Next, using $T_{[1]}\mathcal{G}_- = \mathcal{S}^0_-/\mathbb{C}$ and that $H^0(\mathcal{S}^\bullet, \overline{\partial}_b) \cong \mathrm{Ext}^1_{K(Y)}(\mathcal{E}^\bullet, \mathcal{E}^\bullet) \cong \mathbb{C}$, one sees that $d\tilde{\delta}(\overline{\partial}, 1, 0) : 0 \times (\mathcal{S}^0_-/\mathbb{C}) \times \mathcal{S}^0_+ \to \mathcal{S}^1$ is injective and the image

$$d\tilde{\delta}(\overline{\partial}, 1, 0)(0 \times (\mathcal{S}^0_-/\mathbb{C}) \times \mathcal{S}^0_+) \overset{\subset}{\longrightarrow} \mathcal{S}^1$$

is transversal to $\{x \in \mathcal{S}^1 \mid \overline{\partial}^*_b x = 0\}$. Thus applying the implicit function theorem, we can find a sufficiently small $\epsilon' > 0$ such that there is a holomorphic

$$\zeta : U^{\epsilon'} \to \mathcal{G}_- \times \mathcal{S}^0_+, \quad \zeta(0) = 0,$$

such that $(\tilde{\delta} \circ \zeta)(U^{\epsilon'}) \subset \{x \in \mathcal{S}^1 \mid \overline{\partial}^*_b x = 0\}$.

Because $(\overline{\partial}_b + \tilde{\delta}(z, g, a))^2 = 0$, we have $(\overline{\partial}_b + \tilde{\delta} \circ \zeta(z))^2 = 0$ on $U^{\epsilon'}$. Thus $\tilde{\delta} \circ \zeta : U^{\epsilon'} \to \mathcal{S}^1$ factors through a holomorphic $\rho_1 : U^{\epsilon'} \to \mathfrak{X}_{f^\epsilon}$.

Finally, we calculate the differential $d\rho_1(0)$. By our construction,

$$\tilde{\delta}(z, g, a) = g \circ \xi(z) \circ g^{-1} - \overline{\partial}_b g + \delta(z, g, a).$$

Let $v \in T_0 U^{\epsilon'}$. Because $d\xi(0) = \mathrm{id} : T_0 U^{\epsilon'} \to T_0 \mathcal{S}^1$, we see that

$$d(\tilde{\delta} \circ \zeta)(0)(v) = v + \overline{\partial}_b\big(d(g \circ \zeta)(0)(v)\big) + d(\delta \circ \zeta)(0)(v) = v + \overline{\partial}_b c + \overline{\partial}_b c',$$

for $c \in T_{[\mathrm{id}]}\mathcal{G}_- = \mathcal{S}^0_-/\mathbb{C}$ and $c' \in \mathcal{S}^0_+$. This proves that $d\rho_1(0) : T_0 U^{\epsilon'} \to T\mathfrak{X}_{f^\epsilon}$ is the identity map under the isomorphisms (5.3), (5.4) and (5.5). □

Proposition 18. Let U^ϵ and $\mathfrak{X}_{f^\epsilon}$ be as defined in Proposition 11 and 12. Then there are $\epsilon' > 0$ so that there is holomorphic map of complex spaces $\rho_2 : \mathfrak{X}_{f^\epsilon} \to U^{\epsilon'}$ such that $\rho(\overline{\partial}_b) = \overline{\partial}_b$, and that $d\rho(0) = \mathrm{id}$ under the isomorphisms (5.3), (5.4) and (5.5).

Proof. The proof is a direct application of Lemma 5.3, plus a standard convergence argument. We will omit the details. □

Theorem 19. *Let U^ϵ and $\mathfrak{X}_{f^\epsilon}$ be as defined in Proposition 11 and 12. Then there are $\epsilon' < \epsilon$ so that there is a open embedding of complex spaces $\rho : U^{\epsilon'} \to \mathfrak{X}_{f^\epsilon}$ such that $\rho(0) = 0$, and $d\rho(0) = \mathrm{id}$ under the isomorphisms (5.3), (5.4) and (5.5).*

Proof. For this we only need to show that the holomorphic map ρ constructed in Proposition 17 is an open embedding. Applying Propositions 17 and 18, we can find positive $\epsilon_1 < \epsilon_2 < \epsilon$ so that we have holomorphic maps $\rho : U^{\epsilon_1} \to \mathfrak{X}_{f^{\epsilon_2}}$ and $\rho' : \mathfrak{X}_{f^{\epsilon_2}} \to U^{\epsilon}$ so that $\rho(0) = \rho'(0) = 0$, and $d\rho(0) = d\rho'(0) = \mathrm{id}$. We consider the composite $\rho' \circ \rho : U^{\epsilon_1} \to U^{\epsilon}$. By that both $d\rho(0)$ and $d\rho'(0)$ are identity maps under the isomorphisms (5.3), (5.4) and (5.5), we conclude that $d(\rho' \circ \rho)(0) = \mathrm{id}$. Thus possibly after shrinking ϵ_1 if necessary, $\rho' \circ \rho : U^{\epsilon_1} \to U^{\epsilon}$ is an open embedding. Consequently, $\rho : U^{\epsilon_1} \to \mathfrak{X}_{f^{\epsilon_2}}$ is an open embedding too. This proves the Theorem. □

This proves the existence of CS-charts of the space of small deformations of the complex $\mathcal{E}^\bullet$.

References

1. K. Behrend, Donaldson-Thomas type invariants via microlocal geometry, Ann. of Math. (2) **170** (2009), no. 3, 1307C-1338.
2. K. Behrend and B. Fantechi, The intrinsic normal cone, Invent. Math. **128** (1997), no. 1, 45–88.
3. K. Behrend and E. Getzler, On the holomorphic Chern-Simons functional, Seminar talk, Western Algebraic Geometry Seminar, 2008.
4. C. Brav, V. Bussi, D. Dupont, D. Joyce and B. Szendroi, Symmetries and stabilization for sheaves of vanishing cycles, Preprint, arXiv:1211.3259.
5. V. Bussi, D. Joyce and S. Meinhardt, On motivic vanishing cycles of critical loci, preprint, arXiv:1305.6428.
6. D. Gieseker, A construction of stable bundles on an algebraic surface, J. Differential Geom, **27** (1988), no. 1, 137-C154.
7. R. Gopakumar and C. Vafa, M-Theory and Topological Strings II, ArXiv: 9812127.
8. S. Hosono, M.-H. Saito and A. Takahashi, Relative Lefschetz action and BPS state counting, Internat. Math. Res. Notices (2001), no. 15, 783-816.
9. D. Joyce, A classical model for derived critical loci, preprint, arXiv:1304.4508.
10. D. Joyce and Y. Song, A theory of generalized Donaldson-Thomas invariants, Mem. Amer. Math. Soc. **217** (2012), no. 1020.
11. M. Kashiwara and P. Schapira, Sheaves on manifolds, Grundlehren der Mathematischen Wissenschaften **292**, Springer, 1994.
12. Y-H. Kiem and J. Li, Categorification of Donaldson-Thomas invariants, preprint, arXiv:1212.6444.
13. K. Kodaira, Complex manifolds and deformation of complex structures, Classics in Mathematics. Springer-Verlag, Berlin, 2005.
14. M. Kontsevich and Y. Soibelman, Stability structures, motivic Donaldson-Thomas invariants and cluster transformations, Preprint, arXiv:0811.2435.
15. M. Kontsevich and Y. Soibelman, Motivic Donaldson-Thomas invariants: summary of results, Mirror symmetry and tropical geometry, 55C-89, Contemp. Math., **527**, Amer. Math. Soc., Providence, RI, 2010.

16. M. Kontsevich and Y. Soibelman, Cohomological Hall algebra, exponential Hodge structures and motivic Donaldson-Thomas invariants, Commun. Number Theory Phys. 5 (2011), no. 2, 231C-352.
17. J. Li and G. Tian, Virtual moduli cycles and Gromov-Witten invariants of algebraic varieties, J. Amer. Math. Soc. **11** (1998), no. 1, 119–174.
18. M. Maruyama, Moduli of stable sheaves. II, J. Math. Kyoto Univ. **18** (1978), no. 3, 557C-614.
19. MICA homepage, www.maths.ox.ac.uk/groups/geometry-and-representation-theory/motivic-invariants-and-categorification.
20. K. Miyajima, Kuranishi family of vector bundles and algebraic description of EinsteinCHermitian connections, Publ. RIMS, Kyoto Univ. **25** (1989), 301-C320.
21. R. Narasimhan, Introduction to the theory of analytic spaces, Lecture Notes in Mathematics **25**, Springer-Verlag 1966.
22. M. Saito, On the formalism of mixed sheaves, ArXiv: math/0611597.
23. R. Thomas, A holomorphic Casson invariant for Calabi-Yau 3-folds, and bundles on K3 fibrations, J. Differential Geom. **54** (2000), no. 2, 367-C438.

Exact Boundary Synchronization for a Coupled System of Wave Equations†

Tatsien Li

School of Mathematical Sciences, Fudan University, Shanghai 200433, China; Shanghai Key Laboratory for Contemporary Applied Mathematics; Nonlinear Mathematical Modeling and Methods Laboratory

dqli@fudan.edu.cn

Several kinds of exact synchronizations are introduced for a coupled system of wave equations and these kinds of exact synchronizations can be realized by means of boundary controls.

1. Introduction

Synchronization is a widespread natural phenomenon. Thousands of fireflies may twinkle at the same time; audiences in the theater can applaud with a rhythmic beat; pacemaker cells of the heart function simultaneously; and field crickets give out a unanimous cry — all these are phenomena of synchronization.

In principle, synchronization happens when different individuals possess likeness in nature, that is, they conform essentially to the same governing equation, and meanwhile, the individuals should bear a certain coupled relation.

The phenomenon of synchronization was first observed by Huygens in 1665 [6]. The theoretical research on synchronization phenomena dates back to Fujisaka and Yamada's study of synchronization for coupled equations in 1983 [2]. The previous studies focused on systems described by ODEs such as

$$\frac{\mathrm{d}X_i}{\mathrm{d}t} = f(X_i, t) + \sum_{j=1}^{N} A_{ij} X_j \qquad (i = 1, \cdots, N), \tag{1}$$

†Project supported by the National Basic Research Program of China (No. 2013CB834100) and the National Natural Science Foundation of China (No. 11121101).

where X_i $(i = 1, \cdots, N)$ is an n-dimensional vector, A_{ij} $(i, j = 1, \cdots, N)$ are $n \times n$ matrices, and $f(X, t)$ is an n-dimensional vector function independent of i. The right-hand side of (1) shows that every $X_i (i = 1, \cdots, N)$ possesses two basic features: satisfying a fundamental governing equation and bearing a coupled relation among one another.

Our goal is to synchronize the hyperbolic system through boundary controls. Different from the ODE situation, the coupling of systems can then be fulfilled by coupling of the equations or (and) the boundary conditions, which has richer research implications, moreover, boundary controls can be used to realize our goal in a finite time.

Roughly speaking, we want to find a $T > 0$ and some boundary controls on $[0, T]$, such that from the time $t = T$ on, the system states tend to the same. That is to say, we hope to achieve the synchronization of the system states not only at the moment $t = T$ under the action of boundary controls on $[0, T]$, but also when $t \geq T$ and withdrawing all the controls. This is forever, instead of short-lived, synchronization, as is desired in many actual applications.

Obviously, if the system has the exact boundary null controllability (see [7–8], [13–15]), it must have the exact boundary synchronization, but this is a trivial situation that should be excluded beforehand. The exact boundary null controllability here also means that the system states remain null not only at the moment $t = T$ under the action of boundary controls on $[0, T]$, but also when $t \geq T$ and withdrawing all the controls. However, to exclude in advance the possibility of exact boundary null controllability, we should face the situation of the lack of boundary controls to some extent that frequently occurs in applications.

In this paper, for fixing the idea, we consider only a coupled system of 1-D wave equations in the framework of classical solutions. Similar discussions can be done for a coupled system of higher dimensional wave equations in the framework of weak solutions [9–10]. The results presented here are recently obtained in collaboration with Bopeng Rao, Long Hu and Yimin Wei ([5], [11–12]).

Precisely speaking, we consider the following coupled system of wave equations

$$\frac{\partial^2 U}{\partial t^2} - \frac{\partial^2 U}{\partial x^2} + AU = 0, \tag{2}$$

where

$$U = (u_1, \cdots, u_N)^T \tag{3}$$

is an unknown vector function of (t,x), $A=(a_{ij})$ is a $N\times N$ matrix with constant elements. (2) can be written as

$$\frac{\partial^2 u_i}{\partial t^2}-\frac{\partial^2 u_i}{\partial x^2}+\sum_{j=1}^{N}a_{ij}u_j=0\quad (i=1,\cdots,N). \tag{4}$$

At the end $x=0$, we prescribe anyone of the following boundary conditions:

$$x=0:\ U=H(t)\qquad \text{(Dirichlet type)}, \tag{5a}$$
$$x=0:\ U_x=H(t)\qquad \text{(Newman type)}, \tag{5b}$$
$$x=0:\ U_x-BU=H(t)\qquad \text{(Coupled third type)}, \tag{5c}$$
$$x=0:\ U_x-CU_t=H(t)\qquad \text{(Coupled dissipative type)}, \tag{5d}$$

where $B=(b_{ij})$ and $C=(c_{ij})$ are $N\times N$ matrices with constant elements, and

$$H(t)=(h_1(t),\cdots,h_n(t))^T, \tag{6}$$

the components of which will be totally or partially taken as boundary controls.

Similarly, at the end $x=L$, since no boundary controls are concerned, for one-sided controls we prescribe anyone of the following homogeneous boundary conditions:

$$x=L:\ U=0, \tag{7a}$$
$$x=L:\ U_x=0, \tag{7b}$$
$$x=L:\ U_x+\overline{B}U=0, \tag{7c}$$
$$x=L:\ U_x+\overline{C}U_t=0, \tag{7d}$$

where $\overline{B}=(\bar{b}_{ij})$ and $\overline{C}=(\bar{c}_{ij})$ are $N\times N$ matrices with constant elements.

The initial condition is given by

$$t=0:\ U=\Phi(x),\quad U_t=\Psi(x), \tag{8}$$

where $\Phi(x)=(\varphi_1(x),\cdots,\varphi_N(x))^T$ and $\Psi(x)=(\psi_1(x),\cdots,\phi_N(x))^T$.

For the forward mixed initial-boundary value problem (2), (5), (7) and (8) of the coupled system of wave equations, by means of its reduction to the mixed initial-boundary value problem of a corresponding first order hyperbolic system and the theory of global C^1 solutions for first order hyperbolic systems, we have ([4], [7])

Lemma 1.1 For any given $T > 0$, assume that $\Phi \in (C^2[0,L])^N$, $\Psi \in (C^1[0,L])^N$ and $H \in (C^2[0,T])^N$ (case (5a)) or $(C^1[0,T])^N$ (cases (5b)-(5d)) satisfy the conditions of C^2 compatibility at the points $(t,x) = (0,0)$ and $(0,L)$, respectively. For any given matrices A, B (or C) and $\overline{B}$ (or $\overline{C}$), if

$$\det(C+I) \neq 0 \tag{9}$$

and

$$\det(\overline{C}+I) \neq 0, \tag{10}$$

namely, if -1 is not an eigenvalue of both C and $\overline{C}$, then the mixed initial-boundary value problem (2), (5), (7) and (8) admits a unique C^2 solution $U = U(t,x)$ on the domain $R(T) = \{(t,x)\,|\,0 \leq t \leq T,\ 0 \leq x \leq L\}$.

By Lemma 1.1 and using a constructive method, we can get the following result on the exact boundary null controllability (see [4], [7–8]).

Lemma 1.2 Let

$$T > 2L. \tag{11}$$

For any given initial state $(\Phi, \Psi) \in (C^2[0,L])^N \times (C^1[0,L])^N$, satisfying the conditions of C^2 compatibility at the point $(t,x) = (0,L)$, there exist boundary controls $H \in (C^2[0,T])^N$ (case (5a)) or $(C^1[0,T])^N$ (cases (5b)-(5d)) with support on $[0,T]$ at the end $x = 0$, such that the corresponding mixed initial-boundary value problem (2), (5), (7) and (8) admits a unique C^2 solution $U = U(t,x)$ on $t \geq 0$, $0 \leq x \leq L$, and as $t \geq T$ we have

$$U(t,x) \equiv 0, \quad 0 \leq x \leq L. \tag{12}$$

In what follows, we will use Lemma 1.1 and Lemma 1.2 to consider various kinds of exact boundary synchronization for the coupled system of wave equations (2), (5) and (7).

2. Exact Boundary Synchronization

Definition 2.1 If there exists $T > 0$ such that for any given initial state (Φ, Ψ), we can find some boundary controls with support on $[0,T]$ in $H(t)$ (the precise number of these boundary controls is equal to $N-1$, see the sequel), such that the corresponding mixed initial-boundary value problem (2), (5), (7) and (8) admits a unique C^2 solution

$U = U(t,x) = (u_1(t,x), \cdots, u_N(t,x))^T$ on $t \geq 0$, $0 \leq x \leq L$, and as $t \geq T$ we have

$$u_1(t,x) \equiv \cdots \equiv u_N(t,x) \stackrel{\text{def.}}{=} \tilde{\tilde{u}}(t,x), \quad 0 \leq x \leq L, \tag{13}$$

then the coupled system (2), (5) and (7) is said to possess the **exact boundary synchronization** or is said to be **exactly synchronizable**, and $\tilde{\tilde{u}} = \tilde{\tilde{u}}(t,x)$ given by (13) is called the corresponding **synchronizable state**.

Obviously, if the system possesses the exact boundary null controllability, then it must be exactly synchronizable. However, this trivial situation should be excluded in the discussion.

We can prove

Lemma 2.1 If the coupled system of wave equations (2), (5) and (7) is exactly synchronizable, but not exactly null controllable, then the coupling matrices A, B (or C) and $\overline{B}$ (or $\overline{C}$) should satisfy the following necessary conditions: The values of the sum of every row of these matrices $\sum_{j=1}^{N} a_{ij}$, $\sum_{j=1}^{N} b_{ij}$ (or $\sum_{j=1}^{N} c_{ij}$) and $\sum_{j=1}^{N} \bar{b}_{ij}$ (or $\sum_{j=1}^{N} \bar{c}_{ij}$) are independent of $i = 1, \cdots, N$, namely,

$$\sum_{j=1}^{N} a_{ij} \stackrel{\text{def.}}{=} \tilde{\tilde{a}} \quad (i = 1, \cdots, N), \tag{14}$$

$$\sum_{j=1}^{N} b_{ij} \stackrel{\text{def.}}{=} \tilde{\tilde{b}} \quad \left(or \sum_{j=1}^{N} c_{ij} \stackrel{\text{def.}}{=} \tilde{\tilde{c}} \right) \quad (i = 1, \cdots, N), \tag{15}$$

$$\sum_{j=1}^{N} \bar{b}_{ij} \stackrel{\text{def.}}{=} \tilde{\tilde{\bar{b}}} \quad \left(or \sum_{j=1}^{N} \bar{c}_{ij} \stackrel{\text{def.}}{=} \tilde{\tilde{\bar{c}}} \right) \quad (i = 1, \cdots, N), \tag{16}$$

where $\tilde{\tilde{a}}$, $\tilde{\tilde{b}}$ (or $\tilde{\tilde{c}}$) and $\tilde{\tilde{\bar{b}}}$ (or $\tilde{\tilde{\bar{c}}}$) are constants independent of $i = 1, \cdots, N$.

Remark 2.1 Conditions (14)–(16) mean that all the matrices A, B (or C) and $\overline{B}$ (or $\overline{C}$) have a common right eigenvector $(1, 1, \cdots, 1)^T$ with the corresponding eigenvalues $\tilde{\tilde{a}}$, $\tilde{\tilde{b}}$ (or $\tilde{\tilde{c}}$) and $\tilde{\tilde{\bar{b}}}$ (or $\tilde{\tilde{\bar{c}}}$), respectively. By assumptions (9) and (10). We have $\tilde{\tilde{c}} \neq -1$ and $\tilde{\tilde{\bar{c}}} \neq -1$.

Theorem 2.1 Under assumptions (9) and (10), suppose that (14)–(16)

hold and

$$T > 2L. \tag{17}$$

For any given initial state $(\Phi, \Psi) \in (C^2[0,L])^N \times (C^1[0,L])^N$, satisfying the conditions of C^2 compatibility at the point $(t,x) = (0,L)$, there exist $(N-1)$ boundary controls with support on $[0,T]$ in $H(t)$ (for example, take $h_2(t), \cdots, h_N(t)$ with $h_1(t) \equiv 0$), such that the coupled system (2), (5) and (7) possesses the exact boundary synchronization.

Proof Let

$$w_i = u_i - u_{i+1} \quad (i = 1, \cdots, N-1). \tag{18}$$

It is easily shown that the original mixed initial-boundary value problem (2), (5), (7) and (8) for the variable U can be reduced to a self-closing mixed initial-boundary value problem of the same kind for the variable W as follows:

$$\frac{\partial^2 W}{\partial t^2} - \frac{\partial^2 W}{\partial x^2} + \widetilde{A}W = 0, \tag{19}$$

$$x = 0: \ W = \widetilde{H}(t), \tag{20a}$$

$$x = 0: \ W_x = \widetilde{H}(t), \tag{20b}$$

$$x = 0: \ W_x - \widetilde{B}W = \widetilde{H}(t), \tag{20c}$$

$$x = 0: \ W_x - \widetilde{C}W_t = \widetilde{H}(t), \tag{20d}$$

and

$$x = L: \ W = 0, \tag{21a}$$

$$x = L: \ W_x = 0, \tag{21b}$$

$$x = L: \ W_x + \widetilde{\overline{B}}W = 0, \tag{21c}$$

$$x = L: \ W_x + \widetilde{\overline{C}}W_t = 0, \tag{21d}$$

where $\widetilde{A}$, $\widetilde{B}$ (or $\widetilde{C}$) and $\widetilde{\overline{B}}$ (or $\widetilde{\overline{C}}$) are $(N-1) \times (N-1)$ matrices with constant elements, reduced from A, B (or C) and $\overline{B}$ (or $\overline{C}$), respectively,

$$\widetilde{H}(t) = (\widetilde{h}_1(t), \cdots, \widetilde{h}_{N-1}(t))^T \tag{22}$$

is reduced from $H(t)$.

Thus, according to Lemma 1.2, by means of $(N-1)$ boundary controls $\widetilde{H}(t)$ at the end $x = 0$, the variable W is exactly null controllable, then the variable U is exactly synchronizable. This finished the proof.

Remark 2.2 In the case of coupled dissipative boundary conditions (5d) and (7d), by assumptions (9) and (10), the corresponding coupled matrices $\widetilde{C}$ and $\widetilde{\overline{C}}$ in the reduced coupled dissipative boundary conditions for the variable W also satisfy the assumptions of the same kind, hence, Lemma 1.1 and Lemma 1.2 are always available for the reduced mixed initial-boundary value problem for the variable W.

Remark 2.3 As $t \geq T$, the synchronizable state $\widetilde{\widetilde{u}} = \widetilde{\widetilde{u}}(t,x)$ defined by (13) satisfies the following wave equation

$$\frac{\partial^2 \widetilde{\widetilde{u}}}{\partial t^2} - \frac{\partial^2 \widetilde{\widetilde{u}}}{\partial x^2} + \widetilde{\widetilde{a}}\widetilde{\widetilde{u}} = 0 \tag{23}$$

and the following boundary conditions:

$$x = 0: \ \widetilde{\widetilde{u}} = 0, \tag{24a}$$

$$x = 0: \ \widetilde{\widetilde{u}}_x = 0, \tag{24b}$$

$$x = 0: \ \widetilde{\widetilde{u}}_x - \widetilde{\widetilde{b}}\widetilde{\widetilde{u}} = 0, \tag{24c}$$

$$x = 0: \ \widetilde{\widetilde{u}}_x - \widetilde{\widetilde{c}}\widetilde{\widetilde{u}}_t = 0, \tag{24d}$$

and

$$x = L: \ \widetilde{\widetilde{u}} = 0, \tag{25a}$$

$$x = L: \ \widetilde{\widetilde{u}}_x = 0, \tag{25b}$$

$$x = L: \ \widetilde{\widetilde{u}}_x + \widetilde{\widetilde{\overline{b}}}\widetilde{\widetilde{u}} = 0, \tag{25c}$$

$$x = L: \ \widetilde{\widetilde{u}}_x + \widetilde{\widetilde{\overline{c}}}\widetilde{\widetilde{u}}_t = 0, \tag{25d}$$

where $\widetilde{\widetilde{a}}$, $\widetilde{\widetilde{b}}$ (or $\widetilde{\widetilde{c}}$) and $\widetilde{\widetilde{\overline{b}}}$ (or $\widetilde{\widetilde{\overline{c}}}$) are given by (14)–(16). Moreover, by (9) and (10) we have $\widetilde{\widetilde{c}} \neq -1$ and $\widetilde{\widetilde{\overline{c}}} \neq -1$, then Lemma 1.1 is always available for this system (23)–(25).

Hence, if we know the initial state of $\widetilde{\widetilde{u}} = \widetilde{\widetilde{u}}(t,x)$ at the moment $t = T$:

$$t = T: \ \widetilde{\widetilde{u}} = \varphi(x), \ \widetilde{\widetilde{u}}_t = \psi(x), \quad 0 \leq x \leq L, \tag{26}$$

we can completely determine the evolution of $\widetilde{\widetilde{u}} = \widetilde{\widetilde{u}}(t,x)$ with respect to t.

Remark 2.4 In order to realize the exact boundary synchronization, the $(N-1)$ boundary controls in $H(t)$ can be chosen in infinitely many ways, therefore, the initial state (φ, ψ) in (26) will be not unique. Moreover, under the additional hypotheses $\widetilde{\widetilde{c}} \neq 1$ and $\widetilde{\widetilde{\overline{c}}} \neq 1$ in the cases (24d) and

(25d), it can be shown that the attainable set of all possible initial data of synchronizable state at $t = T$ is the whole space $C^2[0, L] \times C^1[0, L]$ associated with the conditions of C^2 compatibility with the boundary conditions (24) and (25) at $(t, x) = (T, 0)$ and (T, L), respectively.

3. Exact Boundary Synchronization by Groups

The concept of exact boundary synchronization can be generalized to the exact boundary synchronization by groups. Without loss of generality, we consider here the exact boundary synchronization by 2-groups. For this purpose, setting $U = \binom{U^{(1)}}{U^{(2)}}$ with $U^{(1)} = (u_1, \cdots, u_m)^T$ and $U^{(2)} = (u_{m+1}, \cdots, u_N)^T$, we look for some boundary controls in $H(t)$, such that $U^{(1)}$ and $U^{(2)}$ are independently synchronizable.

Definition 3.1 If there exists $T > 0$ such that for any given initial state (Φ, Ψ), we can find some boundary controls with support on $[0, T]$ in $H(t)$ (the precise number of these boundary controls is equal to $(N - 2)$, see the sequel), such that the corresponding mixed initial-boundary value problem (2), (5), (7) and (8) admits a unique C^2 solution $U = U(t, x) = (u_1(t, x), \cdots, u_N(t, x))^T$ on $t \geq 0$, $0 \leq x \leq L$, and for $t \geq T$ we have

$$u_1(t, x) \equiv \cdots \equiv u_m(t, x) \overset{\text{def.}}{=} \tilde{\tilde{u}}^{(1)}(t, x), \quad 0 \leq x \leq L, \tag{27}$$

$$u_{m+1}(t, x) \equiv \cdots \equiv u_N(t, x) \overset{\text{def.}}{=} \tilde{\tilde{u}}^{(2)}(t, x), \quad 0 \leq x \leq L, \tag{28}$$

then the coupled system (2), (5) and (7) is said to possess the **exact boundary synchronization by 2-groups** or is said to be **exactly synchronizable by 2-groups**, and $\tilde{\tilde{U}} = \tilde{\tilde{U}}(t, x) = \binom{\tilde{\tilde{u}}^{(1)}(t, x)}{\tilde{\tilde{u}}^{(2)}(t, x)}$ is called the corresponding **synchronizable state by 2-groups**.

Obviously, if the coupled system with the same boundary controls possesses the exact boundary synchronization, then it must be exactly synchronizable by 2-groups. However, this trivial situation should be excluded in advance. For this purpose we need to impose suitable assumptions on the synchronizable states by 2-groups $\tilde{\tilde{u}}^{(1)}(t, x)$ and $\tilde{\tilde{u}}^{(2)}(t, x)$. We have

Lemma 3.1 Suppose that the coupled system (2), (5) and (7) is exactly synchronizable by 2-groups. Suppose furthermore that at least for an initial state (Φ, Ψ) the synchronizable states by 2-groups $\tilde{\tilde{u}}^{(1)}(t, x)$ and $\tilde{\tilde{u}}^{(2)}(t, x)$ are linearly independent. Then the coupling matrix $A = (a_{ij})$ in the system

(2) of wave equations should satisfy the following necessary conditions:

$$\sum_{j=1}^{m} a_{ij} \overset{\text{def.}}{=} \begin{cases} \widetilde{\widetilde{a}}_{11} & (i = 1, \cdots, m), \\ \widetilde{\widetilde{a}}_{21} & (i = m+1, \cdots, N), \end{cases} \tag{29}$$

$$\sum_{j=m+1}^{N} a_{ij} \overset{\text{def.}}{=} \begin{cases} \widetilde{\widetilde{a}}_{12} & (i = 1, \cdots, m), \\ \widetilde{\widetilde{a}}_{22} & (i = m+1, \cdots, N), \end{cases} \tag{30}$$

where $\widetilde{\widetilde{a}}_{11}$ and $\widetilde{\widetilde{a}}_{12}$ are independent of $i = 1, \cdots, m$, while, $\widetilde{\widetilde{a}}_{21}$ and $\widetilde{\widetilde{a}}_{22}$ are independent of $i = m+1, \cdots, N$.

Based on Lemma 3.1, for the coupling matrices B (or C) and $\overline{B}$ (or $\overline{C}$) in boundary conditions, we impose the following similar assumptions:

$$\begin{aligned} &\sum_{j=1}^{m} b_{ij} \ (\text{resp. } c_{ij}, \bar{b}_{ij} \text{ or } \bar{c}_{ij}) \\ &\quad \overset{\text{def.}}{=} \begin{cases} \widetilde{\widetilde{b}}_{11} \ (\text{resp. } \widetilde{\widetilde{c}}_{11}, \bar{\widetilde{\widetilde{b}}}_{11} \text{ or } \bar{\widetilde{\widetilde{c}}}_{11}) & (i = 1, \cdots, m), \\ \widetilde{\widetilde{b}}_{21} \ (\text{resp. } \widetilde{\widetilde{c}}_{21}, \bar{\widetilde{\widetilde{b}}}_{21} \text{ or } \bar{\widetilde{\widetilde{c}}}_{21}) & (i = m+1, \cdots, N), \end{cases} \end{aligned} \tag{31}$$

$$\begin{aligned} &\sum_{j=m+1}^{N} b_{ij} \ (\text{resp. } c_{ij}, \bar{b}_{ij} \text{ or } \bar{c}_{ij}) \\ &\quad \overset{\text{def.}}{=} \begin{cases} \widetilde{\widetilde{b}}_{12} \ (\text{resp. } \widetilde{\widetilde{c}}_{12}, \bar{\widetilde{\widetilde{b}}}_{12} \text{ or } \bar{\widetilde{\widetilde{c}}}_{12}) & (i = 1, \cdots, m), \\ \widetilde{\widetilde{b}}_{22} \ (\text{resp. } \widetilde{\widetilde{c}}_{22}, \bar{\widetilde{\widetilde{b}}}_{22} \text{ or } \bar{\widetilde{\widetilde{b}}}_{22}) & (i = m+1, \cdots, N). \end{cases} \end{aligned} \tag{32}$$

Thus, we have

Theorem 3.1 Under assumptions (9)-(10), suppose that (29)-(32) hold and $T > 0$ satisfies (17). For any given initial state (Φ, Ψ), there exist $(N-2)$ boundary controls with support on $[0, T]$ in $H(t)$ (for instance, take $h_2(t), \cdots, h_m(t), h_{m+2}(t), \cdots, h_N(t)$ with $h_1(t) \equiv h_{m+1}(t) \equiv 0$), such that the coupled system (2), (5) and (7) possesses the exact boundary synchronization by 2-groups.

Proof Let

$$\begin{cases} w_i = u_i - u_{i+1} & (i = 1, \cdots, m-1), \\ w_i = u_{i+1} - u_{i+2} & (i = m, \cdots, N-2). \end{cases} \tag{33}$$

It is easy to see that under assumptions (29)–(32), the original mixed initial-boundary value problem (2), (5), (7) and (8) for the variable U can be reduced to a self-closing mixed initial-boundary value problem of the same kind for the variable W, see (19)–(21), in which $\widetilde{A}$, $\widetilde{B}$ (or $\widetilde{C}$) and $\widetilde{\widetilde{B}}$ (or $\widetilde{\widetilde{C}}$) are $(N-2)\times(N-2)$ matrices with constant elements and $\widetilde{H}(t) = (\widetilde{h}_1(t), \cdots, \widetilde{h}_{N-2}(t))^T$. Thus, according to Lemma 1.2, by means of $(N-2)$ boundary controls $\widetilde{H}(t)$ at the end $x=0$, the variable W is exactly null controllable, then the variable U is exactly synchronizable by 2-groups. This completes the proof.

Remark 3.1 In the case of coupled dissipative boundary conditions (5d) and (7d), by assumptions (9) and (10), the corresponding coupling matrices $\widetilde{C}$ and $\widetilde{\widetilde{C}}$ in the reduced coupled dissipative boundary conditions for the variable W also satisfy the assumptions of the same kind, hence, Lemma 1.1 and Lemma 1.2 are always available for the reduced mixed initial-boundary value problem for the variable W.

Remark 3.2 In order to realize the exact boundary synchronization by 2-groups, the number of boundary controls is equal to $(N-2)$. Generally speaking, in order to realize the exact boundary synchronization by k-groups, the number of boundary controls is equal to $(N-k)$.

Remark 3.3 As $t \geq T$, the synchronizable state by 2-groups $\widetilde{\widetilde{U}} = \widetilde{\widetilde{U}}(t,x) = \begin{pmatrix} \widetilde{\widetilde{u}}^{(1)}(t,x) \\ \widetilde{\widetilde{u}}^{(2)}(t,x) \end{pmatrix}$ satisfies the following coupled system of wave equations:

$$\frac{\partial^2 \widetilde{\widetilde{U}}}{\partial t^2} - \frac{\partial^2 \widetilde{\widetilde{U}}}{\partial x^2} + \widetilde{\widetilde{A}}\widetilde{\widetilde{U}} = 0, \tag{34}$$

where

$$\widetilde{\widetilde{A}} = \begin{pmatrix} \widetilde{\widetilde{a}}_{11} & \widetilde{\widetilde{a}}_{12} \\ \widetilde{\widetilde{a}}_{21} & \widetilde{\widetilde{a}}_{22} \end{pmatrix} \tag{35}$$

and $\widetilde{\widetilde{a}}_{11}$, $\widetilde{\widetilde{a}}_{12}$, $\widetilde{\widetilde{a}}_{21}$ and $\widetilde{\widetilde{a}}_{21}$ are given by (29)–(30). Moreover, $\widetilde{\widetilde{U}} = \widetilde{\widetilde{U}}(t,x)$ satisfies the following boundary conditions:

$$x = 0: \ \widetilde{\widetilde{U}} = 0, \tag{36a}$$

$$x = 0: \ \widetilde{\widetilde{U}}_x = 0, \tag{36b}$$

$$x = 0: \ \widetilde{\widetilde{U}}_x - \widetilde{\widetilde{B}}\widetilde{\widetilde{U}} = 0, \tag{36c}$$

$$x = 0: \ \widetilde{\widetilde{U}}_x - \widetilde{\widetilde{C}}\widetilde{\widetilde{U}}_t = 0 \tag{36d}$$

and

$$x = L: \ \widetilde{\widetilde{\overline{U}}} = 0, \tag{37a}$$

$$x = L: \ \widetilde{\widetilde{\overline{U}}}_x = 0, \tag{37b}$$

$$x = L: \ \widetilde{\widetilde{\overline{U}}}_x + \overline{\widetilde{\widetilde{\overline{B}}}}\widetilde{\widetilde{\overline{U}}} = 0, \tag{37c}$$

$$x = L: \ \widetilde{\widetilde{\overline{U}}}_x + \overline{\widetilde{\widetilde{\overline{C}}}}\widetilde{\widetilde{\overline{U}}}_t = 0, \tag{37d}$$

where

$$\widetilde{\widetilde{\overline{B}}} = \begin{pmatrix} \widetilde{\widetilde{\overline{b}}}_{11} & \widetilde{\widetilde{\overline{b}}}_{12} \\ \widetilde{\widetilde{\overline{b}}}_{21} & \widetilde{\widetilde{\overline{b}}}_{22} \end{pmatrix}, \quad \widetilde{\widetilde{\overline{C}}} = \begin{pmatrix} \widetilde{\widetilde{\overline{c}}}_{11} & \widetilde{\widetilde{\overline{c}}}_{12} \\ \widetilde{\widetilde{\overline{c}}}_{21} & \widetilde{\widetilde{\overline{c}}}_{22} \end{pmatrix}, \quad \overline{\widetilde{\widetilde{\overline{B}}}} = \begin{pmatrix} \overline{\widetilde{\widetilde{\overline{b}}}}_{11} & \overline{\widetilde{\widetilde{\overline{b}}}}_{12} \\ \overline{\widetilde{\widetilde{\overline{b}}}}_{21} & \overline{\widetilde{\widetilde{\overline{b}}}}_{22} \end{pmatrix}, \quad \overline{\widetilde{\widetilde{\overline{C}}}} = \begin{pmatrix} \overline{\widetilde{\widetilde{\overline{c}}}}_{11} & \overline{\widetilde{\widetilde{\overline{c}}}}_{12} \\ \overline{\widetilde{\widetilde{\overline{c}}}}_{21} & \overline{\widetilde{\widetilde{\overline{c}}}}_{22} \end{pmatrix}, \tag{38}$$

the element of which are provided by (31)–(32). In addition, by assumptions (9) and (10), -1 is not an eigenvalue of 2×2 matrices $\widetilde{\widetilde{\overline{C}}}$ and $\overline{\widetilde{\widetilde{\overline{C}}}}$, then Lemma 1.1 is always available for this system (34) and (36)–(37).

Hence, if we know the initial state of $\widetilde{\widetilde{\overline{U}}} = \widetilde{\widetilde{\overline{U}}}(t,x)$ at the moment $t = T$:

$$t = T: \ \widetilde{\widetilde{\overline{U}}} = (\varphi^{(1)}(x), \varphi^{(2)}(x))^T, \ \widetilde{\widetilde{\overline{U}}}_t = (\psi^{(1)}(x), \psi^{(2)}(x))^T, \quad 0 \le x \le L, \tag{39}$$

then the evolution of $\widetilde{\widetilde{\overline{U}}} = \widetilde{\widetilde{\overline{U}}}(t,x)$ with respect to t can be completely determined.

Remark 3.4 In order to realize the exact boundary synchronization by 2-groups, one has infinitively many ways to choose $(N-2)$ boundary controls at the end $x = 0$. Moreover, as in Remark 2.4, any given state $(\varphi^{(1)}, \varphi^{(2)})^T$ and $(\psi^{(1)}, \psi^{(2)})^T$ in $(C^2[0,L])^2 \times (C^1[0,L])^2$, satisfying the conditions of C^2 compatibility at the points $(t,x) = (T,0)$ and (T,L), respectively, with boundary conditions (36) and (37), belongs to the attainable set of the initial data of synchronizable state by 2-groups at $t = T$.

Thus, at least for some initial states (Φ, Ψ), the synchronizable states by 2-groups $\widetilde{\overline{u}}_1(t,x)$ and $\widetilde{\overline{u}}_2(t,x)$ are linearly independent on $t = T$ then for $t \ge T$. It shows that the corresponding requirement on $\widetilde{\overline{u}}_1$ and $\widetilde{\overline{u}}_2$ given in Lemma 3.1 is reasonable.

4. Exact Boundary Null Controllability and Synchronization by Groups

For the coupled system of wave equations (2), (5) and (7), by Lemma 1.2, we can use N boundary controls to realize the exact boundary null controllability. When the number of boundary controls is less than N, however, generically speaking, it is impossible to realize the same requirement. A natural question is whether we can get the exact boundary null controllability for a part of state variables when the number of boundary controls is less than N. For instance, is it possible or not to realize the exact boundary null controllability for $(N-1)$ state variables by means of only $(N-1)$ boundary controls? Since all the state variables are coupled each other, generally speaking, this requirement is impossible to be realized. However, following the idea given in §3, if we divide the state variables into two groups $U^{(1)} = (u_1, \cdots, u_m)^T$ and $U^{(2)} = (u_{m+1}, \cdots, u_N)^T$ such that $U^{(1)}$ is exactly null controllable, while $U^{(2)}$ is exactly synchronizable, then we can realize the previous requirement in some sense. It leads the following

Definition 4.1 If there exists $T > 0$ such that for any given initial state (Φ, Ψ), we can find some boundary controls with support on $[0, T]$ in $H(t)$ (the precise number of these boundary controls is equal to $(N-1)$, see the sequel), such that the corresponding mixed initial-boundary value problem (2), (5), (7) and (8) admits a unique C^2 solution $U = U(t, x) = (u_1(t, x), \cdots, u_N(t, x))^T$ on $t \geq 0$, $0 \leq x \leq L$, and as $t \geq T$ we have

$$u_1(t, x) \equiv \cdots \equiv u_m(t, x) \equiv 0, \quad 0 \leq x \leq L, \tag{40}$$

$$u_{m+1}(t, x) \equiv \cdots \equiv u_N(t, x) \overset{\text{def.}}{=} \widetilde{\widetilde{u}}(t, x), \quad 0 \leq x \leq L, \tag{41}$$

then the coupled system (2), (5) and (7) is said to possess **the exact boundary null controllability and synchronization by 2-groups** or is said to be **exactly null controllable and synchronizable by 2-groups**, and $\widetilde{\widetilde{u}} = \widetilde{\widetilde{u}}(t, x)$ is called to be the **partially synchronizable state**.

Obviously, if the coupled system (2), (5) and (7) possesses the exact null controllability, then it must satisfy the requirement of Definition 4.1. This situation should be excluded in advance.

Lemma 4.1 Suppose that the couple system (2), (5) and (7) is exactly null controllable and synchronizable by 2-groups, but not exactly null controllable, then the coupling matrix $A = (a_{ij})$ in system (2) must satisfy the following necessary conditions:

$$\sum_{j=m+1}^{N} a_{ij} = 0 \quad (i = 1, \cdots, m), \tag{42}$$

$$\sum_{j=m+1}^{N} a_{ij} \stackrel{\text{def.}}{=} \tilde{\tilde{a}} \quad (i = m+1, \cdots, N), \tag{43}$$

where $\tilde{\tilde{a}}$ is a constant independent of $i = m+1, \cdots, N$. For the coupling matrices $B = (b_{ij})$ (or $C = (c_{ij})$) and $\overline{B} = (\bar{b}_{ij})$ (or $\overline{C} = (\bar{c}_{ij})$) in the boundary conditions (5) and (7), similar necessary conditions hold, namely,

$$\sum_{j=m+1}^{N} b_{ij} \text{ (resp. } c_{ij}, \bar{b}_{ij} \text{ or } \bar{c}_{ij}) = 0 \quad (i = 1, \cdots, m), \tag{44}$$

$$\sum_{j=m+1}^{N} b_{ij} \text{ (resp. } c_{ij}, \bar{b}_{ij} \text{ or } \bar{c}_{ij}) \stackrel{\text{def.}}{=} \tilde{\tilde{b}} \text{ (resp. } \tilde{\tilde{c}},\ \tilde{\tilde{\bar{b}}} \text{ or } \tilde{\tilde{\bar{c}}}) \quad (i = m+1, \cdots, N), \tag{45}$$

where $\tilde{\tilde{b}}$ (or $\tilde{\tilde{c}}$) and $\tilde{\tilde{\bar{b}}}$ (or $\tilde{\tilde{\bar{c}}}$) are constants independent of $i = m+1, \cdots, N$.

Theorem 4.1 Under assumptions (9)–(10), suppose that (42)–(45) hold and $T > 0$ satisfies (17). For any given initial state (Φ, Ψ) we can find $(N-1)$ boundary controls with support on $[0, T]$ in $H(t)$ (for instance, take $h_1(t), \cdots, h_m(t), h_{m+2}(t), \cdots, h_N(t)$ with $h_{m+1}(t) \equiv 0$), such that the coupled system (2), (5) and (7) possesses the exact boundary null controllability and synchronization by 2-groups.

Proof Let

$$\begin{cases} w_i = u_i \quad (i = 1, \cdots, m), \\ w_i = u_i - u_{i+1} \quad (i = m+1, \cdots, N-1). \end{cases} \tag{46}$$

It is easy to see that under assumptions (42)–(45), the original mixed initial-boundary value problem (2), (5), (7) and (8) for the variable U can be reduced to a self-closing mixed initial-boundary value problem of the same kind for the variable $W = (w_1, \cdots, w_{N-1})^T$, see (19)–(21). Thus, according to Lemma 1.2, by means of $(N-1)$ boundary controls $\widetilde{H}(t)$ at the end $x = 0$, the variable W is exactly null controllable, then the variable U is exactly null controllable and synchronizable by 2-groups. This proves Theorem 4.1.

Remark 4.1 In the case of coupled dissipative boundary conditions (5d) and (7d), by assumptions (9) and (10), the corresponding coupling matrices $\widetilde{C}$ and $\widetilde{\widetilde{C}}$ in the reduces coupled dissipative boundary conditions for the variable W also satisfy the assumptions of the same kind, hence Lemma 1.1 and Lemma 1.2 are always available for the reduced initial-boundary value problem for the variable W.

Remark 4.2 As $t \geq T$, the partially synchronizable state $\widetilde{\widetilde{u}} = \widetilde{\widetilde{u}}(t, x)$ satisfies the following wave equation

$$\frac{\partial^2 \widetilde{\widetilde{u}}}{\partial t^2} - \frac{\partial^2 \widetilde{\widetilde{u}}}{\partial x^2} + \widetilde{\widetilde{a}}\widetilde{\widetilde{u}} = 0 \tag{47}$$

and the following boundary conditions

$$x = 0: \ \widetilde{\widetilde{u}} = 0, \tag{48a}$$

$$x = 0: \ \widetilde{\widetilde{u}}_x = 0, \tag{48b}$$

$$x = 0: \ \widetilde{\widetilde{u}}_x - \widetilde{\widetilde{b}}\widetilde{\widetilde{u}} = 0, \tag{48c}$$

$$x = 0: \ \widetilde{\widetilde{u}}_x - \widetilde{\widetilde{c}}\widetilde{\widetilde{u}}_t = 0 \tag{48d}$$

and

$$x = L: \ \widetilde{\widetilde{u}} = 0, \tag{49a}$$

$$x = L: \ \widetilde{\widetilde{u}}_x = 0, \tag{49b}$$

$$x = L: \ \widetilde{\widetilde{u}}_x + \widetilde{\widetilde{\widetilde{b}}}\widetilde{\widetilde{u}} = 0, \tag{49c}$$

$$x = L: \ \widetilde{\widetilde{u}}_x + \widetilde{\widetilde{\widetilde{c}}}\widetilde{\widetilde{u}}_t = 0, \tag{49d}$$

where $\widetilde{\widetilde{a}}$, $\widetilde{\widetilde{b}}$ (or $\widetilde{\widetilde{c}}$) and $\widetilde{\widetilde{\widetilde{b}}}$ (or $\widetilde{\widetilde{\widetilde{c}}}$) are given by (43) and (45). Moreover, by (9) and (10) we have $\widetilde{\widetilde{c}} \neq -1$ and $\widetilde{\widetilde{\widetilde{c}}} \neq -1$, then Lemma 1.1 is always available for this system (47)–(49).

Hence, if we know the initial condition of $\widetilde{\widetilde{u}} = \widetilde{\widetilde{u}}(t, x)$ at the moment $t = T$, then the evolution of the partially synchronizable state $\widetilde{\widetilde{u}} = \widetilde{\widetilde{u}}(t, x)$ with respect to t can be completely determined.

Moreover, any given state (φ, ψ) in $C^2[0, L] \times C^1[0, L]$, satisfying the conditions of C^2 compatibility at the points $(t, x) = (T, 0)$ and (T, L), respectively, with boundary conditions (48) and (49), belongs to the attainable set of the initial data of partially synchronizable state at $t = T$.

Remark 4.3 Taking $m = N-2$ in Theorem 4.1, assumptions (42)–(45) become

$$\sum_{j=N-1}^{N} a_{ij} = 0 \quad (i = 1, \cdots, N-2), \tag{50}$$

$$\sum_{j=N-1}^{N} a_{ij} \overset{\text{def.}}{=} \widetilde{\widetilde{a}} \quad (i = N-1, N) \tag{51}$$

and

$$\sum_{j=N-1}^{N} b_{ij} \text{ (resp. } c_{ij}, \bar{b}_{ij} \text{ or } \bar{c}_{ij}) = 0 \quad (i = 1, \cdots, N-2), \tag{52}$$

$$\sum_{j=N-1}^{N} b_{ij} \text{ (resp. } c_{ij}, \bar{b}_{ij} \text{ or } \bar{c}_{ij}) = \widetilde{\widetilde{b}} \text{ (resp. } \widetilde{\widetilde{c}}, \widetilde{\widetilde{\bar{b}}} \text{ or } \widetilde{\widetilde{\bar{c}}}) \quad (i = N-1, N). \tag{53}$$

Thus, we can use $(N-1)$ (instead of $(N-2)$!) boundary controls to realize the exact boundary null controllability for $(N-2)$ state variables in U.

5. Generalized Exact Boundary Synchronization

We now consider the problem of synchronization from a general mathematical point of view.

Definition 5.1 Let M be an integer with $0 < M < N$. If there exist $T > 0$ and a full row-rank $M \times N$ matrix Ⓗ with constant entries θ_{ij} $(i = 1, \cdots, M; j = 1, \cdots, N)$, such that for any given initial data (8), there exist some boundary controls in $H(t)$ with support on $[0, T]$, such that the C^2 solution $U = U(t, x)$ to the mixed initial-boundary value problem (2), (5), (7) and (8) satisfies

$$t \geq T: \; W(t, x) \overset{\text{def.}}{=} Ⓗ U(t, x) \equiv 0, \tag{54}$$

then the coupled system (2), (5) and (7) of wave equations possesses **the generalized exact boundary synchronization associated with Ⓗ.**

By Definition 5.1, if system (2), (5) and (7) possesses the generalized exact boundary synchronization associated with Ⓗ, then M linearly independent linear combinations, associated with Ⓗ, of the components of U are exactly null controllable. A natural idea is to take a suitable nonsingular linear transformation of the variable U, such that the first M components of the new variable $\widehat{U}$ are exactly null controllable.

Let P be a nonsingular $N \times N$ matrix with constant entries. Taking

$$U = P\widehat{U} \quad \text{or} \quad \widehat{U} = P^{-1}U, \tag{55}$$

the original system (2), (5) and (7) reduces to

$$\widehat{U}_{tt} - \widehat{U}_{xx} + \widehat{A}\widehat{U} = 0, \tag{56}$$

$$x = 0: \ \widehat{U} = \widehat{H}(t), \tag{57a}$$

$$x = 0: \ \widehat{U}_x = \widehat{H}(t), \tag{57b}$$

$$x = 0: \ \widehat{U}_x - \widehat{B}\widehat{U} = \widehat{H}(t), \tag{57c}$$

$$x = 0: \ \widehat{U}_x - \widehat{C}\widehat{U}_t = \widehat{H}(t), \tag{57d}$$

$$x = L: \ \widehat{U} = 0, \tag{58a}$$

$$x = L: \ \widehat{U}_x = 0, \tag{58b}$$

$$x = L: \ \widehat{U}_x + \widehat{\overline{B}}\widehat{U} = 0, \tag{58c}$$

$$x = L: \ \widehat{U}_x + \widehat{\overline{C}}\widehat{U}_t = 0, \tag{58d}$$

where

$$\widehat{A}=P^{-1}AP,\ \widehat{B}=P^{-1}BP \ (\text{or } \widehat{C}=P^{-1}CP),\ \widehat{\overline{B}}=P^{-1}\overline{B}P \ (\text{or } \widehat{\overline{C}}=P^{-1}\overline{C}P) \tag{59}$$

are obtained by operating the same similarity transformation related to P to A, B (or C) and $\overline{B}$ (or $\overline{C}$), respectively, and

$$\widehat{H}(t) = P^{-1}H(t). \tag{60}$$

Under transformation (55), (54) becomes

$$t \geq T: \ W(t,x) = \widehat{\text{Ⓗ}}\, \widehat{U}(t,x) \equiv 0, \tag{61}$$

where

$$\widehat{\text{Ⓗ}} = \text{Ⓗ} P. \tag{62}$$

Since Ⓗ is of full row-rank, taking the Gram–Schmidt orthogonalization method in its columns shows that there exists a nonsingular matrix P such that

$$\text{Ⓗ} P = (I_M, 0)_{M\times N}. \tag{63}$$

For such a matrix P, we have

$$W = (\hat{u}_1, \cdots, \hat{u}_M)^T, \tag{64}$$

then (61)–(62) simply reduce to

$$t \geq T: \ (\hat{u}_1, \cdots, \hat{u}_M)(t, x) \equiv 0, \tag{65}$$

i.e., the first M components of $\widehat{U}(t,x)$ are exactly null controllable. Thus, the generalized exact boundary synchronization is essentially the exact boundary null controllability for a part of new state variables.

If we can get a self-closing system of the same kind for the variable $W = Ⓗ U$ from the original system (2), (5) and (7) for the variable U, then, according to the result on the exact boundary null controllability for the coupled system of wave equations, M boundary controls can be used to realize the exact boundary null controllability for the variable W, and then we get the generalized exact boundary synchronization associated with Ⓗ for the variable U. Thus, if for a nonsingular matrix P satisfying (63), the first M components of the variable $\widehat{U}$, namely, the components of the variable W, constitute a self-closing system, then we can realize the generalized exact boundary synchronization associated with Ⓗ for the variable U. Obviously, this requirement is equivalent to the fact that for such a matrix P, $\widehat{A}$, $\widehat{B}$ (or $\widehat{C}$) and $\widehat{\overline{B}}$ (or $\widehat{\overline{C}}$) given by (59) are all block lower triangular matrices of the following form:

$$\widehat{A} = P^{-1}AP = \begin{pmatrix} \widetilde{A}_{M\times M} & 0_{M\times(N-M)} \\ * & \widetilde{\widetilde{A}}_{(N-M)\times(N-M)} \end{pmatrix}, \tag{66}$$

$$\widehat{B} = P^{-1}BP = \begin{pmatrix} \widetilde{B}_{M\times M} & 0_{M\times(N-M)} \\ * & \widetilde{\widetilde{B}}_{(N-M)\times(N-M)} \end{pmatrix}$$

$$\left(\text{or } \widehat{C} = P^{-1}CP = \begin{pmatrix} \widetilde{C}_{M\times M} & 0_{M\times(N-M)} \\ * & \widetilde{\widetilde{C}}_{(N-M)\times(N-M)} \end{pmatrix}\right), \tag{67}$$

$$\widehat{\overline{B}} = P^{-1}\overline{B}P = \begin{pmatrix} \widetilde{\overline{B}}_{M\times M} & 0_{M\times(N-M)} \\ * & \widetilde{\widetilde{\overline{B}}}_{(N-M)\times(N-M)} \end{pmatrix}$$

$$\left(\text{or } \widehat{\overline{C}} = P^{-1}\overline{C}P = \begin{pmatrix} \widetilde{\overline{C}}_{M\times M} & 0_{M\times(N-M)} \\ * & \widetilde{\widetilde{\overline{C}}}_{(N-M)\times(N-M)} \end{pmatrix}\right), \tag{68}$$

where $\widetilde{A}$, $\widetilde{B}$ (or $\widetilde{C}$) and $\widetilde{\overline{B}}$ (or $\widetilde{\overline{C}}$) are $M \times M$ matrices, $\widetilde{\widetilde{A}}$, $\widetilde{\widetilde{B}}$ (or $\widetilde{\widetilde{C}}$) and $\widetilde{\widetilde{\overline{B}}}$ (or $\widetilde{\widetilde{\overline{C}}}$) are $(N-M)\times(N-M)$ matrices, and $*$ stands for some $(N-M)\times M$ matrices.

In this case, it follows from (56)–(60) that the variable W, the vector composed of the first M components of the variable $\widehat{U}$, satisfies the following self-closing system:

$$W_{tt} - W_{xx} + \widetilde{A}W = 0, \tag{69}$$

$$x = 0: \ W = \widehat{H}_1(t), \tag{70a}$$
$$x = 0: \ W_x = \widehat{H}_1(t), \tag{70b}$$
$$x = 0: \ W_x - \widetilde{B}W = \widehat{H}_1(t), \tag{70c}$$
$$x = 0: \ W_x - \widetilde{C}W_t = \widehat{H}_1(t), \tag{70d}$$

$$x = L: \ W = 0, \tag{71a}$$
$$x = L: \ W_x = 0, \tag{71b}$$
$$x = L: \ W_x + \widetilde{\overline{B}}W = 0, \tag{71c}$$
$$x = L: \ W_x + \widetilde{\overline{C}}W_t = 0, \tag{71d}$$

where $\widetilde{A}$, $\widetilde{B}$ (or $\widetilde{C}$), $\widetilde{\overline{B}}$ (or $\widetilde{\overline{C}}$) are $M \times M$ matrices given by (66)–(68), respectively, and $\widehat{H}_1(t)$ is a vector composed of the first M components of $\widehat{H}(t)$. Moreover, in the case of coupled dissipative boundary conditions (5d) and (7d), by assumptions (9)–(10) and noting (67)–(68), obviously we have

$$\det(\widetilde{C} + I_M) \neq 0 \tag{72}$$

and

$$\det(\widetilde{\overline{C}} + I_M) \neq 0. \tag{73}$$

Hence, the reduced forward mixed problem (69)–(71) with the corresponding initial data for the variable W is always well-posed, and by means of M boundary controls $\widehat{H}_1(t)$, system (69)–(71) can realize the exact boundary null controllability.

Thus, we get

Theorem 5.1 For any given $N \times N$ coupling matrices A, B (or C) and $\overline{B}$ (or $\overline{C}$) with assumptions (9)-(10) and any given full row-rank $M \times N$ matrix Ⓗ, if there exists a nonsingular $N \times N$ matrix P such that (63) and (66)-(68) simultaneously hold, then the coupled system (2), (5) and (7) of wave equations possesses the generalized exact boundary synchronization associated with Ⓗ.

By means of Moore-Penrose generalized inverse (cf. [1]), we can prove the following two theorems.

Theorem 5.2 Suppose that $0 < M < N$. For any given coupling $N \times N$ matrices A, B (or C) and $\overline{B}$ (or $\overline{C}$) and any given $M \times N$ matrix Ⓗ with full row-rank, there exists a nonsingular $N \times N$ matrix P such that (63) and (66)–(68) hold simultaneously, if and only if the null space Ker(Ⓗ) of Ⓗ is a common invariant subspace of A, B (or C) and $\overline{B}$ (or $\overline{C}$), namely,

$$AX, \quad BX \text{ (or } CX), \quad \overline{B}X \text{ (or } \overline{C}X) \in \mathrm{Ker}(Ⓗ), \quad \forall X \in \mathrm{Ker}(Ⓗ). \tag{74}$$

Remark 5.1 By Theorem 5.2 and its proof, the coupling matrices $\widetilde{A}$, $\widetilde{B}$ (or $\widetilde{C}$) and $\widetilde{\overline{B}}$ (or $\widetilde{\overline{C}}$) in the reduced coupled system of wave equations for the variable W are given by

$$\begin{aligned} \widetilde{A} = Ⓗ A Ⓗ^+, \ \widetilde{B} &= Ⓗ B Ⓗ^+ \text{ (or } \widetilde{C} = Ⓗ C Ⓗ^+), \\ \widetilde{\overline{B}} &= Ⓗ \overline{B} Ⓗ^+ \text{ (or } \widetilde{\overline{C}} = Ⓗ \overline{C} Ⓗ^+), \end{aligned} \tag{75}$$

or equivalently by

$$\begin{aligned} Ⓗ A = \widetilde{A} Ⓗ, \quad Ⓗ B &= \widetilde{B} Ⓗ \text{ (or } Ⓗ C = \widetilde{C} Ⓗ), \\ Ⓗ \overline{B} &= \widetilde{\overline{B}} Ⓗ \text{ (or } Ⓗ \overline{C} = \widetilde{\overline{C}} Ⓗ), \end{aligned} \tag{76}$$

where $Ⓗ^+ = Ⓗ^T (Ⓗ Ⓗ^T)^{-1}$ is the Moore-Penrose generalized inverse of Ⓗ.

Theorem 5.3 Suppose that $0 < M < N$. For any given coupling $N \times N$ matrices A, B (or C) and $\overline{B}$ (or $\overline{C}$), there exist a full row-rank $M \times N$ matrix Ⓗ and a nonsingular $N \times N$ matrix P such that (63) and (66)–(68) hold simultaneously, if and only if matrices A, B(or C) and $\overline{B}$(or $\overline{C}$) possess a common invariant subspace $\mathscr{L}$ with dimension $(N - M)$.

Remark 5.2 From the proof of Theorem 5.3, it is easy to see that

$$\mathscr{L} = \mathrm{Ker}(Ⓗ), \tag{77}$$

then $Ⓗ^T$ is a basis matrix of the orthogonal complement $\mathscr{L}^{\perp}$ of $\mathscr{L}$, moreover, the matrix composed of the system of basic solutions to the linear algebraic system

$$Ⓗ X = 0 \tag{78}$$

is a basis matrix of $\mathscr{L}$.

According to Theorem 5.3, if the coupling matrices A, B (or C) and $\overline{B}$ (or $\overline{C}$) possess a common invariant subspace with dimension $(N-M)$, then we can find a full row-rank $M \times N$ matrix Ⓗ such that the coupled system (2), (5) and (7) of wave equations possesses the generalized exact boundary synchronization associated with Ⓗ. In particular, if among A, B (or C) and $\overline{B}$ (or $\overline{C}$), only one matrix (for instance, A) is a really coupling matrix, while other two matrices (for instance, B (or C) and $\overline{B}$ (or $\overline{C}$)) are zero or constant matrices, then once A has an invariant subspace with dimension $(N-M)$, the generalized exact boundary synchronization associated with a corresponding $M \times N$ matrix Ⓗ can be realized. We now consider this possibility more precisely.

For any given $N \times N$ matrix A, by real Schur decomposition (cf. [3]) there exists an $N \times N$ orthogonal matrix Q such that

$$Q^{-1}AQ = \begin{bmatrix} R_{11} & & & & \\ R_{21} & R_{22} & & 0 & \\ \vdots & \vdots & \ddots & & \\ \vdots & \vdots & & \ddots & \\ R_{m1} & R_{m2} & \cdots & & R_{mm} \end{bmatrix}, \tag{79}$$

where R_{kk} $(k = 1, \cdots, m)$ are matrices of order 1 or order 2. When R_{kk} is of order 1, R_{kk} is a real eigenvalue of A; while, when R_{kk} is of order 2, its eigenvalues are composed of a pair of conjugate complex eigenvalues of A, i.e., $R_{kk} = \begin{pmatrix} a & b \\ -b & a \end{pmatrix}$ and $a \pm bi$ are a pair of conjugate complex eigenvalues. Furthermore, through a suitable permutation the order of R_{kk} $(k = 1, \cdots, m)$ can be arbitrarily changed.

Since the orthogonal transformation is a special similarity transformation, and (79) is of the form of block lower triangular matrix, A should have some invariant subspaces, then the generalized exact boundary synchronization associated with corresponding Ⓗ can be always realized, however, the rank M of Ⓗ should be determined according to the structure of (79). For example, when N is even, if A does not have any real eigenvalue, then we can get the generalized exact boundary synchronization only for the even value of M, but not for the odd value of M. While, when N is odd, A has at least one real eigenvalue, then we can always get the generalized exact boundary synchronization for any given value of M with $0 < M < N$.

We now give some applications of the general results mentioned above.

5.1 Exact boundary synchronization

Suppose that $A = (a_{ij})$, $B = (b_{ij})$ (or $C = (c_{ij})$) and $\overline{B} = (\bar{b}_{ij})$ (or $\overline{C} = (\bar{c}_{ij})$) satisfy conditions (14)-(16). In this situation, A, B (or C) and $\overline{B}$ (or $\overline{C}$) possess a common eigenvector $(1, \cdots, 1)^T$ which spans a common one-dimensional invariant subspace of A, B (or C) and $\overline{B}$ (or $\overline{C}$). Then, $M = N - 1$, and by Remark 5.2, Ⓗ^{T} can be taken as a basis matrix of the subspace orthogonal to $(1, \cdots, 1)^T$, essentially we can take

$$\text{Ⓗ} = \begin{pmatrix} 1 & -1 & & & \\ & 1 & -1 & & \\ & & \ddots & \ddots & \\ & & & 1 & -1 \end{pmatrix}_{(N-1)\times N}. \tag{80}$$

This leads to the exact boundary synchronization discussed in §2.

5.2 Exact boundary null controllability and synchronization by 2-groups

Suppose that the coupling matrix A satisfies conditions (42)–(43), and the coupling matrices B (or C) and $\overline{B}$ (or $\overline{C}$) satisfy similar conditions. In this situation, A, B (or C) and $\overline{B}$ (or $\overline{C}$) possess a common eigenvector $(\overbrace{0, \cdots, 0}^{m}, \overbrace{1, \cdots, 1}^{N-m})^T$ which spans a common one-dimensional invariant subspace of A, B (or C) and $\overline{B}$ (or $\overline{C}$). Thus, $M = N - 1$, and by Remark 5.2, Ⓗ^{T} can be taken as a basis matrix of the subspace orthogonal to $(\overbrace{0, \cdots, 0}^{m}, \overbrace{1, \cdots, 1}^{N-m})^T$, essentially we can take

$$\text{Ⓗ} = \begin{pmatrix} \mathrm{I}_m & \\ & \begin{pmatrix} 1 & -1 & & & \\ & 1 & -1 & & \\ & & \ddots & \ddots & \\ & & & 1 & -1 \end{pmatrix}_{(N-m-1)\times(N-m)} \end{pmatrix}. \tag{81}$$

This leads to the exact boundary null controllability and synchronization by 2-groups discussed in §4.

5.3 Exact boundary synchronization by 2-groups

Suppose that A satisfies conditions (29)–(30), and B (or C) and $\overline{B}$ (or $\overline{C}$) satisfy similar conditions. In this situation, the linear space spanned by vectors $(\overbrace{1, \cdots, 1}^{m}, \overbrace{0, \cdots, 0}^{N-m})^T$ and $(\overbrace{0, \cdots, 0}^{m}, \overbrace{1, \cdots, 1}^{N-m})^T$ is a common invariant subspace with dimension 2 of A, B (or C) and $\overline{B}$ (or $\overline{C}$). Thus, $M = N-2$, and by Remark 5.2, Ⓗ^{T} can be taken as a basis matrix of the orthogonal complement of the 2-dimensional invariant subspace, essentially we can take

$$
\text{Ⓗ} = \begin{pmatrix} \begin{pmatrix} 1 & -1 & & & \\ & 1 & -1 & & \\ & & \ddots & \ddots & \\ & & & 1 & -1 \end{pmatrix}_{(m-1)\times m} & \\ & \begin{pmatrix} 1 & -1 & & & \\ & 1 & -1 & & \\ & & \ddots & \ddots & \\ & & & 1 & -1 \end{pmatrix}_{(N-m-1)\times(N-m)} \end{pmatrix}. \tag{82}
$$

This leads to the exact boundary synchronization by 2-groups discussed in §3.

5.4 Similarly, we may consider the exact boundary synchronization and the exact boundary null controllability and synchronization by $k(>1)$-groups.

5.5 A more general example

Let $a = (a_1, \cdots, a_N)^T$ and $b = (b_1, \cdots, b_N)^T$ be two linearly independent vectors. Suppose that the linear space spanned by a and b is a common invariant subspace of A, B (or C) and $\overline{B}$ (or $\overline{C}$). Thus, $M = N-2$, and by Remark 5.2, if Ⓗ^{T} is taken as a basis matrix of the orthogonal complement of this two-dimensional invariant subspace, namely, a matrix composed of the system of basic solutions to the linear algebraic system

$$
\begin{pmatrix} a_1, \cdots, a_N \\ b_1, \cdots, b_N \end{pmatrix} X = 0, \tag{83}
$$

then we can get the corresponding generalized exact boundary synchronization. In this case, Ⓗ does not possess a structure of block diagonal matrix in general.

6. Remarks

6.1. The case of two-sided controls can be similarly discussed.
6.2. A similar treatment can be done for the coupled system of wave equations of dissipative type

$$
\frac{\partial^2 U}{\partial t^2} - \frac{\partial^2 U}{\partial x^2} + \overline{A}\frac{\partial U}{\partial t} = 0. \tag{84}
$$

References

1. A. Ben-Israel, T. N. E. Greville, *Generalized Inverses: Theory and Applications*, 2nd Edition, Springer, 2003.
2. H. Fujisaka, T. Yamada, *Stability theory of synchronized motion in coupled-oscillator systems*, Progress of Theoretical Physics, 69(1983), 32–47.
3. R. A. Horn, C. R. Johnson, *Matrix Analysis*, 2nd Edition, Cambridge University Press, 2013.
4. Long Hu, Fanqiong Ji, Ke Wang, *Exact boundary controllability and exact boundary observability for a coupled system of quasilinear wave equations*, Chin. Ann. Math., 34B(2013), 479–490.
5. Long Hu, Tatsien Li, Bopeng Rao, *Exact boundary synchronization for a coupled system of 1-D wave equations with coupled boundary conditions of dissipative type*, Communications on Pure and Applied Analysis, 13(2014). DOI: 10.3934/cpaa.2014.13.
6. Ch. Huygens, *Œuvres Complètes*, Vol. 15, Swets & Zeitlinger B. V., Amsterdam, 1967.
7. Tatsien Li, *Controllability and Observability for Quasilinear Hyperbolic Systems*, AIMS Series on Applied Mathematics, Vol. 3, AIMS & Higher Education Prcss, 2010.
8. Tatsien Li, Bopeng Rao, *Strong (weak) exact controllability and strong (weak) exact observability for quasilinear hyperbolic systems*, Chin. Ann. Math., 31B(2010), 723–742.
9. Tatsien Li, Bopeng Rao, *Exact synchronization for a coupled system of wave equations with Dirichlet boundary controls*, Chin. Ann. Math., 34B(2013), 139–160.
10. Tatsien Li, Bopeng Rao, *A note on the exact synchronization by groups for a coupled system of wave equations*, To appear in Math. Meth. Appl. Sci. DOI: 10.1002/mma.3062.
11. Tatsien Li, Bopeng Rao, Long Hu, *Exact boundary synchronization for a coupled system of 1-D wave equations*, To appear in ESAIM: COCV. DOI: 10.1051/COCV/2013066.
12. Tatsien Li, Bopeng Rao, Yimin Wei, *Generalized exact boundary synchronization for a coupled system of wave equations*, Discrete and Continuous Dynamical Systems, 34(2014), 2893–2905.
13. J.-L. Lions, *Contrôlabilité Exacte, Perturbations et Stabilization de Systèmes Distribués*, Vol. 1, Masson, 1988.
14. J.-L. Lions, *Exact controllability, stabilization and perturbations for distributed systems*, SIAM Review, 30(1988), 1–68.
15. D. L. Russell, *Controllability and stabilization theory for linear partial differential equations*, Recent progress and open questions, SIAM Review, 20(1978), 639–739.

Scaling Limit for Compressible Viscoelastic Fluids

Xianpeng Hu and Fanghua Lin

Courant Institute of mathematical sciences, New York University, New York, NY 10012, USA

xianpeng@cims.nyu.edu and linf@cims.nyu.edu

The convergence from a sequence of the unique global solutions to the Cauchy problems for compressible viscoelastic fluids to a unique global solution of the incompressible Navier-Stokes equations without external forces is studied for a wide class of initial data as the Mach number and the elastic coefficient go to zero simultaneously. The proofs are based on a set of conservation laws and a list of estimates which are uniform in the scaling parameter as well as a dispersive estimate for the wave equation.

1. Introduction

A multi-dimensional compressible viscoelastic flow is governed by (see Refs. 5, 7, 8, 13, 14):

$$\rho_t + \mathrm{div}(\rho\mathbf{u}) = 0, \tag{1.1a}$$

$$(\rho\mathbf{u})_t + \mathrm{div}\,(\rho\mathbf{u}\otimes\mathbf{u}) - \mu\Delta\mathbf{u} - (\lambda+\mu)\nabla\mathrm{div}\mathbf{u} + \nabla P(\rho) = \mathrm{div}(\rho\mathtt{F}\,\mathtt{F}^\top), \tag{1.1b}$$

$$\mathtt{F}_t + \mathbf{u}\cdot\nabla\mathtt{F} = \nabla\mathbf{u}\,\mathtt{F}, \tag{1.1c}$$

where ρ stands for the density, $\mathbf{u}\in\mathbb{R}^N$ $(N=2,3)$ the velocity, and $\mathtt{F}\in M^{N\times N}$ (the set of $N\times N$ matrices) the deformation gradient. The viscosity coefficients μ,λ are two constants satisfying $\mu>0, 2\mu+N\lambda>0$, which ensures that the operator $-\mu\Delta\mathbf{u}-(\lambda+\mu)\nabla\mathrm{div}\mathbf{u}$ is a strongly elliptic. The pressure $P(\rho)$ is assumed to be an increasing and convex function of ρ for $\rho>0$. The symbol $\otimes$ denotes the Kronecker tensor product, $\mathtt{F}^\top$ means the transpose matrix of $\mathtt{F}$, and the notation $\mathbf{u}\cdot\nabla\mathtt{F}$ is understood to be $(\mathbf{u}\cdot\nabla)\mathtt{F}$. For system (1.1), the corresponding elastic energy is chosen to be the special form of the Hookean linear elasticity:

$$W(\mathtt{F}) = \frac{1}{2}|\mathtt{F}|^2,$$

and for simplicity of our presentation, we assume that $P'(1) = 1$. The arguments and results of this paper can be easily applied to more general cases.

A compressible flow, via physics, behaves asymptotically like an incompressible flow when the density is approximately a constant and the velocity is small in a suitable sense. More precisely, if we scale ρ, $\mathbf{u}$, and $\mathtt{F}$ in the following way:

$$\rho = \rho^\varepsilon(x, \varepsilon t), \quad \mathbf{u} = \varepsilon \mathbf{u}^\varepsilon(x, \varepsilon t), \quad \mathtt{F} = \mathtt{F}^\varepsilon(x, \varepsilon t), \tag{1.2}$$

and we replace μ and λ by the scaled ones as: $\varepsilon\mu$ and $\varepsilon\lambda$.

Where $\varepsilon \in (0, 1)$ is a small parameter and

$\mu > 0$, $2\mu + N\lambda > 0$. Such a scaling ensures in particular that the limit equation as $\varepsilon \to 0$ is not an Euler type system. Under these scalings, system (1.1) becomes

$$\begin{cases} \rho^\varepsilon_t + \operatorname{div}(\rho^\varepsilon \mathbf{u}^\varepsilon) = 0, \\ (\rho^\varepsilon \mathbf{u}^\varepsilon)_t + \operatorname{div}(\rho^\varepsilon \mathbf{u}^\varepsilon \otimes \mathbf{u}^\varepsilon) - \mu\Delta\mathbf{u}^\varepsilon - (\lambda + \mu)\nabla \operatorname{div}\mathbf{u}^\varepsilon + \frac{\nabla P(\rho^\varepsilon)}{\varepsilon^2} \\ \qquad = \frac{1}{\varepsilon^2}\operatorname{div}(\rho^\varepsilon\, \mathtt{F}^\varepsilon\, (\mathtt{F}^\varepsilon)^\top), \\ \mathtt{F}^\varepsilon_t + \mathbf{u}^\varepsilon \cdot \nabla \mathtt{F}^\varepsilon = \nabla \mathbf{u}^\varepsilon\, \mathtt{F}^\varepsilon. \end{cases} \tag{CV^ε}$$

If ε is fixed, the global existence and uniqueness of strong solution had been considered in Refs. 7, 13 in the critical spaces (see the definition below) when the fluid variables are close to the equilibrium. Various conservation laws played key roles in the arguments, see for example Proposition 1.1 below. Using conservation laws (quantities) for global existence of strong solution to (CV^ε) as well as their convergence are, of course, not superficial. They are part of the intrinsic properties of viscoelastic fluids. Formally, we observe that when $\rho \to 1$, the first equation in (CV^ε) leads to: $\operatorname{div}\mathbf{u} = 0$, which is the incompressibility of a flow, and the first two terms in the second equation of (CV^ε) induce the term

$$\mathbf{u}_t + \operatorname{div}(\mathbf{u} \otimes \mathbf{u}) = \mathbf{u}_t + \mathbf{u} \cdot \nabla \mathbf{u}.$$

On the other hand, the incompressible Navier-Stokes equations without external forces is giving by

$$\begin{cases} v_t + (v \cdot \nabla)v - \mu\Delta v + \nabla\Pi = 0, \\ \operatorname{div} v = 0. \end{cases} \tag{INS}$$

Thus one would expect that strong solutions of (CV^ε) converge in suitable functional spaces to the strong solutions of (INS) when ρ goes to a constant say 1 as ε goes to 0 provided that one can verify the right hand side of the second equation of (CV^ε) goes to the gradient of a pressue due to internal elastic deformations. In other words, in such a convergence, one part of the hydrostatic pressure Π in (INS) is the "limit" of $(P(\rho)-P(1))/\varepsilon^2$ in (CV^ε), and the another contribution of the pressue Π comes from (the leading part of) the elastic force in this asymptotic limit.

This paper is devoted to the rigorous justification of the above formal reasoning of the scaling limit for global strong solutions of the compressible viscoelastic flows for a more general class of the initial data than the ones considered earlier, e.g., Ref. 9. In fact, we shall assume $\rho_0^\varepsilon = 1 + \varepsilon b_0^\varepsilon$ with $(b_0^\varepsilon, \mathbf{u}_0^\varepsilon, \mathtt{F}_0^\varepsilon)$ uniformly bounded in a convenient functional space as ε goes to 0. Furthermore, when

$$\rho^\varepsilon = 1 + \varepsilon b^\varepsilon,$$

we are led to study

$$\begin{cases} b_t^\varepsilon + \frac{\mathrm{div}\mathbf{u}^\varepsilon}{\varepsilon} = -\mathrm{div}(b^\varepsilon \mathbf{u}^\varepsilon), \\ (\mathbf{u}_i^\varepsilon)_t + \mathbf{u}^\varepsilon \cdot \nabla \mathbf{u}_i^\varepsilon - \frac{\mu\Delta \mathbf{u}_i^\varepsilon + (\lambda+\mu)\partial_i \mathrm{div}\mathbf{u}^\varepsilon}{1+\varepsilon b^\varepsilon} + \frac{P'(1+\varepsilon b^\varepsilon)}{1+\varepsilon b^\varepsilon}\frac{\partial_i b^\varepsilon}{\varepsilon} = \frac{1}{\varepsilon^2}\mathtt{F}_{jk}^\varepsilon \partial_j \mathtt{F}_{ik}^\varepsilon, \\ \mathtt{F}_t^\varepsilon + \mathbf{u}^\varepsilon \cdot \nabla \mathtt{F}^\varepsilon = \nabla \mathbf{u}^\varepsilon\, \mathtt{F}^\varepsilon, \\ (b^\varepsilon, \mathbf{u}^\varepsilon, \mathtt{F}^\varepsilon)|_{t=0} = (b_0^\varepsilon, \mathbf{u}_0^\varepsilon, \mathtt{F}_0^\varepsilon). \end{cases} \tag{1.3}$$

Notice that, here we have used the fact $\mathrm{div}(\rho^\varepsilon (\mathtt{F}^\varepsilon)^\top) = 0$ for all positive time, which is ensured by (CV^ε) and the assumption $\mathrm{div}(\rho_0^\varepsilon (\mathtt{F}_0^\varepsilon)^\top) = 0$ (see Ref. 13 or Lemma 6.1 of Ref. 7). We wish to show $\mathbf{u}^\varepsilon$ to tend to v which solves (INS). One of the dfficulties is that we need to deal with the propagation of acoustic waves with the speed ε^{-1}.

Despite the fact that we can eventually prove the convergence of $\mathbf{u}^\varepsilon$ to v, it is not clear that what would be the limiting equation for the third equation in (1.3) when the parameter ε goes to zero. Though the third equation in (1.3) plays no role in the corrsponding (INS) or that it can be derived from the fluid particle trajectory equation as in Ref. 10, it would be interesting to understand in the present case if there are any missing informations coded in possible limiting equations. In this paper, we will focus on the case of the whole space $\mathbb{R}^N$. The periodic domain case may be different due no dispersive estimate for the wave equations and it will be studied later.

A huge literature has been devoted to the existence of solutions for compressible fluids, (CV^ε), (INS) and to the convergence of $(\rho^\varepsilon, \mathbf{u}^\varepsilon, \mathtt{F}^\varepsilon)$ when ε goes to zero. Roughly speaking, two different heuristics have been introduced. The first one is for the class of *well-prepared* data. For example, $\rho_0^\varepsilon = 1 + \mathcal{O}(\varepsilon^2)$, $\mathrm{div}\mathbf{u}_0^\varepsilon = \mathcal{O}(\varepsilon)$, $\mathtt{F}_0^\varepsilon = \mathtt{F} + \mathcal{O}(\varepsilon)$ for compressible viscoelastic fluds as in[9]). In this case, the time derivative of the sequence of the density at $t = 0$ is uniformly bounded and the usual energy method works well. The problem in this case is of parabolic and dissipative nature. The second heuristics is for a larger class of the initial data which are not so well-prepared. This is the one we considered here. Here certain dispersive estimates for wave equations may play a crucial role. And the problem is of mixed parabolic dissipative and hyperbolic dispersive nature. In this paper, we consider classical strong solutions. Hence we would need to work in critical functional spaces. To explain critical spaces, we observe that (CV^ε) is invariant under the following transformation

$$\begin{cases}(\rho_0(x), \mathbf{u}_0(x), \mathtt{F}_0(x)) \to (\rho_0(lx), l\mathbf{u}_0(lx), \mathtt{F}_0(lx)),\\ (\rho(t,x), \mathbf{u}(t,x), \mathtt{F}(t,x)) \to (\rho(l^2t, lx), l\mathbf{u}(l^2t, lx), \mathtt{F}(l^2t, lx))\end{cases} \tag{1.4}$$

for $l > 0$ and with changes of the pressure law P into l^2P. This suggests the following definition: *A functional space* $\mathfrak{A} \subset \mathcal{S}'(\mathbb{R}^N) \times (\mathcal{S}'(\mathbb{R}^N))^N \times (\mathcal{S}'(\mathbb{R}^N))^{N\times N}$ *is called a critical space if the associated norm is invariant under the transformation* $(\rho, \mathbf{u}, \mathtt{F}) \to (\rho(l\cdot), l\mathbf{u}(l\cdot), \mathtt{F}(l\cdot))$ *(up to a constant independent of* l*), where* $\mathcal{S}'$ *is the space of tempered distributions, i.e., the dual of the Schwartz space* $\mathcal{S}$. According to this definition, $B^{\frac{N}{2}} \times (B^{\frac{N}{2}-1})^N \times B^{\frac{N}{2}}$ (see Section 2 for the definition of $B^s := \dot{B}^s_{2,1}(\mathbb{R}^N)$) is a critical space. Our motivations to use the homogeneous Besov space B^s with the derivative index $\frac{N}{2}$ including following two: first, $B^{\frac{N}{2}}$ is an algebra embedded in L^∞, which allows us to control the density and the deformation gradient from below and from above without requiring more regularity on derivatives of ρ and $\mathtt{F}$; second, the product is continuous from $B^{\frac{N}{2}-\alpha} \times B^{\frac{N}{2}}$ to $B^{\frac{N}{2}-\alpha}$ for $0 \le \alpha < N$.

In Refs. 7 and 13, authors showed that (1.1) is well-posed for initial data

$$(b_0, \mathbf{u}_0, \mathtt{F}_0 - I) \in \tilde{B}^{\frac{N}{2}-1,\frac{N}{2}} \times \left(B^{\frac{N}{2}-1}\right)^N \times \left(\tilde{B}^{\frac{N}{2}-1,\frac{N}{2}}\right)^{N\times N},$$

where $\tilde{B}^{\frac{N}{2}-1,\frac{N}{2}} = B^{\frac{N}{2}-1} \cap B^{\frac{N}{2}}$. For convieneces, we summarize the main result of Refs. 7 and 13 as

Proposition 1.1. *There exists two positive constants* γ *and* Γ*, such that, if* $\rho_0 - 1 \in \tilde{B}^{\frac{N}{2}-1,\frac{N}{2}}$, $\mathbf{u}_0 \in B^{\frac{N}{2}-1}$, $\mathtt{F}_0 - I \in \tilde{B}^{\frac{N}{2}-1,\frac{N}{2}}$ *satisfy*

- $\|\rho_0 - 1\|_{\tilde{B}^{\frac{N}{2}-1,\frac{N}{2}}} + \|\mathbf{u}_0\|_{B^{\frac{N}{2}-1}} + \|\mathtt{F}_0 - I\|_{\tilde{B}^{\frac{N}{2}-1,\frac{N}{2}}} \le \gamma$;
- $\operatorname{div}(\rho_0 \mathtt{F}_0^{\top}) = 0$;
- $\mathtt{F}_{lk}(0)\nabla_{x_l}\mathtt{F}_{ij}(0) = \mathtt{F}_{lj}(0)\nabla_{x_l}\mathtt{F}_{ik}(0)$,

then system (1.1) *has a solution* $(\rho,\ \mathbf{u},\ \mathtt{F})$ *with* $(\rho-1, \mathbf{u},\ \mathtt{F}-I)$ *in* $\mathfrak{B}^{\frac{N}{2}}$ *satisfying*

$$\|(\rho-1, \mathbf{u}, \mathtt{F}-I)\|_{\mathfrak{B}^{\frac{N}{2}}} \le \Gamma\Big(\|\rho_0-1\|_{\tilde{B}^{\frac{N}{2}-1,\frac{N}{2}}} + \|\mathbf{u}_0\|_{B^{\frac{N}{2}-1}} + \|\mathtt{F}_0-I\|_{\tilde{B}^{\frac{N}{2}-1,\frac{N}{2}}}\Big).$$

In the above statement, we used the following notation.

Notation. *For* $T > 0$, *and* $s \in \mathbb{R}$, *we denote*

$$\begin{aligned}\mathfrak{B}^s_T = \Big\{(\rho, \mathbf{u}, E) \in \Big(&L^1(0,T;\tilde{B}^{s+1,s}) \cap C([0,T];\tilde{B}^{s-1,s})\Big) \\ &\times \big(L^1(0,T;B^{s+1}) \cap C([0,T];B^{s-1})\big)^N \\ &\times \Big(L^1(0,T;\tilde{B}^{s+1,s}) \cap C([0,T];\tilde{B}^{s-1,s})\Big)^{N\times N}\Big\}\end{aligned}$$

and

$$\begin{aligned}\|(\rho, \mathbf{u}, E)\|_{\mathfrak{B}^s_T} = &\|\rho\|_{L^\infty_T(\tilde{B}^{s-1,s})} + \|\mathbf{u}\|_{L^\infty_T(B^{s-1,s})} + \|E\|_{L^\infty_T(\tilde{B}^{s-1,s})} \\ &+ \|\rho\|_{L^1_T(\tilde{B}^{s+1,s})} + \|\mathbf{u}\|_{L^1_T(B^{s+1})} + \|E\|_{L^1_T(\tilde{B}^{s+1,s})}.\end{aligned}$$

We use the notation $\mathfrak{B}^s$ *if* $T = +\infty$ *by changing the interval* $[0,T]$ *into* $[0,\infty)$ *in the notation above.*

For the incompressible Navier-Stokes equation, the system is also invariant under a similar scaling (1.4). The global existence for (INS) had been obtained by H. Fujita and T.Kato in Ref. 6 and it can be restated as

Theorem 1.1 (H. Fujita and T.Kato). *There exists a constant* $c > 0$ *such that if* $v_0 \in \dot{H}^{\frac{N}{2}-1}$ *satisfies* $\|v_0\|_{\dot{H}^{\frac{N}{2}-1}} \le c$, *then (INS) has a unique solution in* $L^\infty(\mathbb{R}^+;\dot{H}^{\frac{N}{2}-1}) \cap L^2(\mathbb{R}^+;\dot{H}^{\frac{N}{2}})$.

This paper is organized as follows. In Section 2, we describe a few definitions related to Besov spaces (including homogeneous and hybird Besov spaces) and recall a couple basic properties that will be needed later on. In Section 3, we state the main convergence result, Theorem 3.1, and give a sketch of the proof of Theorem 3.1. The final Section 4 is devoted to the details of the rigorous proof of Theorem 3.1.

2. Besov Spaces and Basics

Throughout this paper, we use C for a generic constant, and denote $A \leq CB$ by $A \lesssim B$. Also we use $(\alpha_q)_{q\in\mathbb{Z}}$ to denote a sequence such that $\sum_{q\in\mathbb{Z}} \alpha_q \leq 1$. $(f|g)$ denotes the inner product of two functions f, g in $L^2(\mathbb{R}^N)$. The standard summation notation over the repeated indices is adopted.

The definition of homogeneous Besov spaces is built on an homogeneous Littlewood–Paley decomposition. First, we introduce a function $\psi \in C^\infty(\mathbb{R}^N)$, supported in the shell

$$\mathcal{C} = \{\xi \in \mathbb{R}^N : \frac{5}{6} \leq |\xi| \leq \frac{12}{5}\},$$

such that

$$\sum_{q\in\mathbb{Z}} \psi(2^{-q}\xi) = 1, \text{ if } \xi \neq 0.$$

Denoting by $h := \mathcal{F}^{-1}\psi$ the inverse Fourier transform of ψ, we define the dyadic blocks as follows:

$$\Delta_q f = \psi(2^{-q}D)f = 2^{qN} \int_{\mathbb{R}^N} h(2^q y) f(x-y)dy,$$

and

$$S_q f = \sum_{p\leq q-1} \Delta_p f,$$

where D is the first order differential operator. The formal decomposition

$$f = \sum_{q\in\mathbb{Z}} \Delta_q f \tag{2.1}$$

is called the homogeneous Littlewood-Paley decomposition. For $s \in \mathbb{R}$ and $f \in \mathcal{S}'(\mathbb{R}^N)$, we denote

$$\|f\|_{B^s} := \sum_{q\in\mathbb{Z}} 2^{sq} \|\Delta_q f\|_{L^2}.$$

The homogeneous Besov spaces are defined as:

Definition 2.1. Let $s \in \mathbb{R}$ and $m = -\left[\frac{N}{2} + 1 - s\right]$. If $m < 0$, we set

$$B^s = \left\{ f \in \mathcal{S}'(\mathbb{R}^N) : \|f\|_{B^s} < \infty \text{ and } f = \sum_{q\in\mathbb{Z}} \Delta_q f \text{ in } \mathcal{S}'(\mathbb{R}^N) \right\}.$$

If $m \geq 0$, we denote by $\mathcal{P}_m$ the set of polynomials with N variables of degree $\leq m$ and define

$$B^s = \left\{ f \in \mathcal{S}'(\mathbb{R}^N)/\mathcal{P}_m : \|f\|_{B^s} < \infty \text{ and } f = \sum_{q\in\mathbb{Z}} \Delta_q f \text{ in } \mathcal{S}'(\mathbb{R}^N)/\mathcal{P}_m \right\}.$$

Functions in B^s have many good properties (for example, see Proposition 2.5 in Ref. 2). A variant of the homogeneous Besov space, so called "hybird Besov space" was proposed to deal with the different regularity for low and high frequencies (see Refs. 2 and 3). The definition of the hybird Besov spaces are given as follows (see Definition 2.8 in Refs. 2 or 3).

Definition 2.2. Let $s, t \in \mathbb{R}$. We set

$$\|f\|_{\tilde{B}^{s,t}} = \sum_{q\leq 0} 2^{qs}\|\Delta_q f\|_{L^2} + \sum_{q>0} 2^{qt}\|\Delta_q f\|_{L^2}.$$

Denoting $m = -\left[\frac{N}{2} + 1 - s\right]$, we define

$$\tilde{B}^{s,t} = \left\{ f \in \mathcal{S}'(\mathbb{R}^N) : \|f\|_{\tilde{B}^{s,t}} < \infty \right\} \text{ if } m < 0,$$

$$\tilde{B}^{s,t} = \left\{ f \in \mathcal{S}'(\mathbb{R}^N)/\mathcal{P}_m : \|f\|_{\tilde{B}^{s,t}} < \infty \right\} \text{ if } m \geq 0.$$

Remark 2.1. Some remarks about the hybrid Besov spaces are in order:

- $\tilde{B}^{s,s} = B^s$;
- If $s \leq t$, then $\tilde{B}^{s,t} = B^s \cap B^t$. Otherwise, $\tilde{B}^{s,t} = B^s + B^t$. In particular, $\tilde{B}^{s,\frac{N}{2}} \hookrightarrow L^\infty$ as $s \leq \frac{N}{2}$;
- The space $\tilde{B}^{0,s}$ coincides with the usual nonhomogeneous Besov space

$$\left\{ f \in \mathcal{S}'(\mathbb{R}^N) : \|\chi(D)f\|_{L^2} + \sum_{q\geq 0} 2^{qs}\|\Delta_q f\|_{L^2} < \infty \right\},$$

  where $\chi(\xi) = 1 - \sum_{q\geq 0} \phi(2^{-q}\xi)$;
- If $s_1 \leq s_2$ and $t_1 \geq t_2$, then $\tilde{B}^{s_1,t_1} \hookrightarrow \tilde{B}^{s_2,t_2}$.

For products of functions in hybrid Besov spaces, we have (see Proposition 1.4 in Ref. 1 and Proposition 2.10 in Ref. 2):

Proposition 2.1. *Given $s_1, s_2, t_1, t_2 \in \mathbb{R}$.*

- *For all $s_1, s_2 > 0$,*

$$\|fg\|_{\tilde{B}^{s_1,s_2}} \lesssim \|f\|_{L^\infty}\|g\|_{\tilde{B}^{s_1,s_2}} + \|g\|_{L^\infty}\|f\|_{\tilde{B}^{s_1,s_2}}.$$

- *If* $1 \le p_1 \le p_2 \le \infty$, $s_1 \le \frac{N}{p_1}$, $s_2 \le \frac{N}{p_2}$ *and* $s_1 + s_2 \ge 0$, *then*

$$fg\|_{B_{p_2}^{s_1+s_2-\frac{N}{p_1}}} \lesssim \|f\|_{B_{p_1}^{s_1}} \|g\|_{B_{p_2}^{s_2}}.$$

In order to state our existence result, we let $T > 0$, $r \in [0, \infty]$ and X be a Banach space. Denote by $\mathcal{M}(0, T; X)$ the set of measurable functions on $(0, T)$ valued in X. For $f \in \mathcal{M}(0, T; X)$, we define

$$\|f\|_{L_T^r(X)} = \left(\int_0^T \|f(\tau)\|_X^r d\tau \right)^{\frac{1}{r}} \quad \text{if } r < \infty,$$

$$\|f\|_{L_T^\infty(X)} = \sup \mathrm{ess}_{\tau \in (0,T)} \|f(\tau)\|_X.$$

Denote

$$L^r(0, T; X) = \{f \in \mathcal{M}(0, T; X) : \|f\|_{L_T^r(X)} < \infty\}.$$

If $T = \infty$, we use $L^r(\mathbb{R}^+; X)$ and $\|f\|_{L^r(X)}$ to denote the corresponding spaces and norms. Also we let $C([0, T], X)$ (or $C(\mathbb{R}^+, X)$) the set of continuous X-valued functions on $[0, T]$ (resp. $\mathbb{R}^+$). We shall further use $C_b(\mathbb{R}^+; X)$ for the set of bounded continuous X-valued functions.

Remark 2.2. If $f(t, x) \in L^r(0, T; B_p^s)$, then the following equivalence holds

$$\|w(\theta \cdot)\|_{B_p^s} \approx \theta^{s - \frac{N}{p}} \|w\|_{B_p^s} \quad \text{for all} \quad \theta > 0, \tag{2.2}$$

$$\|f(\theta^a \cdot, \theta^b \cdot)\|_{L^r(0,\infty; B_p^s)} \approx \theta^{b\left(s - \frac{N}{p}\right) - \frac{a}{r}} \|f\|_{L^r(0,\infty; B_p^s)}. \tag{2.3}$$

3. Main Result and Sketch of Proof

In this section, we state the main convergence result as the parameter tends to zero and outline the sketch of the idea.

To begin with, our main convergence result as $\varepsilon \to 0$ can be described as

Theorem 3.1. *Let* γ *and* Γ *be same as those in Proposition 1.1. Assume that* $\rho_0^\varepsilon = 1 + \varepsilon b_0$ *with* $b_0^\varepsilon \in \tilde{B}^{\frac{N}{2}-1, \frac{N}{2}}$, $\mathbf{u}_0^\varepsilon \in B^{\frac{N}{2}-1}$, $\frac{\mathrm{F}_0^\varepsilon - I}{\varepsilon} \in \tilde{B}^{\frac{N}{2}-1, \frac{N}{2}}$ *satisfy that for all* $0 < \varepsilon$

$$C_0^\varepsilon \stackrel{def}{=} \|b_0^\varepsilon\|_{B^{\frac{N}{2}-1}} + \varepsilon \|b_0^\varepsilon\|_{B^{\frac{N}{2}}} + \|\mathbf{u}_0^\varepsilon\|_{B^{\frac{N}{2}-1}} + \left\| \frac{\mathrm{F}_0^\varepsilon - I}{\varepsilon} \right\|_{B^{\frac{N}{2}-1}} + \varepsilon \left\| \frac{\mathrm{F}_0^\varepsilon - I}{\varepsilon} \right\|_{B^{\frac{N}{2}}}$$
$$\le \gamma,$$

and $\mathcal{P}\mathbf{u}_0^\varepsilon$ *converges to* v_0 *with* $\mathrm{div} v_0 = 0$ *and* $\|v_0\|_{B^{\frac{N}{2}-1}} \le \gamma$. *Then there is a unique global solution* $(\rho^\varepsilon, \mathbf{u}^\varepsilon, \mathrm{F}^\varepsilon)$ *of* (CV^ε) *such that*

$$C^\varepsilon \stackrel{def}{=} \|b^\varepsilon\|_{B^{\frac{N}{2}-1}} + \varepsilon\|b^\varepsilon\|_{B^{\frac{N}{2}}} + \|\mathbf{u}^\varepsilon\|_{B^{\frac{N}{2}-1}} + \left\|\frac{\mathrm{F}^\varepsilon - I}{\varepsilon}\right\|_{B^{\frac{N}{2}-1}} + \varepsilon\left\|\frac{\mathrm{F}^\varepsilon - I}{\varepsilon}\right\|_{B^{\frac{N}{2}}}$$
$$\le \Gamma\gamma.$$

Moreover the density ρ^ε, *the velocity* $\mathbf{u}^\varepsilon$, *and the variance of the deformation gradient* $\frac{\mathrm{F}^\varepsilon - I}{\varepsilon}$ *converge to* 1, v *and* 0, *respectively, where* v *is the unique solution to (INS) with initial data* v_0.

Remark 3.1. We note that if $(\overline{b}, \overline{u}, \overline{\mathrm{F}})$ is a solution of (1.1), then $b^\varepsilon = \varepsilon^{-1}\overline{b}(x/\varepsilon, t/\varepsilon^2)$, $u^\varepsilon = \varepsilon^{-1}\overline{\mathbf{u}}(x/\varepsilon, t/\varepsilon^2)$ and $F^\varepsilon = \overline{\mathrm{F}}(x/\varepsilon, t/\varepsilon^2)$ is a solution of (CV^ε). Moreover, the norms C_0^ε and C^ε correspond exactly to the norm defined in Prop. 1.1. Hence the global existence described in the theorem 3.1 follows from Prop. 1.1, see also section 4.1 below for details.

Theorem 3.1 implies that the first two equations in (CV^ε) will converge to their couterparts in incompressible Navier-Stokes equations, while the variance of the deformation gradient from the identity matrix will converges to zero. However, the kind of convergences we can establish here does not ensure what would be the limit of the third equation in (CV^ε). It is partially due to the high singular term $\frac{\nabla \mathbf{u}}{\varepsilon}$ in the equations and the lack of *a priori* estimates on the time derivative of $\frac{\mathrm{F}^\varepsilon - I}{\varepsilon}$.

Remark 3.2. In the case of initial-boundary value problems, the rigorously verification will depend on the estimates near the spatial boundary that will be discussed elsewhere. One can also obtain various convergences and verify a similar result as Theorem 3.1 in the case of periodic domain, except that we can only obtain the weak convergence of the incompressible part of the velocity since we can not apply the dispersive estimate for wave equations, see also Ref. 1.

Remark 3.3. The system (CV^ε) is similar to the sigular limit of compressible magnetohydrodynamic fluids (MHD) with zero magnetic diffusivity in Ref. 12 as the Mach number and the Alfven number go to zero simultenously. As discussed in Ref. 12, when the Alfven number goes to zero, one can obtain a uniform bound for the Lorentz forces, but can not guarantee *a priori* convergence of the Loerentz forces. Here we are able to verify the convergence of the elastic forces as the parameter goes to zero. What has played key roles in our arguments is a set of local conservation laws(quantities), see Proposition 3.1. The extension of our result to MHD may become possible now because of the recent global well-posedness result for the incompressible MHD equations with zero magnetic diffusivity and small initial data established in Ref. 11.

We now outline the proofs of the Theorem 3.1. First of all, we recall conservation laws or local conserved quantities for compressible viscoelastic fluids (see, for example, Refs. 7 and 13).

Proposition 3.1. *Assume that* $(\rho^\varepsilon, \mathbf{u}^\varepsilon, \mathtt{F}^\varepsilon)$ *is a solution of the system* (CV^ε)*. Then the following identities*

$$\mathrm{div}(\rho^\varepsilon(\mathtt{F}^\varepsilon)^\top) = 0, \tag{3.1}$$

and

$$(\mathtt{F}^\varepsilon)_{lk}\nabla_l(\mathtt{F}^\varepsilon)_{ij} = (\mathtt{F}^\varepsilon)_{lj}\nabla_l(\mathtt{F}^\varepsilon)_{ik} \tag{3.2}$$

hold for all time $t > 0$ *if they are true initially.*

Thanks to Proposition 3.1, we shall reformulate (CV^ε). First, one notices that

$$\mathtt{F}^\varepsilon_{jk}\partial_j\mathtt{F}^\varepsilon_{ik} = \mathrm{div}\mathtt{F}^\varepsilon + (\mathtt{F}^\varepsilon_{jk} - \delta_{jk})\nabla_j\mathtt{F}^\varepsilon_{ik} \tag{3.3}$$

where

$$\delta_{ij} = \begin{cases} 0, & \text{if} \quad i \neq j; \\ 1, & \text{if} \quad i = j. \end{cases}$$

Since $\mathrm{divcurl} f = 0$ for any vector-valued function f, one has

$$(-\Delta)^{-1}\mathrm{divcurldiv}(\mathtt{F}^\varepsilon - I) = 0.$$

On the other hand, we have

$$\begin{aligned}
&(-\Delta)^{-1}\mathrm{divcurldiv}(\mathtt{F}^\varepsilon - I)\\
&= (-\Delta)^{-1}\mathrm{div}(\partial_{x_k}\partial_{x_j}(\mathtt{F}^\varepsilon_{ij} - \delta_{ij}) - \partial_{x_i}\partial_{x_j}(\mathtt{F}^\varepsilon_{kj} - \delta_{kj}))\\
&\overset{(3.2)}{=} (-\Delta)^{-1}\mathrm{div}(\partial_{x_j}\partial_{x_j}(\mathtt{F}^\varepsilon_{ik} - \delta_{ik}) - \partial_{x_j}\partial_{x_j}(\mathtt{F}^\varepsilon_{ki} - \delta_{ki}))\\
&\quad + (-\Delta)^{-1}\mathrm{div}\partial_{x_j}\Big((\mathtt{F}^\varepsilon_{lj} - \delta_{lj})\nabla_l\mathtt{F}^\varepsilon_{ik} - (\mathtt{F}^\varepsilon_{lk} - \delta_{lk})\nabla_l\mathtt{F}^\varepsilon_{ij}\\
&\quad - (\mathtt{F}^\varepsilon_{lj} - \delta_{lj})\nabla_l\mathtt{F}^\varepsilon_{ki} + (\mathtt{F}^\varepsilon_{li} - \delta_{li})\nabla_l\mathtt{F}^\varepsilon_{kj}\Big)\\
&= \mathrm{div}(\mathtt{F}^\varepsilon - I) - \mathrm{div}((\mathtt{F}^\varepsilon)^\top - I) + \mathcal{N}^\varepsilon
\end{aligned}$$

with

$$\mathcal{N}^\varepsilon = (-\Delta)^{-1}\mathrm{div}\partial_{x_j}\Big((\mathrm{F}^\varepsilon_{lj} - \delta_{lj})\nabla_l \mathrm{F}^\varepsilon_{ik} - (\mathrm{F}^\varepsilon_{lk} - \delta_{lk})\nabla_l \mathrm{F}^\varepsilon_{ij} \\ - (\mathrm{F}^\varepsilon_{lj} - \delta_{lj})\nabla_l \mathrm{F}^\varepsilon_{ki} + (\mathrm{F}^\varepsilon_{li} - \delta_{li})\nabla_l \mathrm{F}^\varepsilon_{kj})\Big).$$

Thus we obtain

$$\mathrm{div}(\mathrm{F}^\varepsilon - I) = \mathrm{div}((\mathrm{F}^\varepsilon)^\top - I) - \mathcal{N}^\varepsilon. \tag{3.4}$$

By (3.1), one has

$$\mathrm{div}((\mathrm{F}^\varepsilon)^\top - I) = -\varepsilon\nabla b^\varepsilon - \varepsilon\mathrm{div}(((\mathrm{F}^\varepsilon)^\top - I)b^\varepsilon). \tag{3.5}$$

Thanks to (3.3), (3.4), and (3.5), we can rewrite (1.3) as

$$\begin{cases} b^\varepsilon_t + \frac{\mathrm{div}\mathbf{u}^\varepsilon}{\varepsilon} = -\mathrm{div}(b^\varepsilon \mathbf{u}^\varepsilon), \\ (\mathbf{u}^\varepsilon_i)_t + \mathbf{u}^\varepsilon\cdot\nabla\mathbf{u}^\varepsilon_i - \frac{\mu\Delta\mathbf{u}^\varepsilon_i + (\lambda+\mu)\partial_i\mathrm{div}\mathbf{u}^\varepsilon}{1+\varepsilon b^\varepsilon} + \left(1 + \frac{P'(1+\varepsilon b^\varepsilon)}{1+\varepsilon b^\varepsilon}\right)\frac{\partial_i b^\varepsilon}{\varepsilon} = \mathcal{M}^\varepsilon_i, \\ \mathrm{F}^\varepsilon_t + \mathbf{u}^\varepsilon\cdot\nabla\mathrm{F}^\varepsilon = \nabla\mathbf{u}^\varepsilon\,\mathrm{F}^\varepsilon, \\ (b^\varepsilon, \mathbf{u}^\varepsilon, \mathrm{F}^\varepsilon)|_{t=0} = (b^\varepsilon_0, \mathbf{u}^\varepsilon_0, \mathrm{F}^\varepsilon_0) \end{cases} \tag{3.6}$$

with

$$\mathcal{M}^\varepsilon_i = \frac{1}{\varepsilon^2}\left((\mathrm{F}^\varepsilon_{jk} - \delta_{jk})\nabla_j \mathrm{F}^\varepsilon_{ik} - \varepsilon\mathrm{div}(((\mathrm{F}^\varepsilon)^\top - I)b^\varepsilon) - \mathcal{N}^\varepsilon\right).$$

To obtain uniform estimates, let us split the velocity into a divergence-free part $\mathcal{P}\mathbf{u}^\varepsilon$ and a gradient part $\mathcal{Q}\mathbf{u}^\varepsilon$. The operator $\mathcal{P}$ is the projection to the divergence-free vectors, and the operator $\mathcal{Q}$ is the projection to its orthogonal compliment, that is, the curl-free vectors. They can be defined as

$$\mathcal{P} = I - \nabla\Delta^{-1}\mathrm{div}, \quad \text{and} \quad \mathcal{Q} = I - \mathcal{P},$$

where I is the identity operator. The equation (CV^ε) can then be split according to this decomposition of the velocity:

$$\partial_t b^\varepsilon + \frac{\mathrm{div}\mathcal{Q}\mathbf{u}^\varepsilon}{\varepsilon} = -\mathrm{div}(b^\varepsilon\mathbf{u}^\varepsilon), \tag{3.7a}$$

$$\partial_t\mathcal{Q}\mathbf{u}^\varepsilon - \nu\Delta\mathcal{Q}\mathbf{u}^\varepsilon + 2\frac{\nabla b^\varepsilon}{\varepsilon} = \mathcal{Q}\left(\mathcal{M} - \mathbf{u}^\varepsilon\cdot\nabla\mathbf{u}^\varepsilon - \frac{\varepsilon b^\varepsilon}{1+\varepsilon b^\varepsilon}\mathcal{A}\mathbf{u}^\varepsilon - K(\varepsilon b^\varepsilon)\frac{\nabla b^\varepsilon}{\varepsilon}\right), \tag{3.7b}$$

$$\partial_t\mathcal{P}\mathbf{u}^\varepsilon - \nu\Delta\mathcal{P}\mathbf{u}^\varepsilon = \mathcal{P}\left(\mathcal{M} - \mathbf{u}^\varepsilon\cdot\nabla\mathbf{u}^\varepsilon - \frac{\varepsilon b^\varepsilon}{1+\varepsilon b^\varepsilon}\mathcal{A}\mathbf{u}^\varepsilon\right), \tag{3.7c}$$

$$\mathrm{F}^\varepsilon_t + \mathbf{u}^\varepsilon\cdot\nabla\mathrm{F}^\varepsilon = \nabla\mathbf{u}^\varepsilon\,\mathrm{F}^\varepsilon, \tag{3.7d}$$

$$(b^\varepsilon, \mathbf{u}^\varepsilon, \mathrm{F}^\varepsilon)|_{t=0} = (b^\varepsilon_0, \mathbf{u}^\varepsilon_0, \mathrm{F}^\varepsilon_0), \tag{3.7e}$$

where the vector $\mathcal{M}$ and the function K are defined as

$$\mathcal{M}^\varepsilon \stackrel{def}{=} (\mathcal{M}_1^\varepsilon, \mathcal{M}_2^\varepsilon, \mathcal{M}_3^\varepsilon),$$

$$K(z) \stackrel{def}{=} \frac{P'(1+z)}{1+z} - 1$$

with $K(0) = 0$ and $\mathcal{A} = \mu\Delta + (\lambda+\mu)\nabla\mathrm{div}$. Here the implicit summation over repeated indices is assumed throughout this paper.

We note that in the equations for the incompressible part $\mathcal{P}\mathbf{u}^\varepsilon$, there is no linear coupling with other three equations in the system (3.7), and hence it is reasonable to believe that one may obtain a control on $\mathcal{P}\mathbf{u}^\varepsilon$ in view of the standard theory for parabolic equations. On the other hand, there is a strong coupling between other three equations in the system (3.7). To overcome the difficulties, we will use in particular some dispersive estimates for the wave equations (the so-called *Strichartz* estimates, see Ref. 1 and references therein). More precisely, the following lemma will be needed in our arguments. One may find its proof in Proposition 7.1 in Ref. 1.

Lemma 3.1 (*Strichartz* estimates). *Let (b,d) be a solution of the following system*

$$\begin{cases} \partial_t b + \frac{\Lambda d}{\varepsilon} = J, \\ \partial_t d - \frac{\Lambda b}{\varepsilon} = K. \end{cases}$$

Then for any $s \in \mathbb{R}$ and positive time T (possible infinite), the following estimate holds

$$\|(b,d)\|_{L^r_T(B_p^{s+N(1/p-1/2)+1/r})} \lesssim \varepsilon^{1/r}(\|(b_0,d_0)\|_{B_2^s} + \|(J,K)\|_{L^1_T(B_2^s)})$$

with $p \geq 2$, $2/r \leq \min(1,(N-1)(1/2-1/p))$ and $(r,p,N) \neq (2,\infty,3)$.

Strichartz estimates for wave equations along with some estimates for parabolic equations will give us a list of uniform estimates on $(b^\varepsilon, \mathbf{u}^\varepsilon, \frac{\mathrm{F}^\varepsilon - I}{\varepsilon})$. The latter would be important to derive convergences of various quantities. Indeed our main result will follow from the following Proposition.

Proposition 3.2. *Under assumptions as in Theorem 3.1, the following results hold:*

- *Existence: For all $0 < \varepsilon$, system (CV^ε) has a unique solution $(b^\varepsilon, \mathbf{u}^\varepsilon, \mathrm{F}^\varepsilon)$ in $\mathfrak{B}^{\frac{N}{2}}$ such that The norm as defined in Theorem 3.1,*

C^ε is bounded by $\Gamma\gamma$. (INS) has a unique solution $v \in \mathfrak{B}^{\frac{N}{2}}$ such that

$$\|v\|_{L^1(B^{\frac{N}{2}+1})} + \|v\|_{L^\infty(B^{\frac{N}{2}-1})} \leq \Gamma \|v_0\|_{B^{\frac{N}{2}-1}}.$$

- *Convergence: If*

$$\begin{cases} \|\mathcal{P}\mathbf{u}_0^\varepsilon - v_0\|_{\tilde{B}_p^{\frac{3}{p}-1,\frac{2}{p}-\frac{3}{2}}} \lesssim \varepsilon, & \text{as} \quad N=3, \\ \|\mathcal{P}\mathbf{u}_0^\varepsilon - v_0\|_{\tilde{B}_p^{\frac{2}{p}-1,\frac{3}{2p}-\frac{7}{4}}} \lesssim \varepsilon, & \text{as} \quad N=2, \end{cases}$$

then

- *If $N=3$, for all $p \in [2,\infty)$ and $q \in (1,2)$,*

$$\|\mathcal{P}\mathbf{u}^\varepsilon - v\|_{L^1(\tilde{B}_p^{\frac{3}{p}+1,\frac{2}{p}+\frac{1}{2}})} + \|\mathcal{P}\mathbf{u}^\varepsilon - v\|_{L^\infty(\tilde{B}_p^{\frac{3}{p}-1,\frac{2}{p}-\frac{3}{2}})} + \|(b^\varepsilon, \mathcal{Q}\mathbf{u}^\varepsilon)\|_{L^{\frac{2p}{p-2}}(B_p^{\frac{2}{p}-\frac{1}{2}})} \lesssim \varepsilon^{\frac{1}{2}-\frac{1}{p}},$$

and

$$\left\| \frac{\mathrm{F}^\varepsilon - I}{\varepsilon} \right\|_{L^q_{loc}(\tilde{B}_p^{\frac{3}{p},\frac{2}{p}-\frac{1}{2}})} \lesssim \varepsilon^{\min\{\frac{2-q}{q},\frac{1}{2}-\frac{1}{p}\}}.$$

- *If $N=2$, for all $p \in [2,6]$ and $q \in (1,2)$,*

$$\|\mathcal{P}\mathbf{u}^\varepsilon - v\|_{L^1(\tilde{B}_p^{\frac{2}{p}+1,\frac{3}{2p}+\frac{1}{4}})} + \|\mathcal{P}\mathbf{u}^\varepsilon - v\|_{L^\infty(\tilde{B}_p^{\frac{2}{p}-1,\frac{3}{2p}-\frac{7}{4}})} + \|(b^\varepsilon, \mathcal{Q}\mathbf{u}^\varepsilon)\|_{L^{\frac{4p}{p-2}}(B_p^{\frac{3}{2p}-\frac{3}{4}})} \lesssim \varepsilon^{\frac{1}{4}-\frac{1}{2p}},$$

and

$$\left\| \frac{\mathrm{F}^\varepsilon - I}{\varepsilon} \right\|_{L^q_{loc}(\tilde{B}_p^{\frac{2}{p},\frac{3}{2p}-\frac{3}{4}})} \lesssim \varepsilon^{\min\{\frac{2-q}{q},\frac{1}{4}-\frac{1}{2p}\}}.$$

To prove Proposition 3.2, we first use an appropriate change of variables which enables us to apply Proposition 1.1. Under the smallness assumptions, we get in particular global solutions $(b^\varepsilon, \mathbf{u}^\varepsilon, \mathrm{F}^\varepsilon)$ in $\mathfrak{B}^{\frac{N}{2}}$ with a list of estimates which are uniform in ε. Next, as these uniform estimates will lead to uniform bounds in $L^1(\mathbb{R}^+; B_2^{\frac{N}{2}-1})$ for the right-hand side of the first two equations in (3.7), and for $\nu\Delta\mathcal{Q}\mathbf{u}^\varepsilon$, then we use Lemma 3.1 to conclude that $(b^\varepsilon, \mathcal{Q}\mathbf{u}^\varepsilon)$ converges to zero. Also, these uniform estimates enable us to prove the nonlinear part $\mathcal{M}$ will converge to 0 as $\varepsilon \to 0$. Finally, the estimates for heat equations allow us to verify that $\mathcal{P}\mathbf{u}^\varepsilon \to v$ in a suitable functional space. Then the strong convergence will be followed by a classical interpolation method with the aid of these uniform estimates.

4. Proof of Proposition 3.2

The proof will be divided into several steps.

4.1. *Existence of global solution for* (CV^ε) *and uniform estimates*

To this end, let us introduce the following change of unknowns and change of variables:

$$\overline{\rho} \stackrel{def}{=} \varepsilon b(\varepsilon^2 t, \varepsilon x), \quad \overline{\mathbf{u}} \stackrel{def}{=} \varepsilon \mathbf{u}(\varepsilon^2 t, \varepsilon x) \quad \text{and} \quad \overline{\mathtt{F}} \stackrel{def}{=} \mathtt{F}(\varepsilon^2 t, \varepsilon x).$$

Then $(b, \mathbf{u}, \mathtt{F})$ solves (1.3) if and only if $(\overline{\rho}, \overline{\mathbf{u}}, \overline{\mathtt{F}})$ solves

$$\begin{cases} \overline{\rho}_t + \mathrm{div}\overline{\mathbf{u}} = -\mathrm{div}(\overline{\rho}\overline{\mathbf{u}}), \\ (\overline{\mathbf{u}}_i)_t + \overline{\mathbf{u}} \cdot \nabla \overline{\mathbf{u}}_i - \frac{\mu\Delta\overline{\mathbf{u}}_i + (\lambda+\mu)\partial_i \mathrm{div}\overline{\mathbf{u}}}{1+\overline{\rho}} + \frac{P'(1+\overline{\rho})}{1+\overline{\rho}}\partial_i\overline{\rho} = \alpha\overline{\mathtt{F}}_{jk}\partial_j\overline{\mathtt{F}}_{ik}, \\ \overline{\mathtt{F}}_t + \overline{\mathbf{u}} \cdot \nabla\overline{\mathtt{F}} = \nabla\overline{\mathbf{u}}\,\overline{\mathtt{F}}, \\ (\overline{\rho}, \overline{\mathbf{u}}, \overline{\mathtt{F}})|_{t=0} = (\overline{\rho}_0, \overline{\mathbf{u}}_0, \overline{\mathtt{F}}_0). \end{cases} \tag{4.1}$$

According to Proposition 1.1, there exist two positive constants

$$\gamma = \gamma(N, \mu, \lambda, P) \quad \text{and} \quad \Gamma = \Gamma(N, \mu, \lambda, P)$$

such that (4.1) has a unique solution in $\mathfrak{B}^{\frac{N}{2}}$ satisfying

$$\|(\overline{\rho}, \overline{\mathbf{u}}, \overline{\mathtt{F}})\|_{\mathfrak{B}^{\frac{N}{2}}} \le \Gamma\|(\overline{\rho}_0, \overline{\mathbf{u}}_0, \overline{\mathtt{F}}_0)\|_{\mathfrak{B}^{\frac{N}{2}}}, \tag{4.2}$$

provided that

$$C_0^\varepsilon \stackrel{def}{=} \|\overline{\rho}_0\|_{\tilde{B}^{\frac{N}{2}-1,\frac{N}{2}}} + \|\overline{\mathbf{u}}_0\|_{B^{\frac{N}{2}-1}} + \|\overline{\mathtt{F}}_0 - I\|_{\tilde{B}^{\frac{N}{2}-1,\frac{N}{2}}} \le \gamma. \tag{4.3}$$

Due to the scaling properties (2.2), (2.3), we easily obtain

$$\begin{aligned} &\|\overline{\rho}_0\|_{\tilde{B}^{\frac{N}{2}-1,\frac{N}{2}}} + \|\overline{\mathbf{u}}_0\|_{B^{\frac{N}{2}-1}} + \|\overline{\mathtt{F}}_0 - I\|_{\tilde{B}^{\frac{N}{2}-1,\frac{N}{2}}} \\ &\quad = \|b_0\|_{B^{\frac{N}{2}-1}} + \varepsilon\|b_0\|_{B^{\frac{N}{2}}} + \|\mathbf{u}_0\|_{B^{\frac{N}{2}-1}} + \varepsilon\left\|\frac{\mathtt{F}_0 - I}{\varepsilon}\right\|_{B^{\frac{N}{2}}} + \left\|\frac{\mathtt{F}_0 - I}{\varepsilon}\right\|_{B^{\frac{N}{2}-1}}, \end{aligned} \tag{4.4}$$

and

$$\begin{aligned} &\|(\overline{\rho}, \overline{\mathbf{u}}, \overline{\mathtt{F}})\|_{\mathfrak{B}^{\frac{N}{2}}} \\ &\quad \ge \varepsilon\|b^\varepsilon\|_{L^\infty(0,\infty;B^{\frac{N}{2}})} + \|b^\varepsilon\|_{L^\infty(0,\infty;B^{\frac{N}{2}-1})} + \varepsilon\left\|\frac{\mathtt{F}^\varepsilon - I}{\varepsilon}\right\|_{L^\infty(0,\infty;B^{\frac{N}{2}})} \\ &\quad + \left\|\frac{\mathtt{F}^\varepsilon - I}{\varepsilon}\right\|_{L^\infty(0,\infty;B^{\frac{N}{2}-1})} + \|\mathbf{u}^\varepsilon\|_{L^\infty(0,\infty;B^{\frac{N}{2}-1})} + \|\mathbf{u}^\varepsilon\|_{L^1(0,\infty;B^{\frac{N}{2}+1})}. \end{aligned} \tag{4.5}$$

Moreover, by the interpolation

$$\|w\|_{B^s} \le \|w\|^{\frac{1}{2}}_{\tilde{B}^{s-1,s}} \|w\|^{\frac{1}{2}}_{\tilde{B}^{s+1,s}},$$

one has

$$\|w\|_{L^2(0,\infty;B^s)} \le \|w\|^{\frac{1}{2}}_{L^\infty(0,\infty;\tilde{B}^{s-1,s})} \|w\|^{\frac{1}{2}}_{L^1(0,\infty;\tilde{B}^{s+1,s})}.$$

Hence, from (4.2) and the definition of $\mathfrak{B}^{\frac{N}{2}}$, we obtain that

$$\|\overline{\rho}\|_{L^2(0,\infty;B^{\frac{N}{2}})} + \|\overline{\mathtt{F}} - I\|_{L^2(0,\infty;B^{\frac{N}{2}})} \le C\Gamma\gamma.$$

The latter implies that

$$\|b^\varepsilon\|_{L^2(0,\infty;B^{\frac{N}{2}})} + \left\|\frac{\mathtt{F}^\varepsilon - I}{\varepsilon}\right\|_{L^2(0,\infty;B^{\frac{N}{2}})} \le C\Gamma\gamma. \tag{4.6}$$

Based on these uniform estimates, one can show that

Lemma 4.1. *For these solutions, $\mathcal{M}^\varepsilon$ are uniformly bounded in $L^1(B^{\frac{N}{2}-1})$. More precisely, we have*

$$\|\mathcal{M}^\varepsilon\|_{L^1(B^{\frac{N}{2}-1})} \lesssim C\Gamma^2\gamma^2.$$

Proof. Recall that

$$\mathcal{M}^\varepsilon_i = \frac{1}{\varepsilon^2}\left((\mathtt{F}^\varepsilon_{jk} - \delta_{jk})\nabla_j \mathtt{F}^\varepsilon_{ik} - \varepsilon \mathrm{div}(((\mathtt{F}^\varepsilon)^\top - I)b^\varepsilon) - \mathcal{N}^\varepsilon\right).$$

To prove this lemma, we estimate $\mathcal{M}^\varepsilon$ term by term. First, for $\frac{1}{\varepsilon^2}\left((\mathtt{F}^\varepsilon_{jk} - \delta_{jk})\nabla_j \mathtt{F}^\varepsilon_{ik}\right)$, we have, by (4.6)

$$\begin{aligned}
&\left\|\frac{1}{\varepsilon^2}\left((\mathtt{F}^\varepsilon_{jk} - \delta_{jk})\nabla_j \mathtt{F}^\varepsilon_{ik}\right)\right\|_{L^1(B^{\frac{N}{2}-1})} \\
&= \left\|\frac{1}{\varepsilon^2}\left((\mathtt{F}^\varepsilon_{jk} - \delta_{jk})\nabla_j (\mathtt{F}^\varepsilon_{ik} - \delta_{ik})\right)\right\|_{L^1(B^{\frac{N}{2}-1})} \\
&\lesssim \left\|\frac{\mathtt{F}^\varepsilon - I}{\varepsilon}\right\|_{L^2(B^{\frac{N}{2}})} \left\|\nabla\left(\frac{\mathtt{F}^\varepsilon - I}{\varepsilon}\right)\right\|_{L^2(B^{\frac{N}{2}-1})} \\
&\lesssim \left\|\frac{\mathtt{F}^\varepsilon - I}{\varepsilon}\right\|^2_{L^2(B^{\frac{N}{2}})} \le C\Gamma^2\gamma^2.
\end{aligned} \tag{4.7}$$

Secondly, for the term $\frac{1}{\varepsilon}\mathrm{div}(((\mathtt{F}^\varepsilon)^\top - I)b^\varepsilon)$, we observe that, by (4.6)

$$\begin{aligned}\left\|\frac{1}{\varepsilon}\mathrm{div}\left(((\mathtt{F}^\varepsilon)^\top - I)\,b^\varepsilon\right)\right\|_{L^1(B^{\frac{N}{2}-1})} &\lesssim \left\|\frac{\mathtt{F}^\varepsilon - I}{\varepsilon}b^\varepsilon\right\|_{L^1(B^{\frac{N}{2}})}\\ &\lesssim \left\|\frac{\mathtt{F}^\varepsilon - I}{\varepsilon}\right\|_{L^2(B^{\frac{N}{2}})}\|b^\varepsilon\|_{L^2(B^{\frac{N}{2}})} \le C\Gamma^2\gamma^2.\end{aligned} \tag{4.8}$$

Finally, for $\frac{1}{\varepsilon^2}\mathcal{N}^\varepsilon$, we have, by (4.6)

$$\begin{aligned}\left\|\frac{1}{\varepsilon^2}\mathcal{N}^\varepsilon\right\|_{L^1(B^{\frac{N}{2}-1})} &\lesssim \left\|\frac{\mathtt{F}^\varepsilon - I}{\varepsilon}\nabla\left(\frac{\mathtt{F}^\varepsilon - I}{\varepsilon}\right)\right\|_{L^1(B^{\frac{N}{2}-1})}\\ &\lesssim \left\|\frac{\mathtt{F}^\varepsilon - I}{\varepsilon}\right\|_{L^2(B^{\frac{N}{2}})}\left\|\nabla\left(\frac{\mathtt{F}^\varepsilon - I}{\varepsilon}\right)\right\|_{L^2(B^{\frac{N}{2}-1})}\\ &\lesssim \left\|\frac{\mathtt{F}^\varepsilon - I}{\varepsilon}\right\|^2_{L^2(B^{\frac{N}{2}})} \le C\Gamma^2\gamma^2.\end{aligned} \tag{4.9}$$

This completes the proof. □

From the bounds of $\mathcal{M}$ and $\mathbf{u}^\varepsilon$ in $L^1(B^{\frac{N}{2}-1})$, equation (3.7c) implies that $\partial_t\mathcal{P}\mathbf{u}^\varepsilon$ is also bounded in $L^1(B^{\frac{N}{2}-1})$. Also, as an easy consequence of the uniform bound on $\frac{\mathcal{N}^\varepsilon}{\varepsilon^2}$, one obtains the uniform bound on the divergece free part of $\frac{1}{\varepsilon^2}\mathrm{div}(\mathtt{F}^\varepsilon - I)$, $\mathcal{P}\frac{1}{\varepsilon^2}\mathrm{div}(\mathtt{F}^\varepsilon - I)$, in $L^1(B^{\frac{N}{2}-1})$. More precisely, we have

Corollary 4.1. *For the sequence $\{\mathtt{F}^\varepsilon\}_{\{\varepsilon>0\}}$, the following bound holds*

$$\left\|\frac{1}{\varepsilon}\mathcal{P}\mathrm{div}\left(\frac{\mathtt{F}^\varepsilon - I}{\varepsilon}\right)\right\|_{L^1(B^{\frac{N}{2}-1})} \le C\Gamma^2\gamma^2.$$

Proof. Indeed, from (3.4) and (3.5), one deduces that

$$\mathrm{div}(\mathtt{F}^\varepsilon - I) = -\varepsilon\nabla b^\varepsilon - \varepsilon\mathrm{div}(((\mathtt{F}^\varepsilon)^\top - I)b^\varepsilon) - \mathcal{N}^\varepsilon,$$

and hence

$$\frac{1}{\varepsilon}\mathcal{P}\mathrm{div}\left(\frac{\mathtt{F}^\varepsilon - I}{\varepsilon}\right) = -\mathcal{P}\mathrm{div}\left(\frac{(\mathtt{F}^\varepsilon)^\top - I}{\varepsilon}b^\varepsilon\right) - \mathcal{P}\frac{\mathcal{N}^\varepsilon}{\varepsilon^2}.$$

Inequalities (4.8) and (4.9) implies that the right-hand side of the above identity is bounded by $C\Gamma^2\gamma^2$ in $L^1(B^{\frac{N}{2}-1})$, and thus

$$\left\|\frac{1}{\varepsilon}\mathcal{P}\mathrm{div}\left(\frac{\mathtt{F}^\varepsilon - I}{\varepsilon}\right)\right\|_{L^1(B^{\frac{N}{2}-1})} \le C\Gamma^2\gamma^2.$$

This finishes the proof. □

4.2. *Convergence of b^ε and $\mathcal{Q}\mathbf{u}^\varepsilon$*

The convergence of b^ε and $\mathcal{Q}\mathbf{u}^\varepsilon$ are based on the following lemma. In the rest of this paper, the space $B^s_{p,1}$ will be denoted by B^s_p for convenience.

Lemma 4.2. *Let C_0^ε is defined as* (4.3). *The solutions obtained in the previous subsection satisfy*

- *If $N = 3$: For all $2 \le p < \infty$,*

$$\|(b^\varepsilon, \mathcal{Q}\mathbf{u}^\varepsilon)\|_{L^{\frac{2p}{p-2}}(B_p^{\frac{2}{p}-\frac{1}{2}})} \lesssim \varepsilon^{\frac{1}{2}-\frac{1}{p}};$$

- *If $N = 2$: For all $2 \le p \le \infty$,*

$$\|(b^\varepsilon, \mathcal{Q}\mathbf{u}^\varepsilon)\|_{L^{\frac{4p}{p-2}}(B_p^{\frac{3}{2p}-\frac{3}{4}})} \lesssim \varepsilon^{\frac{1}{4}-\frac{1}{2p}}.$$

Proof. Let

$$d^\varepsilon \stackrel{def}{=} \Lambda^{-1}\mathrm{div}\mathcal{Q}\mathbf{u}^\varepsilon,$$

we have, according to (3.7)

$$\begin{cases} \partial_t b^\varepsilon + \frac{\Lambda d^\varepsilon}{\varepsilon} = J, \\ \partial_t d^\varepsilon - 2\frac{\Lambda b^\varepsilon}{\varepsilon} = K, \\ (b^\varepsilon, d^\varepsilon)|_{t=0} = (b_0^\varepsilon, \Lambda^{-1}\mathrm{div}\mathcal{Q}\mathbf{u}_0^\varepsilon), \end{cases} \tag{4.10}$$

with

$$J \stackrel{def}{=} -\mathrm{div}(b^\varepsilon \mathbf{u}^\varepsilon),$$

and

$$K \stackrel{def}{=} \nu\Delta d^\varepsilon - \Lambda^{-1}\mathrm{div}\mathcal{Q}\left(\mathbf{u}^\varepsilon\cdot\nabla\mathbf{u}^\varepsilon + \frac{\varepsilon b^\varepsilon}{1+\varepsilon b^\varepsilon}\mathcal{A}\mathbf{u}^\varepsilon + \frac{K(\varepsilon b^\varepsilon)\nabla b^\varepsilon}{\varepsilon} - \mathcal{M}^\varepsilon\right).$$

One notices that $\mathcal{Q}\mathbf{u}^\varepsilon = -\nabla\Lambda^{-1}d^\varepsilon$, and hence estimations for $\mathcal{Q}\mathbf{u}^\varepsilon$ or d^ε are equivalent (up to harmless constants). Applying Lemma 3.1 to (4.10) with $s = \frac{N}{2} - 1$, $2 \le p < \infty$ and $r = \frac{2p}{p-2}$ if $N = 3$ or $2 \le p \le \infty$ and $r = \frac{4p}{p-2}$ if $N = 2$, we readily obtain estimates in Lemma 4.2, whenever

$$\|(J, K)\|_{L^1(B^{\frac{N}{2}-1})} \lesssim C\Gamma^2\gamma^2, \tag{4.11}$$

is valid. Here we have observed that by (4.4), one has

$$\|(b_0^\varepsilon, d_0^\varepsilon)\|_{B^{\frac{N}{2}-1}} \lesssim C\Gamma^2\gamma^2.$$

To verify (4.11), we need to use uniform estimates obtained in the previous subsection. Indeed, one can derive by using (4.4)–(4.6) that

$$\begin{aligned}\|J\|_{L^1(B^{\frac{N}{2}-1})} &\lesssim \|b^\varepsilon \mathbf{u}^\varepsilon\|_{L^1(B^{\frac{N}{2}})} \lesssim \|b^\varepsilon\|_{L^2(B^{\frac{N}{2}})}\|\mathbf{u}^\varepsilon\|_{L^2(B^{\frac{N}{2}})}\\ &\lesssim \|b^\varepsilon\|_{L^2(B^{\frac{N}{2}})}\|\mathbf{u}^\varepsilon\|^{\frac{1}{2}}_{L^\infty(B^{\frac{N}{2}-1})}\|\mathbf{u}^\varepsilon\|^{\frac{1}{2}}_{L^1(B^{\frac{N}{2}+1})} \lesssim C\Gamma^2\gamma^2,\end{aligned}$$

$$\|\Delta d^\varepsilon\|_{L^1(B^{\frac{N}{2}-1})} \lesssim \|d^\varepsilon\|_{L^1(B^{\frac{N}{2}+1})} \lesssim \|\mathbf{u}^\varepsilon\|_{L^1(B^{\frac{N}{2}+1})} \lesssim C\Gamma^2\gamma^2,$$

$$\|\mathbf{u}^\varepsilon\cdot\nabla\mathbf{u}^\varepsilon\|_{L^1(B^{\frac{N}{2}-1})} \lesssim \|\mathbf{u}^\varepsilon\|_{L^2(B^{\frac{N}{2}})}\|\nabla\mathbf{u}^\varepsilon\|_{L^2(B^{\frac{N}{2}-1})} \lesssim C\Gamma^2\gamma^2,$$

$$\begin{aligned}\left\|\frac{\varepsilon b^\varepsilon}{1+\varepsilon b^\varepsilon}\mathcal{A}\mathbf{u}^\varepsilon\right\|_{L^1(B^{\frac{N}{2}-1})} &\lesssim \varepsilon\|b^\varepsilon\|_{L^\infty(B^{\frac{N}{2}})}\|\mathcal{A}\mathbf{u}^\varepsilon\|_{L^1(B^{\frac{N}{2}-1})}\\ &\lesssim \varepsilon\|b^\varepsilon\|_{L^\infty(B^{\frac{N}{2}})}\|\mathbf{u}^\varepsilon\|_{L^1(B^{\frac{N}{2}+1})}\\ &\lesssim C\Gamma^2\gamma^2,\end{aligned}$$

$$\left\|\frac{K(\varepsilon b^\varepsilon)\nabla b^\varepsilon}{\varepsilon}\right\|_{L^1(B^{\frac{N}{2}-1})} \lesssim \|b^\varepsilon\|_{L^2(B^{\frac{N}{2}})}\|\nabla b^\varepsilon\|_{L^2(B^{\frac{N}{2}-1})} \lesssim C\Gamma^2\gamma^2.$$

Combining above estimates with Lemma 4.1, one concludes (4.11). This completes the proof. □

4.3. *Convergence of* $\frac{\mathtt{F}^\varepsilon - I}{\varepsilon}$

Based on the previous bounds on b^ε and estimates (4.8), (4.9) and (3.2), one can obtain the following bound for $\frac{\mathtt{F}^\varepsilon-I}{\varepsilon}$.

Lemma 4.3. *Let Γ, γ be defined as in Theorem 3.1. For $1 < q < 2$ and $p \geq 2$, the solution obtained in the previous subsection satisfies that if $N = 3$*

$$\left\|\frac{\mathtt{F}^\varepsilon - I}{\varepsilon}\right\|_{L^q_{loc}(\tilde{B}_p^{\frac{3}{p},\frac{2}{p}-\frac{1}{2}})} \lesssim \varepsilon^{\min\{\frac{2-q}{q},\frac{1}{2}-\frac{1}{p}\}},$$

and if $N = 2$

$$\left\|\frac{\mathtt{F}^\varepsilon - I}{\varepsilon}\right\|_{L^q_{loc}(\tilde{B}_p^{\frac{2}{p},\frac{3}{2p}-\frac{3}{4}})} \lesssim \varepsilon^{\min\{\frac{2-q}{q},\frac{1}{4}-\frac{1}{2p}\}}.$$

Therefore, $\frac{\mathtt{F}^\varepsilon-I}{\varepsilon}$ converges to 0 in coresponding spaces as $\varepsilon \to 0$.

Proof. In fact, from (3.4) and (3.5), one has

$$\mathrm{div}\left(\frac{\mathrm{F}^\varepsilon - I}{\varepsilon}\right) = -\nabla b^\varepsilon - \varepsilon \mathrm{div}\left(\left(\frac{(\mathrm{F}^\varepsilon)^\top - I}{\varepsilon}\right) b^\varepsilon\right) - \frac{1}{\varepsilon}\mathcal{N}^\varepsilon.$$

Notice that estimates (4.8) and (4.9) imply

$$\varepsilon \left\| \mathrm{div}\left(\left(\frac{(\mathrm{F}^\varepsilon)^\top - I}{\varepsilon}\right) b^\varepsilon\right)\right\|_{L^1(B^{\frac{N}{2}-1})} \lesssim \varepsilon \Gamma^2 \gamma^2,$$

and

$$\left\| \frac{1}{\varepsilon}\mathcal{N}^\varepsilon \right\|_{L^1(B^{\frac{N}{2}-1})} \lesssim \varepsilon \Gamma^2 \gamma^2.$$

Hence, we have

$$\left\| \mathrm{div}\left(\frac{\mathrm{F}^\varepsilon - I}{\varepsilon} + b^\varepsilon I\right)\right\|_{L^1(B^{\frac{N}{2}-1})} \lesssim \varepsilon. \tag{4.12}$$

Notice that if $2 \le p < \infty$, one has, see Proposition 1.3.5 in Ref. 4

$$\begin{cases} B^{\frac{1}{2}} \hookrightarrow B_p^{\frac{3}{p}-1} \hookrightarrow \tilde{B}_p^{\frac{3}{p}-1,\frac{2}{p}-\frac{3}{2}}, & \text{if} \quad N = 3; \\ B^0 \hookrightarrow B_p^{\frac{2}{p}-1} \hookrightarrow \tilde{B}_p^{\frac{2}{p}-1,\frac{3}{2p}-\frac{3}{4}}, & \text{if} \quad N = 2. \end{cases}$$

Therefore, if $N = 3$, using (4.6) and (4.12), we have, as $1 < q < 2$

$$\begin{aligned}
&\left\| \mathrm{div}\left(\frac{\mathrm{F}^\varepsilon - I}{\varepsilon} + b^\varepsilon I\right)\right\|_{L^q(\tilde{B}_p^{\frac{3}{p}-1,\frac{2}{p}-\frac{3}{2}})} \\
&\quad \lesssim \left\| \mathrm{div}\left(\frac{\mathrm{F}^\varepsilon - I}{\varepsilon} + b^\varepsilon I\right)\right\|_{L^q(B^{\frac{1}{2}})} \\
&\quad \lesssim \left\| \mathrm{div}\left(\frac{\mathrm{F}^\varepsilon - I}{\varepsilon} + b^\varepsilon I\right)\right\|_{L^1(B^{\frac{1}{2}})}^{\frac{2-q}{q}} \left\| \mathrm{div}\left(\frac{\mathrm{F}^\varepsilon - I}{\varepsilon} + b^\varepsilon I\right)\right\|_{L^2(B^{\frac{1}{2}})}^{\frac{2q-2}{q}} \\
&\quad \lesssim \left\| \mathrm{div}\left(\frac{\mathrm{F}^\varepsilon - I}{\varepsilon} + b^\varepsilon I\right)\right\|_{L^1(B^{\frac{1}{2}})}^{\frac{2-q}{q}} \left\| \frac{\mathrm{F}^\varepsilon - I}{\varepsilon} + b^\varepsilon I\right\|_{L^2(B^{\frac{3}{2}})}^{\frac{2q-2}{q}} \\
&\quad \lesssim \left\| \mathrm{div}\left(\frac{\mathrm{F}^\varepsilon - I}{\varepsilon} + b^\varepsilon I\right)\right\|_{L^1(B^{\frac{1}{2}})}^{\frac{2-q}{q}} (C\Gamma^2\gamma^2)^{\frac{2q-2}{q}} \\
&\quad \lesssim \varepsilon^{\frac{2-q}{q}}.
\end{aligned}$$

Similarly, if $N = 2$, we have, as $1 < q < 2$

$$\left\| \mathrm{div}\left(\frac{\mathrm{F}^\varepsilon - I}{\varepsilon} + b^\varepsilon I\right)\right\|_{L^q(\tilde{B}_p^{\frac{2}{p}-1,\frac{3}{2p}-\frac{7}{4}})} \lesssim \varepsilon^{\frac{2-q}{q}}.$$

If $2 < p < \infty$, Proposition 1.3.5 in Ref. 4 implies

$$\begin{cases} B_p^{\frac{2}{p}-\frac{3}{2}} \hookrightarrow \tilde{B}_p^{\frac{3}{p}-1,\frac{2}{p}-\frac{3}{2}}, & \text{if} \quad N=3; \\ B_p^{\frac{3}{2p}-\frac{7}{4}} \hookrightarrow \tilde{B}_p^{\frac{2}{p}-1,\frac{3}{2p}-\frac{7}{4}}, & \text{if} \quad N=2. \end{cases}$$

Hence, combined with Lemma 4.2, we have if $N = 3$

$$\begin{aligned} &\left\| \operatorname{div}\left(\frac{\mathrm{F}^\varepsilon - I}{\varepsilon}\right) \right\|_{L^q_{loc}(\tilde{B}_p^{\frac{3}{p}-1,\frac{2}{p}-\frac{3}{2}})} \\ &\quad \le \left\| \operatorname{div}\left(\frac{\mathrm{F}^\varepsilon - I}{\varepsilon} + b^\varepsilon I\right) \right\|_{L^q(\tilde{B}_p^{\frac{3}{p}-1,\frac{2}{p}-\frac{3}{2}})} + \|\nabla b^\varepsilon\|_{L^q_{loc}(\tilde{B}_p^{\frac{3}{p}-1,\frac{2}{p}-\frac{3}{2}})} \\ &\quad \lesssim \varepsilon^{\frac{2-q}{q}} + \|\nabla b^\varepsilon\|_{L_{loc}^{\frac{2p}{p-2}}(B_p^{\frac{2}{p}-\frac{3}{2}})} \\ &\quad \lesssim \varepsilon^{\frac{2-q}{q}} + \|b^\varepsilon\|_{L^{\frac{2p}{p-2}}(B_p^{\frac{2}{p}-\frac{1}{2}})} \\ &\quad \lesssim \varepsilon^{\frac{2-q}{q}} + \varepsilon^{\frac{1}{2}-\frac{1}{p}} \lesssim \varepsilon^{\min\{\frac{2-q}{q},\frac{1}{2}-\frac{1}{p}\}}, \end{aligned} \tag{4.13}$$

and similarly if $N = 2$,

$$\left\| \operatorname{div}\left(\frac{\mathrm{F}^\varepsilon - I}{\varepsilon}\right) \right\|_{L^q_{loc}(\tilde{B}_p^{\frac{2}{p}-1,\frac{3}{2p}-\frac{7}{4}})} \lesssim \varepsilon^{\min\{\frac{2-q}{q},\frac{1}{4}-\frac{1}{2p}\}}. \tag{4.14}$$

On the other hand, from (3.2), we have

$$\begin{aligned} &\operatorname{curl}\left(\frac{\mathrm{F}^\varepsilon - I}{\varepsilon}\right) \\ &\quad = \varepsilon\left(\left(\frac{\mathrm{F}^\varepsilon_{lj} - \delta_{lj}}{\varepsilon}\right)\nabla_l\left(\frac{\mathrm{F}^\varepsilon_{ik} - \delta_{ik}}{\varepsilon}\right) - \left(\frac{\mathrm{F}^\varepsilon_{lk} - \delta_{lk}}{\varepsilon}\right)\nabla_l\left(\frac{\mathrm{F}^\varepsilon_{ij} - \delta_{ij}}{\varepsilon}\right)\right). \end{aligned}$$

As for (4.7), it is easy to check that the right hand side of the above identity is bounded by $C\Gamma^2\gamma^2\varepsilon$ in $L^1(B^{\frac{N}{2}-1})$. Hence, one further concludes

$$\left\| \operatorname{curl}\left(\frac{\mathrm{F}^\varepsilon - I}{\varepsilon}\right) \right\|_{L^1(B^{\frac{N}{2}-1})} \lesssim \varepsilon. \tag{4.15}$$

Therefore, we have if $N=3$

$$\begin{aligned}
&\left\|\operatorname{curl}\left(\frac{\mathrm{F}^{\varepsilon}-I}{\varepsilon}\right)\right\|_{L^q(\tilde{B}_p^{\frac{3}{p}-1,\frac{2}{p}-\frac{3}{2}})}\\
&\leq\left\|\operatorname{curl}\left(\frac{\mathrm{F}^{\varepsilon}-I}{\varepsilon}\right)\right\|_{L^q(B^{\frac{1}{2}})}\\
&\leq\left\|\operatorname{curl}\left(\frac{\mathrm{F}^{\varepsilon}-I}{\varepsilon}\right)\right\|_{L^1(B^{\frac{1}{2}})}^{\frac{2-q}{q}}\left\|\operatorname{curl}\left(\frac{\mathrm{F}^{\varepsilon}-I}{\varepsilon}\right)\right\|_{L^2(B^{\frac{1}{2}})}^{\frac{2q-2}{q}}\\
&\leq\left\|\operatorname{curl}\left(\frac{\mathrm{F}^{\varepsilon}-I}{\varepsilon}\right)\right\|_{L^1(B^{\frac{1}{2}})}^{\frac{2-q}{q}}\left\|\frac{\mathrm{F}^{\varepsilon}-I}{\varepsilon}\right\|_{L^2(B^{\frac{3}{2}})}^{\frac{2q-2}{q}}\\
&\lesssim\varepsilon^{\frac{2-q}{q}},
\end{aligned}\tag{4.16}$$

and similarly if $N=2$

$$\left\|\operatorname{curl}\left(\frac{\mathrm{F}^{\varepsilon}-I}{\varepsilon}\right)\right\|_{L^q(\tilde{B}_p^{\frac{2}{p}-1,\frac{3}{2p}-\frac{7}{4}})}\lesssim\varepsilon^{\frac{2-q}{q}}.\tag{4.17}$$

Hence, estimates (4.13) (4.14), (4.16), (4.17) and Bernstein's inequality imply that if $N=3$

$$\left\|\frac{\mathrm{F}^{\varepsilon}-I}{\varepsilon}\right\|_{L^q_{loc}(\tilde{B}_p^{\frac{3}{p},\frac{2}{p}-\frac{1}{2}})}\lesssim\varepsilon^{\min\{\frac{2-q}{q},\frac{1}{2}-\frac{1}{p}\}},$$

and if $N=2$

$$\left\|\frac{\mathrm{F}^{\varepsilon}-I}{\varepsilon}\right\|_{L^q_{loc}(\tilde{B}_p^{\frac{2}{p},\frac{3}{2p}-\frac{3}{4}})}\lesssim\varepsilon^{\min\{\frac{2-q}{q},\frac{1}{4}-\frac{1}{2p}\}}.$$

This finishes the proof. □

Remark 4.1. There is an alternative proof of the convergence of $\frac{\mathrm{F}^{\varepsilon}-I}{\varepsilon}$ to 0. Indeed, from Corollary 4.1, we know that

$$\left\|\mathcal{P}\operatorname{div}\left(\frac{\mathrm{F}^{\varepsilon}-I}{\varepsilon}\right)\right\|_{L^1(B^{\frac{N}{2}-1})}\lesssim\varepsilon.$$

From (3.4) and (3.5), one deduce that

$$\mathcal{Q}\operatorname{div}\left(\frac{\mathrm{F}^{\varepsilon}-I}{\varepsilon}\right)=-\nabla b^{\varepsilon}-\mathcal{Q}\operatorname{div}(((\mathrm{F}^{\varepsilon})^{\top}-I)b^{\varepsilon})-\mathcal{Q}\frac{\mathcal{N}^{\varepsilon}}{\varepsilon}.$$

It can be easily shown that

$$\left\|-\mathcal{Q}\operatorname{div}(((\mathrm{F}^{\varepsilon})^{\top}-I)b^{\varepsilon})-\mathcal{Q}\frac{\mathcal{N}^{\varepsilon}}{\varepsilon}\right\|_{L^1(B^{\frac{N}{2}-1})}\lesssim\varepsilon,$$

and hence in view of Lemma 4.2, the quantity $\mathcal{Q}\mathrm{div}\left(\frac{\mathtt{F}^\varepsilon - I}{\varepsilon}\right)$ is bounded in a suitable space by a positive power of ε. Therefore, the quantity div $\left(\frac{\mathtt{F}^\varepsilon - I}{\varepsilon}\right)$ is bounded in a suitable space by a positive power of ε.

Also, as in (4.15), we know that curl $\left(\frac{\mathtt{F}^\varepsilon - I}{\varepsilon}\right)$ is bounded by ε in $L^1(B^{\frac{N}{2}-1})$. Thus, the quantity $\frac{\mathtt{F}^\varepsilon - I}{\varepsilon}$ is bounded in a suitable space by a positive power of ε, and thus converges to 0 in that space as $\varepsilon \to 0$.

The strong convergence in Lemma 4.3, the uniform bound in $L^\infty(\mathbb{R}^+, B^{\frac{N}{2}-1})$, and the interpolations imply that $\frac{\mathtt{F}^\varepsilon - I}{\varepsilon}$ converges to 0 in the intermediate spaces between $L^\infty(\mathbb{R}^+, B^{\frac{N}{2}-1})$ and these spaces in Lemma 4.3. Moreover, as a consequence of the convergence of $\frac{\mathtt{F}^\varepsilon - I}{\varepsilon}$, the nonlinear terms $\mathcal{M}^\varepsilon$ and $\mathcal{N}^\varepsilon$ also converge to zero. In fact, we have

Lemma 4.4. *Let $\mathcal{M}^\varepsilon$ and $\mathcal{N}^\varepsilon$ be defined as before. As $\varepsilon \to 0$, for any $1 < q < 2$, if $N = 3$,*

$$\mathcal{M}^\varepsilon, \mathcal{N}^\varepsilon \to 0 \ in \ L^q_{loc}(\tilde{B}_p^{\frac{3}{p}-1, \frac{2}{p}-\frac{3}{2}}),$$

while if $N = 2$

$$\mathcal{M}^\varepsilon, \mathcal{N}^\varepsilon \to 0 \ in \ L^q_{loc}(\tilde{B}^{\frac{2}{p}-1, \frac{3}{2p}-\frac{7}{4}}).$$

Proof. We only prove the case as $N = 3$, and the case as $N = 2$ can be treated similarly. By the definitions of $\mathcal{M}^\varepsilon$ and $\mathcal{N}^\varepsilon$, it suffices to prove

$$\frac{1}{\varepsilon^2}\left((\mathtt{F}^\varepsilon_{jk} - \delta_{jk})\nabla_j(\mathtt{F}^\varepsilon_{ik} - \delta_{ik}\right) \to 0, \tag{4.18}$$

and

$$\frac{1}{\varepsilon}\mathrm{div}(((\mathtt{F}^\varepsilon)^\top - I)b^\varepsilon) \to 0 \tag{4.19}$$

in $L^q_{loc}(\tilde{B}_p^{\frac{3}{p}-1, \frac{2}{p}-\frac{3}{2}})$ as $\varepsilon \to 0$.

To see (4.18), we note that, using Proposition 2.1 and Lemma 4.3

$$\begin{aligned}
&\left\| \frac{1}{\varepsilon^2}\left((\mathtt{F}^\varepsilon_{jk} - \delta_{jk})\nabla_j(\mathtt{F}^\varepsilon_{ik} - \delta_{ik}\right) \right\|_{L^q_{loc}(\tilde{B}_p^{\frac{3}{p}-1, \frac{2}{p}-\frac{3}{2}})} \\
&\le \left\| \frac{\mathtt{F}^\varepsilon - I}{\varepsilon} \right\|_{L^{2q-2}_{loc}(\tilde{B}_p^{\frac{3}{p}, \frac{2}{p}-\frac{1}{2}})} \left\| \nabla \frac{\mathtt{F}^\varepsilon - I}{\varepsilon} \right\|_{L^2(B^{\frac{1}{2}})} \\
&\lesssim \varepsilon^{\min\{\frac{2-q}{q-1}, \frac{1}{2}-\frac{1}{p}\}} \left\| \frac{\mathtt{F}^\varepsilon - I}{\varepsilon} \right\|_{L^2(B^{\frac{3}{2}})} \\
&\lesssim \varepsilon^{\min\{\frac{2-q}{q-1}, \frac{1}{2}-\frac{1}{p}\}},
\end{aligned}$$

and hence

$$\frac{1}{\varepsilon^2}\left((\mathtt{F}^\varepsilon_{jk}-\delta_{jk})\nabla_j(\mathtt{F}^\varepsilon_{ik}-\delta_{ik}\right)\to 0 \text{ in } L^q_{loc}(\tilde{B}_p^{\frac{3}{p}-1,\frac{2}{p}-\frac{3}{2}})$$

as $\varepsilon\to 0$.

Similarly, for (4.19), we have, using Proposition 2.1 and Lemma 4.3

$$\begin{aligned}\left\|\frac{1}{\varepsilon}\mathrm{div}(((\mathtt{F}^\varepsilon)^\top-I)b^\varepsilon))\right\|_{L^q_{loc}(\tilde{B}_p^{\frac{3}{p}-1,\frac{2}{p}-\frac{3}{2}})} &\lesssim \left\|\frac{1}{\varepsilon}((\mathtt{F}^\varepsilon)^\top-I)b^\varepsilon\right\|_{L^q_{loc}(\tilde{B}_p^{\frac{3}{p},\frac{2}{p}-\frac{1}{2}})}\\ &\lesssim \left\|\frac{\mathtt{F}^\varepsilon-I}{\varepsilon}\right\|_{L^{2q-2}_{loc}(\tilde{B}_p^{\frac{3}{p},\frac{2}{p}-\frac{1}{2}})}\|b^\varepsilon\|_{L^2(B^{\frac{3}{2}})}\\ &\lesssim \varepsilon^{\min\{\frac{2-q}{q-1},\frac{1}{2}-\frac{1}{p}\}},\end{aligned}$$

and hence

$$\frac{1}{\varepsilon}\mathrm{div}(((\mathtt{F}^\varepsilon)^\top-I)b^\varepsilon)\to 0 \text{ in } L^q_{loc}(\tilde{B}_p^{\frac{3}{p}-1,\frac{2}{p}-\frac{3}{2}})$$

as $\varepsilon\to 0$.

This finishes the proof. □

Remark 4.2. Actually, from Lemma 4.4, the quantities $\frac{1}{\varepsilon}\mathcal{P}\mathrm{div}\left(\frac{\mathtt{F}^\varepsilon-I}{\varepsilon}\right)$ are not only bounded in $L^1(B^{\frac{N}{2}-1})$, but also converge to zero in $L^q_{loc}(\tilde{B}_p^{\frac{3}{p}-1,\frac{2}{p}-\frac{3}{2}})$ if $N=3$, and $L^q_{loc}(\tilde{B}_p^{\frac{2}{p}-1,\frac{3}{2p}-\frac{7}{4}})$ if $N=2$ as $\varepsilon\to 0$.

4.4. *Convergence of $\mathcal{P}\mathbf{u}^\varepsilon$*

To this end, we denote

$$w^\varepsilon=\mathcal{P}\mathbf{u}^\varepsilon-v.$$

Applying Leray Projector $\mathcal{P}$ to the second equation of (CV^ε) and subtracting (INS) from it yields a heat equation for w^ε

$$\begin{cases}\partial_t w^\varepsilon-\mu\Delta w^\varepsilon=H^\varepsilon,\\ w^\varepsilon|_{t=0}=0,\end{cases}\tag{4.20}$$

where H^ε is given by

$$\begin{aligned}H^\varepsilon \stackrel{def}{=} -\mathcal{P}\Big(&w^\varepsilon\cdot\nabla v+\mathbf{u}^\varepsilon\cdot\nabla w^\varepsilon+\mathcal{Q}\mathbf{u}^\varepsilon\cdot\nabla v+\mathbf{u}^\varepsilon\cdot\nabla\mathcal{Q}\mathbf{u}^\varepsilon\\ &+\frac{\varepsilon b^\varepsilon}{1+\varepsilon b^\varepsilon}\mathcal{A}\mathbf{u}^\varepsilon+\mathcal{M}^\varepsilon\Big).\end{aligned}$$

Case 1: $N=3$. For $2\le p<\infty$, we have, by interpolation

$$L^2\left(0,\infty;B^{\frac{14-k}{2k+4}}_{\frac{k+2}{2}}\right)=\left[L^1(0,\infty;B_2^{\frac{5}{2}});L^{\frac{2k}{k-2}}(0,\infty;B_k^{-\frac{1}{2}+\frac{2}{k}})\right]_{\frac{2}{k+2}}.$$

Due to (4.5) and Lemma 4.2, we obtain

$$\|\mathcal{Q}\mathbf{u}^\varepsilon\|_{L^2\left(\tilde{B}_p^{\frac{3}{p},\frac{2}{p}-\frac{1}{2}}\right)} \lesssim \|\mathcal{Q}\mathbf{u}^\varepsilon\|_{L^2\left(B_p^{\frac{4}{p}-\frac{1}{2}}\right)} \lesssim \varepsilon^{\frac{1}{2}-\frac{1}{p}} \quad \text{for all } 2 \le p < \infty. \tag{4.21}$$

Next, we claim that w^ε converges to 0 in $L^q_{loc}(\tilde{B}_p^{\frac{3}{p}+1,\frac{2}{p}+\frac{1}{2}}) \cap C\left(0,\infty;\tilde{B}_p^{\frac{3}{p}-1,\frac{2}{p}-\frac{3}{2}}\right)$ $(2 \le p < \infty)$. In fact, we only need to prove

$$Y_p \stackrel{def}{=} \|w^\varepsilon\|_{L^1\left(0,\infty;\tilde{B}_p^{\frac{3}{p}+1,\frac{2}{p}+\frac{1}{2}}\right)} + \|w^\varepsilon\|_{L^\infty\left(0,\infty;\tilde{B}_p^{\frac{3}{p}-1,\frac{2}{p}-\frac{3}{2}}\right)} \lesssim \varepsilon^{\frac{1}{2}-\frac{1}{p}}. \tag{4.22}$$

In order to verify (4.22), we will apply Proposition 7.3 in Ref. 1 and estimates in two previous subsections. Actually, we have

$$\|\mathcal{P}(w^\varepsilon\cdot\nabla v)\|_{L^1(\tilde{B}_p^{\frac{3}{p}-1,\frac{2}{p}-\frac{3}{2}})} \lesssim \|\nabla v\|_{L^2(B^{\frac{1}{2}})}\|w^\varepsilon\|_{L^2(\tilde{B}_p^{\frac{3}{p},\frac{2}{p}-\frac{1}{2}})},$$

$$\|\mathcal{P}(\mathbf{u}^\varepsilon\cdot\nabla w^\varepsilon)\|_{L^1(\tilde{B}_p^{\frac{3}{p}-1,\frac{2}{p}-\frac{3}{2}})} \lesssim \|\mathbf{u}^\varepsilon\|_{L^2(B^{\frac{3}{2}})}\|\nabla w^\varepsilon\|_{L^2(\tilde{B}_p^{\frac{3}{p}-1,\frac{2}{p}-\frac{3}{2}})},$$

$$\begin{aligned}\|\mathcal{P}(\mathcal{Q}\mathbf{u}^\varepsilon\cdot\nabla v)\|_{L^1(\tilde{B}_p^{\frac{3}{p}-1,\frac{2}{p}-\frac{3}{2}})} &\lesssim \|\nabla v\|_{L^2(B^{\frac{1}{2}})}\|\mathcal{Q}\mathbf{u}^\varepsilon\|_{L^2(\tilde{B}_p^{\frac{3}{p},\frac{2}{p}-\frac{1}{2}})}\\ &\lesssim \varepsilon^{\frac{1}{2}-\frac{1}{p}},\end{aligned}$$

$$\begin{aligned}\|\mathcal{P}(\mathbf{u}^\varepsilon\cdot\nabla\mathcal{Q}\mathbf{u}^\varepsilon)\|_{L^1(\tilde{B}_p^{\frac{3}{p}-1,\frac{2}{p}-\frac{3}{2}})} &\lesssim \|\mathbf{u}^\varepsilon\|_{L^2(B^{\frac{3}{2}})}\|\nabla\mathcal{Q}\mathbf{u}^\varepsilon\|_{L^2(\tilde{B}_p^{\frac{3}{p}-1,\frac{2}{p}-\frac{3}{2}})}\\ &\lesssim \varepsilon^{\frac{1}{2}-\frac{1}{p}},\end{aligned}$$

$$\|\mathcal{P}\mathcal{M}\|_{L^1(\tilde{B}_p^{\frac{3}{p}-1,\frac{2}{p}-\frac{3}{2}})} \lesssim \varepsilon^{\frac{1}{2}-\frac{1}{p}}.$$

Thanks to the embedding $B_2^{\frac{1}{p}} \hookrightarrow \tilde{B}_p^{\frac{3}{p}-1,\frac{2}{p}-\frac{3}{2}}$ and

$$L^{\frac{4p}{3p+2}}(B_2^{\frac{1}{p}+2}) = \left[L^1(B^{\frac{5}{2}}), L^\infty(B^{\frac{1}{2}})\right]_{\frac{3p+2}{4p}},$$

we further have, using (4.5)

$$\begin{aligned}\left\|\mathcal{P}\left(\frac{\varepsilon b^\varepsilon}{1+\varepsilon b^\varepsilon}\mathcal{A}\mathbf{u}^\varepsilon\right)\right\|_{L^1(\tilde{B}_p^{\frac{3}{p}-1,\frac{2}{p}-\frac{3}{2}})} &\lesssim \|\varepsilon b^\varepsilon\|_{L^{\frac{4p}{p-2}}(B^{\frac{3}{2}})}\|\mathcal{A}\mathbf{u}^\varepsilon\|_{L^{\frac{4p}{3p+2}}(\tilde{B}_p^{\frac{3}{p}-1,\frac{2}{p}-\frac{3}{2}})}\\ &\lesssim \|\varepsilon b^\varepsilon\|_{L^{\frac{4p}{p-2}}(B^{\frac{3}{2}})}\|\mathcal{A}\mathbf{u}^\varepsilon\|_{L^{\frac{4p}{3p+2}}(B^{\frac{1}{p}})}\\ &\lesssim \varepsilon\|b^\varepsilon\|^{\frac{p-2}{2p}}_{L^2(B^{\frac{3}{2}})}\|b^\varepsilon\|^{\frac{p+2}{2p}}_{L^\infty(B^{\frac{3}{2}})}\|\mathbf{u}^\varepsilon\|_{L^{\frac{4p}{3p+2}}(B^{\frac{1}{p}+2})}\\ &\lesssim \varepsilon\left(\frac{1}{\varepsilon}\right)^{\frac{1}{2}+\frac{1}{p}} \lesssim \varepsilon^{\frac{1}{2}-\frac{1}{p}}.\end{aligned}$$

Combining these estimates together, we get

$$\begin{aligned} Y_p &\le C\varepsilon^{\frac{1}{2}-\frac{1}{p}} + C\gamma\|w^\varepsilon\|_{L^2(\tilde{B}_p^{\frac{3}{p},\frac{2}{p}-\frac{1}{2}})} \\ &\le C\varepsilon^{\frac{1}{2}-\frac{1}{p}} + C\gamma Y_p, \end{aligned}$$

and hence (4.22) follows by choosing γ sufficiently small.

Case 2: $N = 2$**.** In this case, again by the interpolation, we have

$$L^2\left(0,\infty; B_{\frac{6k+4}{k+6}}^{\frac{14+k}{6k+4}}\right) = \left[L^1(0,\infty;B^2); L^{\frac{4k}{k-2}}(0,\infty;B_k^{-\frac{3}{4}+\frac{3}{2k}})\right]_{\frac{k+2}{3k+2}}.$$

Hence, by Lemma 4.2, we obtain

$$\|\mathcal{Q}\mathbf{u}^\varepsilon\|_{L^2(\tilde{B}_p^{\frac{2}{p},\frac{3}{2p}-\frac{3}{4}})} \lesssim \|\mathcal{Q}\mathbf{u}^\varepsilon\|_{L^2(B_p^{\frac{5}{2p}-\frac{1}{4}})} \lesssim \varepsilon^{\frac{1}{4}-\frac{1}{2p}} \text{ for all } 2 \le p \le 6. \quad (4.23)$$

Now we claim that

$$Y_p \stackrel{def}{=} \|w\|_{L^1(\tilde{B}_p^{\frac{2}{p}+1,\frac{3}{2p}+\frac{1}{4}})} + \|w^\varepsilon\|_{L^\infty(\tilde{B}_p^{\frac{2}{p}-1,\frac{3}{2p}-\frac{7}{4}})} \lesssim \varepsilon^{\frac{1}{4}-\frac{1}{2p}}.$$

By the estimates for the product, we have

$$\|\mathcal{P}(w^\varepsilon\cdot\nabla v)\|_{L^1(\tilde{B}_p^{\frac{2}{p}-1,\frac{3}{2p}-\frac{7}{4}})} \lesssim \|\nabla v\|_{L^2(B^0)}\|w^\varepsilon\|_{L^2(\tilde{B}_p^{\frac{2}{p},\frac{3}{2p}-\frac{3}{4}})},$$

$$\|\mathcal{P}(\mathbf{u}^\varepsilon\cdot\nabla w^\varepsilon)\|_{L^1(\tilde{B}_p^{\frac{2}{p}-1,\frac{3}{2p}-\frac{7}{4}})} \lesssim \|\mathbf{u}^\varepsilon\|_{L^2(B^1)}\|\nabla w^\varepsilon\|_{L^2(\tilde{B}_p^{\frac{2}{p}-1,\frac{3}{2p}-\frac{7}{4}})},$$

$$\begin{aligned} \|\mathcal{P}(\mathcal{Q}\mathbf{u}^\varepsilon\cdot\nabla v)\|_{L^1(\tilde{B}_p^{\frac{2}{p}-1,\frac{3}{2p}-\frac{7}{4}})} &\lesssim \|\nabla v\|_{L^2(B^0)}\|\mathcal{Q}\mathbf{u}^\varepsilon\|_{L^2(\tilde{B}_p^{\frac{2}{p},\frac{3}{2p}-\frac{3}{4}})} \\ &\lesssim \varepsilon^{\frac{1}{4}-\frac{1}{2p}}, \end{aligned}$$

$$\begin{aligned} \|\mathcal{P}(\mathbf{u}^\varepsilon\cdot\nabla\mathcal{Q}\mathbf{u}^\varepsilon)\|_{L^1(\tilde{B}_p^{\frac{2}{p}-1,\frac{3}{2p}-\frac{7}{4}})} &\lesssim \|\mathbf{u}^\varepsilon\|_{L^2(B^1)}\|\nabla\mathcal{Q}\mathbf{u}^\varepsilon\|_{L^2(\tilde{B}_p^{\frac{2}{p}-1,\frac{3}{2p}-\frac{7}{4}})} \\ &\lesssim \varepsilon^{\frac{1}{4}-\frac{1}{2p}}, \end{aligned}$$

$$\|\mathcal{P}\mathcal{M}\|_{L^1(\tilde{B}_p^{\frac{2}{p}-1,\frac{3}{2p}-\frac{7}{4}})} \lesssim \varepsilon^{\frac{1}{4}-\frac{1}{2p}}.$$

On the other hand, note that as $p \le 6$, $\frac{5}{2p} - \frac{1}{4} > 0$. Moreover, thanks to the embedding $B^{\frac{1}{2p}-\frac{1}{4}} \hookrightarrow \tilde{B}_p^{\frac{2}{p}-1,\frac{3}{2p}-\frac{7}{4}}$ and

$$L^{\frac{8p}{7p+2}}(B^{\frac{1}{2p}+\frac{7}{4}}) = \left[L^1(B^2), L^\infty(B^0)\right]_{\frac{7p+2}{8p}},$$

we also have

$$
\begin{aligned}
&\left\|\mathcal{P}\left(\frac{\varepsilon b^\varepsilon}{1+\varepsilon b^\varepsilon}\mathcal{A}\mathbf{u}\right)\right\|_{L^1(\tilde{B}_p^{\frac{2}{p}-1,\frac{3}{2p}-\frac{7}{4}})}\\
&\lesssim \|\varepsilon b^\varepsilon\|_{L^{\frac{8p}{p-2}}(B^1)}\|\mathcal{A}\mathbf{u}\|_{L^{\frac{8p}{7p+2}}(\tilde{B}_p^{\frac{2}{p}-1,\frac{3}{2p}-\frac{7}{4}})}\\
&\lesssim \|\varepsilon b^\varepsilon\|_{L^{\frac{8p}{p-2}}(B^1)}\|\mathcal{A}\mathbf{u}\|_{L^{\frac{8p}{7p+2}}(B^{\frac{1}{2p}-\frac{1}{4}})}\\
&\lesssim \varepsilon\|b^\varepsilon\|^{\frac{p-2}{4p}}_{L^{\frac{8p}{p-2}}(B^1)}\|b^\varepsilon\|^{\frac{3p+2}{4p}}_{L^\infty(B^1)}\|\mathbf{u}^\varepsilon\|_{L^{\frac{8p}{7p+2}}\left(B^{\frac{1}{2p}+\frac{7}{4}}\right)}\\
&\lesssim \varepsilon\left(\frac{1}{\varepsilon}\right)^{\frac{3p+2}{4p}}\lesssim \varepsilon^{\frac{1}{4}-\frac{1}{2p}}.
\end{aligned}
$$

As in the case $N = 3$, we can also obtain the desired inequality.

References

1. Danchin, R., *Zero Mach number limit in critical spaces for compressible Navier-Stokes equations.* Ann. Sci. cole Norm. Sup. 35 (2002), 27–75.
2. Danchin, R., *Global existence in critical spaces for flows of compressible viscous and heat-conductive gases.* Arch. Ration. Mech. Anal. 160 (2001), 1–39.
3. Danchin, R., *Global existence in critical spaces for compressible Navier-Stokes equations.* Invent. Math. 141 (2000), 579–614.
4. Danchin, R., *Fourier analysis methods for PDE's.*
5. Dafermos, C. M., *Hyperbolic conservation laws in continuum physics.* Second edition. Grundlehren der Mathematischen Wissenschaften, 325. Springer-Verlag, Berlin, 2005.
6. Fujita, H.; Kato, T. *On the Navier-Stokes initial value problem. I.* Arch. Rational Mech. Anal. 16 (1964) 269–315.
7. Hu, X., Wang, D.,*Global existence for the multi-dimensional compressible viscoelastic flows.* J. Differential Equations 250 (2011) 1200–1231.
8. Gurtin, M. E., *An introduction to Continuum Mechanics.* Mathematics in Science and Engineering, 158. Academic Press, New York-London, 1981.
9. Lei, Z.; Zhou, Y., *Global existence of classical solutions for the two-dimensional Oldroyd model via the incompressible limit.* SIAM J. Math. Anal. 37 (2005), 797–814.
10. Lin, F; Liu, C.; Zhang, P., *On hydrodynamics of viscoelastic fluids.* Comm. Pure Appl. Math. 58 (2005), 1437–1471.
11. Lin,F; Xu,L.; Zhang,P., *Global small solutions to 2-D incompressible MHD system.* Preprint(2012).
12. Klainerman, S.; Majda, A., *Singular limits of quasilinear hyperbolic systems with large parameters and the incompressible limit of compressible fluids.* Comm. Pure Appl. Math. 34 (1981), 481–524.

13. Qian, J.; Zhang, Z., *Global well-posedness for compressible viscoelastic fluids near equilibrium.* Arch. Ration. Mech. Anal. 198 (2010), 835–868.
14. Renardy, M.; Hrusa, W. J.; Nohel, J. A., *Mathematical Problems in Viscoelasticity.* Longman Scientific and Technical; copublished in the US with John Wiley, New York, 1987.

Uniqueness Modulo Reduction of Bergman Meromorphic Compactifications of Canonically Embeddable Bergman Manifolds

Ngaiming Mok[1]

The University of Hong Kong, Department of Mathematics,
Pokfulam Road, Hong Kong
nmok@hku.hk

A complex manifold is said to be a Bergman manifold if the Bergman kernel form induces in the standard way a Kähler metric on the manifold. A Bergman manifold is said to be canonically embeddable if the canonical map into a possibly infinite-dimensional projective space defined using the Hilbert space of square-integrable holomorphic n-forms is a holomorphic embedding. In this article we define for a canonically embeddable Bergman manifold X the notion of Bergman meromorphic compactifications $i : X \hookrightarrow Z$ into compact complex manifolds Z characterized in terms of extension properties concerning the Bergman kernel form on X, and define the notion of minimal elements among such compactifications. We prove that such a compact complex manifold Z is necessarily Moishezon. When X is given, assuming the existence of Bergman meromorphic compactifications $i : X \hookrightarrow Z$ we prove the existence of a minimal element among them. More precisely, starting with any Bergman meromorphic compactification $i : X \hookrightarrow Z$ we construct reductions of the compactification, and show that any reduction necessarily defines a minimal element. We show that up to a certain natural equivalence relation the minimal Bergman meromorphic compactification is unique. Examples of such compactifications include Borel embeddings of bounded symmetric domains into their compact dual manifolds and also those arising from canonical realizations of bounded homogeneous domains as Siegel domains or as bounded domains on Euclidean spaces and hence as domains on projective spaces.

Motivated by the result of Clozel–Ullmo [CU, 2003] concerning germs of holomorphic isometries of the Poincaré disk into polydisks arising from a problem in Arithmetic Dynamics, in Mok [Mo2, 2012] the author launched

[1]Research partially supported by the GRF 7032/08P of the HKRGC, Hong Kong.

a systematic study of holomorphic isometries up to normalizing constants first of all between bounded domains of Euclidean spaces. Extension results obtained for such germs of holomorphic maps break down into two types, viz., interior extension results which recover results of Calabi [Ca, 1953] in these cases by a different method, and boundary extension results, which concern properties of extensions of graphs of such germs of holomorphic maps beyond boundaries of the bounded domains. The latter type of results are not accessible by the method of Calabi [Ca, *loc. cit.*] since the boundary of a bounded domain may completely disappear once the domain is embedded into the infinite-dimensional complex projective space $\mathbb{P}^\infty$ by means of an orthonormal basis of the Hilbert space of square-integrable holomorphic functions. Both interior and boundary extension results were extended to the more general context of relatively compact domains of complex manifolds.

A complex manifold is said to be a Bergman manifold if the Bergman kernel form induces in the standard way a Kähler metric on the manifold. A Bergman manifold X is said to be canonically embeddable if the canonical map $\Phi_X : X \to \mathbb{P}^\infty$ defined using any orthonormal basis of Hilbert space of square-integrable holomorphic n-forms is a holomorphic embedding. Of particular interest in this article is the case where a canonically embeddable Bergman manifold X is realized as a domain on a compact complex manifold M in such a way that the Bergman kernel form $\mathcal{K}_X(z,w)$ extends meromorphically in $(z,\overline{w})$ to M, as exemplified by the case of a bounded symmetric domain realized as an open subset of its compact dual manifold by means of the Borel embedding. Since biholomorphisms between Bergman manifolds induce holomorphic isometries with respect to Bergman metrics, given any two compactifications $i_1 : X \hookrightarrow M_1$ and $i_2 : X \hookrightarrow M_2$ satisfying the afore-said extension property on Bergman kernels, it follows readily from Mok [Mo2] that the identity map on X extends to a meromorphic correspondence between M_1 and M_2. In this article we are interested in such compactifications, and more generally on compactifications $i : X \hookrightarrow M$, called Bergman meromorphic compactifications, which satisfy slightly weaker extension properties concerning the Bergman kernel forms which are nonetheless strong enough for meromorphic extendibility to remain valid (cf. §2, especially Definition 2.1 and Corollary 2.2 for details). We prove that any Bergman meromorphic compactication M of X is necessarily Moishezon, and moreover that there exists a minimal element M_0 among such compactifications, in the sense that any Bergman meromorphic compactification M of X dominates such a minimal element via a finite meromorphic map which extends the identity map on X. We call this the minimal Bergman meromorphic compactification, which is uniquely determined up to a bimeromorphic map which is biholomorphic on X. In order to prove the existence of a minimal element, we introduce a procedure of reduction starting from any given Bergman meromorphic compactication $i : X \hookrightarrow M$, and show that a reduction of the latter compactification is necessarily minimal in the sense we described.

Borel embeddings of bounded symmetric domains into their compact dual manifolds are minimal Bergman meromorphic compactifications. We give further examples of such compactifications given by canonical realizations of bounded homogeneous domains as Siegel domains (cf. Pyatetskii-Shapiro [Py, 1969]) or as bounded domains on Euclidean spaces (cf. Xu [Xu, 2005]) and hence as domains on projective spaces.

For an introduction to results on bounded domains and more generally Bergman manifolds revolving around holomorphic isometries and related notions including holomorphic measure-preserving maps, we refer the reader to the survey Mok [Mo1, 2011].

1. Summary of Background and Relevant Results

We recall first of all the class of Bergman manifolds among complex manifolds and also the subclass of canonically embeddable Bergman manifolds among them. For the definitions recall first of all that the Bergman kernel form on an n-dimensional complex manifold (assumed connected by convention) is defined in the standard way using any orthonormal basis of the Hilbert space of square-integrable holomorphic n-forms. We have the following definition from Mok [Mo2, Definition (2.2.1)].

Definition 1.1. *Let X be a complex manifold and denote by ω_X its canonical line bundle. Suppose the Hilbert space $H^2(X, \omega_X)$ of square-integrable holomorphic n-forms on X has no base points, and denote by $\mathcal{K}_X(z, w)$ the Bergman kernel form on X. Regarding $\mathcal{K}_X(z, z)$ as a Hermitian metric h on the anti-canonical line bundle $\omega_X^\star$, we denote by $\beta_X \geq 0$ the curvature form of the dual metric h^* on ω_X, and write ds_X^2 for the corresponding semi-Kähler metric on X. We say that (X, ds_X^2) is a Bergman manifold whenever ds_X^2 and equivalently β_X are positive definite. We call (X, ds_X^2) a canonically embeddable Bergman manifold if furthermore the canonical map $\Phi_X : X \to \mathbb{P}(H^2(X, \omega_X)^\star)$ is a holomorphic embedding.*

Here $H^2(X, \omega_X)$ is said to have no base points if and only if the common zeros of all $\nu \in H^2(X, \omega_X)$ is the empty set. The advantage of using the Bergman kernel form $\mathcal{K}_X(z,w)$ lies in the fact that the latter form is defined independent of the choice of local holomorphic coordinates, given that the norm of a holomorphic n-form ν is simply the square-root of $\int_X (\sqrt{-1})^{n^2} \nu \wedge \overline{\nu}$. Bounded domains in the Euclidean space and more generally bounded domains in a Stein manifold furnish examples of canonically embeddable Bergman manifolds. In the case where $X = D \Subset \mathbb{C}^n$ is a bounded domain it is customary to define the Bergman kernel $K_D(z,w)$ in terms of the space $H^2(D)$ of square-integrable holomorphic functions. In terms of the Euclidean coordinates $(z_1, \ldots, z_n)$, for the Bergman kernel form we have $\mathcal{K}_D(z,w) = K_D(z,w)\big(\frac{\sqrt{-1}}{2} dz^1 \wedge \overline{dz^1}\big) \wedge \cdots \wedge \big(\frac{\sqrt{-1}}{2} dz^n \wedge \overline{dz^n}\big)$. By Mok [Mo2, Theorem (2.2.1)] we have the following general extension result for germs of holomorphic isometries up to normalizing constants between canonically embeddable Bergman manifolds.

Theorem 1.1. *Let M and Q be complex manifolds and let $D \Subset M$ resp. $\Omega \Subset Q$ be a relatively compact subdomain in M* resp. *Q. Suppose D and Ω are canonically embeddable Bergman manifolds. Let $x_0 \in D$, λ be a positive real number and $f : (D, \lambda\, ds^2_D; x_0) \to (\Omega, ds^2_\Omega; f(x_0))$ be a germ of holomorphic isometry. Suppose furthermore that $\mathcal{K}_D(z,w)$ extends meromorphically in $(z, \overline{w})$ to $M \times D$ and likewise the Bergman kernel form $\mathcal{K}_\Omega(\zeta, \xi)$ extends meromorphically in $(\zeta, \overline{\xi})$ to $Q \times \Omega$. Then, the germ* Graph$(f) \subset D \times \Omega$ *at $(x_0, f(x_0))$ extends to an irreducible complex-analytic subvariety $S^\sharp \subset M \times Q$. If in addition (Ω, ds^2_Ω) is complete as a Kähler manifold, then $S := S^\sharp \cap (D \times \Omega)$ is the graph of a holomorphic isometric embedding $F : (D, \lambda\, ds^2_D) \to (\Omega, ds^2_\Omega)$. If (D, ds^2_D) is furthermore assumed to be complete as a Kähler manifold, then $F : D \to \Omega$ is proper.*

2. Bergman Meromorphic Compactifications

Let (X, ds^2_X) be a canonically embeddable Bergman manifold. We consider first of all open embeddings $i : X \hookrightarrow Z$ into compact complex manifolds Z for which the Bergman kernel form $\mathcal{K}_X(z,w)$ extends meromorphically in $(z, \overline{w})$ to $Z \times Z$. As an application of Theorem 1.1, we have

Corollary 2.1. *Let (X, ds^2_X) be a canonically embeddable Bergman manifold. For $k = 1, 2$ let $i_k : X \hookrightarrow Z_k$ be an open embedding of X into a compact complex manifold Z_k such that, identifying X with $X_k := i_k(X)$,* *the Bergman kernel form $\mathcal{K}_X(z,w)$ extends meromorphically in $(z,\overline{w})$ to $Z_k \times Z_k$. Then, the identity map $id_X : X \to X$ extends to a correspondence between Z_1 and Z_2.*

Proof. Since the identity map $id_X : X \to X$ is a holomorphic isometry with respect to the Bergman metric, by Theorem 1.1, Graph$(id_X) \subset X \times X$ extends to an irreducible complex-analytic subvariety $S \subset Z_1 \times Z_2$. The canonical projections $\pi_i : S \to X_i$; $i = 1, 2$; are generically finite maps since S contains Graph(id_X), i.e., the diagonal of X, as an open subset. In other words, $S \subset Z_1 \times Z_2$ is a correspondence, as desired. □

Let (X, ds_X^2) be an n-dimensional Bergman manifold. At a point $(x, x') \in X \times X$, in terms of holomorphic coordinates (z_i) on a neighborhood U of x on X and holomorphic coordinates (w_j) on a neighborhood U' of x' on X, we have $\mathcal{K}_X(z,w) = s(z,w)dz^1 \wedge \cdots \wedge dz^n \wedge \overline{dw^1} \wedge \cdots \wedge \overline{dw^n}$. On $X \times \{x'\}$ we can write $\mathcal{K}_X(z,x') = \sigma_{x'}(z) \wedge \overline{dw^1} \wedge \cdots \wedge \overline{dw^n}$, where σ is a holomorphic n-form on X, and $\sigma_{x'}(z) = s(z,x')dz^1 \wedge \cdots \wedge dz^n$ on a neighborhood of x. We may say that $\mathcal{K}_{X,x'}$ is uniquely determined modulo a choice of normalization at x', more precisely modulo a choice of an ordered basis for $T^*_{x'}(X)$, normally given by the ordered basis at x' defined by the differentials of a choice of holomorphic coordinates at x'. Writing now $w \in X$ (in place of $x' \in X$) for a variable point on X, we have on $X \times \{w\}$ a holomorphic n-form $\mathcal{K}_{X,w} := \sigma_w$ on X which is uniquely determined up to a non-zero multiplicative constant. As will be obvious in the ensuing discussion the statements concerning $\mathcal{K}_{X,w}$ will be independent of the choices made. We are now ready to define the notion of a Bergman meromorphic compactification.

Definition 2.1. *Let (X, ds_X^2) be an n-dimensional canonically embeddable Bergman manifold, and $i : X \hookrightarrow Z$ be an open embedding of X into a compact complex manifold Z. Choose any base point $x_0 \in X$ and define $\sigma_0 := \mathcal{K}_{X,x_0}$, which is uniquely determined up to a non-zero multiplicative constant. Writing $\mathcal{K}_X(z,w) = K_X^\flat(z,w)\Big(\big(\sqrt{-1}\big)^{n^2}\sigma_0(z) \wedge \overline{\sigma_0(w)}\Big)$ on X, we say that $i : X \hookrightarrow Z$ is a Bergman meromorphic compactification if and only if (a) the function $K_X^\flat(z,w)$ extends meromorphically in $(z,\overline{w})$ from $X \times X$ to $Z \times Z$; and (b) there exists an open embedding $i : X \hookrightarrow Z'$ into a compact complex manifold Z' such that the identity map id_X extends to a (possibly) branched covering $\xi : Z' \to Z$ and such that $\xi^\star(\sigma_0)$ extends meromorphically to Z'.*

Suppose $i : X \hookrightarrow Z$ is a Bergman meromorphic compactification in the sense of Definition 2.1 with respect to the choice of a base point $x_0 \in X$. Replacing x_0 by $x_1 \in X$ and defining $\sigma_1 := \mathcal{K}_{X,x_1}$, we have

$$K^\flat_X(z,w)\big(\sigma_0(z) \wedge \overline{\sigma_0(w)}\big) = K^\flat_X(z,w)\frac{\sigma_0(z)}{\sigma_1(z)}\overline{\left(\frac{\sigma_0(w)}{\sigma_1(w)}\right)}\big(\sigma_1(z) \wedge \overline{\sigma_1(w)}\big)\,.$$

Noting that from the choice of σ_0, $K^\flat_X(z,x_0)$ is from Definition 1.1 a non-zero constant function, we have $\frac{\sigma_0(z)}{\sigma_1(z)} = \frac{1}{K^\flat_X(z,x_1)}$ up to a non-zero multiplicative constant and we see that $K^\flat_X(z,w)$ is replaced by $\frac{K^\flat_X(z,w)}{K^\flat_X(z,x_1)K^\flat_X(x_1,w)}$ up to a non-zero multiplicative constant. Thus, the assumption (a) that $K^\flat_X(z,w)$ extends meromorphically in $(z,\overline{w})$ from $X\times X$ to $Z\times Z$ is independent of the choice of a base point $x_0 \in X$. Assuming (a) the condition (b) is also independent of the choice of the base point $x_0 \in X$. In fact, replacing the base point x_0 by $x_1 \in X$, σ_0 is replaced by some $\sigma_1 \in H^2(X,\omega_X)$ such that $\sigma_1 = h\sigma_0$ for some meromorphic function h on X which by (a) extends meromorphically to Z, so that $\xi^\star\sigma_1 = (h\circ\xi)(\xi^\star\sigma_0)$ extends meromorphically to Z'.

Observe also that $i : X \hookrightarrow Z$ is a Bergman meromorphic compactification whenever the differential (n,n)-form $\mathcal{K}_X(z,w)$ extends meromorphically in $(z,\overline{w})$ from $X \times X$ to $Z \times Z$. Moreover, from the proof of the extension theorem given in Mok [Mo2, Theorem 2.2.1] (Theorem 1.1 of the current article), the starting point is the functional identity there given by

$$-\sqrt{-1}\partial\overline{\partial}\log K^\flat_\Omega(f(z),f(z)) = -\lambda\sqrt{-1}\partial\overline{\partial}\log K^\flat_D(z,z)\ ;$$

$$\log K^\flat_\Omega(f(z),f(z)) = \lambda\log K^\flat_D(z,z)\ .$$

As is evident from the arguments in Mok [Mo2], imposing the weaker requirements on the meromorphic extendibility of $K^\flat_D(z,w)$ to $M\times M$ in $(z,\overline{w})$ in place of the same on $\mathcal{K}_D(z,w)$, together with the meromorphic extendibility of $K^\flat_\Omega(\zeta,\xi)$ to $Q\times Q$ in $(\zeta,\overline{\xi})$ in place of the same on $\mathcal{K}_\Omega(\zeta,\xi)$, we can still derive the theorem basing on the functional identity in the above. (The meanings of the functions $K^\flat_D(z,w)$ and $K^\flat_\Omega(\zeta,\xi)$ are analogous to that of $K^\flat_X(z,w)$ as given in Definition 2.1 here.) In particular, in the equidimensional case we are considering, where $D = \Omega = X$ and $f : D \to \Omega$ is the identity map id_X on X, and $M = Z_1, Q = Z_2$ are compact complex manifolds, we have the following strengthened version of Corollary 2.1.

Corollary 2.2. *Let (X, ds^2_X) be a canonically embeddable Bergman manifold. For $k = 1, 2$ let $i_k : X \hookrightarrow Z_k$ be Bergman meromorphic compactifications. Then, the identity map $id_X : X \to X$ extends to a correspondence between Z_1 and Z_2.*

As an example of a Bergman meromorphic compactification let X be the underlying complex manifold of an n-dimensional Hermitian symmetric manifold of the noncompact type. Then, X is biholomorphic to a bounded symmetric domain $D \Subset \mathbb{C}^n$ by means of the Harish-Chandra embedding. Let Z be the compact dual manifold of X and $i : X \hookrightarrow Z$ be the Borel embedding. (For example, $X = B^n$ is the n-dimensional complex unit ball, $Z = \mathbb{P}^n$ is the n-dimensional projective space, and $i : B^n \hookrightarrow \mathbb{P}^n$ is given by the standard embedding $B^n \Subset \mathbb{C}^n$ and the standard compactification $\mathbb{C}^n \subset \mathbb{P}^n$.) Then, the Bergman kernel form $\mathcal{K}_X(z, w)$ extends meromorphically in $(z, \overline{w})$ to $Z \times Z$ (cf. §5). Let $\mu : Z' \to Z$ be a finite ramified covering from a compact complex manifold Z' onto Z, such that, for some connected component $X' \subset Z'$ of $\mu^{-1}(X)$, the map $\mu|_{X'}$ maps X' bihiolomorphically onto X. Z' can be regarded as a Bergman meromorphic compactification of X when we identify X with X' by means of $(\mu|_{X'})^{-1}$, noting that the Bergman kernel form on X' can be obtained by pulling back the Bergman kernel form $\mathcal{K}_X$ on X by μ. Take two ramified covers $\mu_1 : Z_1 \to Z$ and $\mu_2 : Z_2 \to Z$ of compact complex manifolds Z_1 and Z_2 branched outside of $X \subset Z$, and define $S_0 \subset Z_1 \times Z_2$ by $S_0 := \{(z_1, z_2) : \mu_1(z_1) = \mu_2(z_2)\}$. Then, $X \subset Z_1$ and $X \subset Z_2$ are Bergman meromorphic compactifications and the correspondence $S \subset Z_1 \times Z_2$ as given in Theorem 1.1 is simply the irreducible component of S_0 containing Graph(id_X).

Remark If in the definition of Bergman meromorphic compactifications $i : X \hookrightarrow Z$ we dropped the requirement (b), viz., that there exists an open embedding $i : X \hookrightarrow Z'$ into a compact complex manifold Z' such that the identity map id_X extends to a (possibly) branched covering $\xi : Z' \to Z$ and such that $\xi^\star(\sigma_0)$ extends meromorphically to Z', Corollary (2.2) would still hold true. We choose nonetheless to introduce the current definition for two reasons. On the one hand, as will be seen in §3 and §4, for the purpose of constructing a minimal Bergman meromorphic compactification by a reduction process, it is necessary from the methods of proofs to extend the class of compactifications $i : X \hookrightarrow Z$ considered beyond those for which the Bergman kernel form $\mathcal{K}_X(z, w)$ extends meromorphically in $(z, \overline{w})$ to $Z \times Z$, since it is not clear that the latter class is preserved when passing to desingularized models of quotient spaces. On the other hand, adding (b) implies that the Bergman kernels extend at least as multi-valued sections of the ambient manifold, so that the requirements on the Bergman kernel may be said to be algebraic, at least when Z is a projective manifold. (As will be proven in Corollary 3.1, Z is in general always Moishezon, i.e., bimeromorphic to a projective manifold.)

3. Reduction of Bergman Meromorphic Compactifications

Corollary 2.2 is a consequence of the proof of Theorem 1.1 in the special case of a biholomorphism between two complex manifolds. In this case, we are going to show that the multivalence of the extended map arises in general exactly as in the example in the last paragraph of §2.

First of all we give a reduction result for Bergman meromorphic compactifications. Let (X, ds^2_X) be an n-dimensional canonically embeddable Bergman manifold, and $i : X \hookrightarrow Z$ be a Bergman meromorphic compactification of X. Given any finite set of distinct points $\{x_0, \cdots, x_m\}$ on X we have a meromorphic map $\Psi_m : X \dashrightarrow \mathbb{P}^m$ given by $\Psi_m(z) = [\mathcal{K}_{X,x_0}(z), \cdots, \mathcal{K}_{X,x_m}(z)]$. Recall that for $0 \le i \le m$ the holomorphic n-form $\mathcal{K}_{X,x_0}(z)$ on X is uniquely determined only up to a non-zero multiplicative constant. Hence, $\Psi_m : X \dashrightarrow \mathbb{P}^m$ is well-defined only up to projective linear transformations on $\mathbb{P}^m$ of a special form. We define now the notion of a reduced meromorphic compactification.

Definition 3.1. *The map $i : X \hookrightarrow Z$ is said to be a reduced Bergman meromorphic compactification if and only if there exists a finite number of points $x_i \in X, 0 \le i \le m$, such that the meromorphic map $\Psi_m : X \dashrightarrow \mathbb{P}^m$ extends to a generically injective meromorphic map $\Psi^\sharp_m : Z \dashrightarrow \mathbb{P}^m$.*

The generic injectivity of $\Psi^\sharp_m$ is satisfied if and only if, writing $E \subset Z$ for the set of indeterminacy of $\Psi^\sharp_m$ and defining $Y := \overline{\Psi^\sharp_m(Z - E)}$, $\Psi^\sharp_m : Z \dashrightarrow Y$ is a bimeromorphic map. Next, we establish a reduction result for Bergman meromorphic compactifications.

Proposition 3.1. *Let (X, ds^2_X) be a canonically embeddable Bergman manifold, and $i : X \hookrightarrow Z$ be a Bergman meromorphic compactification. Then, there exists a reduced Bergman meromorphic compactification $i^\flat : X \hookrightarrow Z^\flat$, a meromorphic map $\mu : Z \to Z^\flat$ which maps $i(X) \subset Z$ biholomorphically onto $i^\flat(X) \subset Z^\flat$, such that $i^\flat = \mu \circ i$. Equivalently, there exists a reduced*

Bergman meromorphic compactification $i^\flat : X \hookrightarrow Z^\flat$, a smooth modification $\rho : \widehat{Z} \to Z$ with the blow-up locus $E \subset Z$ lying outside of X, and a ramified covering $\nu : \widehat{Z} \to Z^\flat$ such that $i^\flat = \nu \circ \widehat{i}$, where $\widehat{i} : X \hookrightarrow \widehat{Z}$ is an open holomorphic embedding such that $i = \rho \circ \widehat{i}$.

Proof. Assume for the time being that the Bergman kernel form $\mathcal{K}_X(z,w)$ extends meromorphically in $(z,\overline{w})$ from $X \times X$ to $Z \times Z$. Let $(x_i)_{i=0}^\infty$ be a dense sequence of points on X. Consider $\sigma_i := \mathcal{K}_{X,x_i}$. By the reproducing property of $\mathcal{K}_X(z,w)$, for $x \in X$, a square-integrable holomorphic n-form $\nu \in H^2(X,\omega_X)$ is orthogonal to $\mathcal{K}_{X,x}$ if and only if $\nu(x) = 0$. Thus, any ν in the orthogonal complement of the linear span of $\sigma_i, 0 \le i < \infty$, must vanish on the dense set $(x_i)_{i=0}^\infty$ and hence identically on X. In other words, $H^2(X,\omega_X)$ is the topological closure of the linear span of $\sigma_i, 0 \le i < \infty$. By the Gram-Schmidt process, we obtain from $\{\sigma_i\}_{i=0}^\infty$ an orthonormal basis $\{\tau_i\}_{i=0}^\infty$ of $H^2(X,\omega_X)$. (Note that for $k \ge 1$ it is possible that σ_k is linearly dependent on $\sigma_0, \cdots, \sigma_{k-1}$.) Since $\mathcal{K}_X(z,w)$ extends to $Z \times Z$ as a function meromorphic in $(z,\overline{w})$, each $\sigma_i = \mathcal{K}_{X,x_i}$, $0 \le i < \infty$, extends meromorphically from X to Z. Furthermore, each $\tau_i, 0 \le i < \infty$, is expressed by the Gram-Schmidt process as a linear combination of a finite number of σ_j, and as such each τ_i extends meromorphically to Z. By assumption, the canonical map $\Phi_X : X \to \mathbb{P}\left(H^2\left(X,\omega_X\right)^\star\right)$ is a holomorphic embedding. For each integer $m \ge 1$, let $\Phi_m : X \dashrightarrow \mathbb{P}^m$ be the meromorphic mapping defined by $\Phi_m(z) = [\tau_0(z)), \tau_1(z), \cdots, \tau_m(z)]$. Let $A_m \subset X$ be the base locus of Φ_m. We have $A_1 \supset A_2 \cdots \supset A_m \supset \cdots$. Since by assumption $H^2(X,\omega_X)$ has no base locus on X, we have $\bigcap_{m=1}^\infty A_m = \emptyset$. □

By the meromorphic extension of each τ_i to Z we see that $A_m = V_m \cap X$, where $V_m \subset Z$ is a complex-analytic subvariety. Since each complex-analytic subvariety of the compact complex manifold Z has at most a finite number of irreducible branches, it follows that by adjoining a finite number of elements τ_i, we have $A_m = \emptyset$ for m sufficiently large. Since the canonical map $\Phi_X : X \to \mathbb{P}\left(H^2\left(X,\omega_X\right)^\star\right)$ is a holomorphic embedding, using the same argument one deduces that for m sufficiently large, $\Phi_m : X \to \mathbb{P}^m$ is a holomorphic embedding. Choose such a positive integer m and denote by $\Phi_m^\sharp : Z \dashrightarrow \mathbb{P}^m$ the meromorphic extension of Φ_m from X to the compact complex manifold Z. Write $E \subset Z$ for the set of indeterminacies of $\Phi_m^\sharp$, where $E \cap X = \emptyset$, and define $Y = \overline{\Phi_m^\sharp(Z - E)}$. Let $\alpha : Y' \to Y$ be a normalization of Y. The holomorphic map $\Phi_m : X \to Y$ lifts to $\Phi_m' : X \to Y'$, mapping X biholomorphically onto an open subset X' of Y' (noting that Y' is locally irreducible). Denote by $\gamma : Z^\flat \to Y'$ a desingularization of Y' such that γ is unramified over the smooth part $\mathrm{Reg}(Y')$, so that in particular $\gamma|_{\gamma^{-1}(X')} : \gamma^{-1}(X') \to X'$ is a biholomorphism. Identifying now X naturally with $\gamma^{-1}(X')$, we have an open embedding $i^\flat : X \hookrightarrow Z^\flat$. By definition, identifying X with $i^\flat(X)$, the function $K^\flat_X(z,w)$ on X extends meromorphically in $(z,\overline{w})$ from $X \times X$ to $Z^\flat \times Z^\flat$. Since $\Phi^\sharp_m : Z \dashrightarrow Y$ is bimeromorphic, replacing Z by a smooth modification $\rho : \widehat{Z} \to Z$ with the blow-up locus $E \subset Z$ lying outside of X, and denoting by $\widehat{i} : X \hookrightarrow \widehat{Z}$ the canonical lifting of $i : X \hookrightarrow Z$, we have a lifting of $\Phi^\sharp_m$ to a ramified covering $\nu : \widehat{Z} \to Z^\flat$ such that $i^\flat = \nu \circ \widehat{i}$. To verify that the open embedding $i^\flat : X \hookrightarrow Z^\flat$ is a Bergman meromorphic compactification it remains to check condition (b) in Definition 3.1. Using the (possibly) branched covering $\nu : \widehat{Z} \to Z^\flat$, and denoting by $\sigma^\flat_0$ the holomorphic n-form on $i^\flat(X) \subset Z^\flat$ corresponding to the holomorphic n-form σ_0 on $i(X) \subset Z$, $\nu^*(\sigma^\flat_0) = \rho^*(\sigma_0)$ extends meromorphically to $\widehat{Z}$. The proof of Proposition 3.1 is thus complete under the extra assumption that $\mathcal{K}_X(z,w)$ extends meromorphically to $Z \times Z$. Tautologically the Bergman meromorphic compactification $i^\flat : X \hookrightarrow Z^\flat$ is reduced.

In general, by assumption there exists an open embedding $i' : X \hookrightarrow Z'$ into a compact complex manifold Z', such that id_X extends to a (possibly) branched covering $\xi : Z' \to Z$ and such that $\xi^*(\sigma_0)$ extends meromorphically to Z'. Then, obviously the Bergman kernel $\mathcal{K}_X(z,w)$ extends meromorphically to Z'. The preceding arguments then apply to give a reduction $i'^\flat : Z' \to Z'^\flat$ and a meromorphic map $\mu' : Z' \to Z'^\flat$ with the desired properties. Suppose z'_1, z'_2 are unramified points of ξ such that $\xi(z'_1) = \xi(z'_2)$ and such that μ' is holomorphic at z'_1 and z'_2. It follows from the definition of μ', which can be equivalently defined by pull-backs of certain meromorphic functions on Z (serving as inhomogeneous coordinates for the image of $\Phi^\sharp_m$), that we must have $\mu'(z'_1) = \mu'(z'_2)$, and hence $\mu' : Z' \to Z'^\flat$ descends to $\mu : Z \to Z'^\flat$. It suffices now to take $Z^\flat$ to be $Z'^\flat$ to complete the proof of Proposition 3.1. □

From the existence of the meromorphic map $\Phi^\sharp_m : Z \dashrightarrow Y$ of maximal rank $n = \dim(X)$ over X we deduce readily

Corollary 3.1. *Let (X, ds^2_X) be a canonically embeddable Bergman manifold, and $i : X \hookrightarrow Z$ be a Bergman meromorphic compactification of X. Then, Z is a Moishezon manifold.*

We define now a natural equivalence relation among Bergman meromorphic compactifications of a given canonically embeddable Bergman manifold, as follows.

Definition 3.2. *Let X be a canonically embeddable Bergman manifold. We say that two Bergman meromorphic compactifications $i_1 : X \hookrightarrow Z_1$ and $i_2 : X \hookrightarrow Z_2$ are equivalent to each other if and only if, identifying X with an open subset of Z_1, resp. Z_2, the identity map id_X extends to a bimeromorphic map between Z_1 and Z_2.*

Starting with a Bergman meromorphic compactification $i : X \hookrightarrow Z$ we have constructed in Proposition 3.1 a reduced Bergman meromorphic compactification $i^\flat : X \hookrightarrow Z^\flat$, which is well-defined up to equivalence. *A priori* the latter depends on the choice of a dense sequence $(x_i)_{i=0}^{\infty}$ on X and the choice of a positive integer m such that $\Psi_m : X \to \mathbb{P}^m$ is an embedding. We will call $i^\flat : X \hookrightarrow Z^\flat$ a reduction of $i : X \hookrightarrow Z$. Next, we introduce the notion of minimal Bergman meromorphic compactifications.

4. Minimality of Reduced Bergman Meromorphic Compactifications

Given a canonically embeddable Bergman manifold X, there is a natural partial ordering among its Bergman meromorphic compactifications $i : X \hookrightarrow Z$, where $i_1 : X \hookrightarrow Z_1$ is said to dominate $i_2 : X \hookrightarrow Z_2$ if and only if $i_2 = \rho \circ i_1$ for some meromorphic mapping $\rho : Z_1 \dashrightarrow Z_2$ where ρ restricts to a biholomorphic map from $i_1(X) \subset Z_1$ onto $i_2(X) \subset Z_2$. A minimal element among Bergman meromorphic compactifications of $i : X \hookrightarrow Z$ with respect to this partial ordering will be called a minimal Bergman meromorphic compactification. In other words, we have

Definition 4.1. *Fixing a canonically embeddable Bergman manifold X, a Bergman meromorphic compactification $i_0 : X \hookrightarrow Z_0$ is said to be minimal if and only if, given any Bergman meromorphic compactification $i : X \hookrightarrow Z$, the biholomorphism $h : i(X) \to i_0(X)$ corresponding to the identity map id_X extends to a meromorphic map $\eta : Z \to Z_0$.*

For a given canonically embeddable Bergman manifold X, we now relate the reduction of its Bergman meromorphic compactifications to the notion of minimality in Definition 4.1. When a single Bergman meromorphic compactification $i : X \hookrightarrow Z$ is given, Proposition 3.1 gives a reduction

$i^\flat : X \hookrightarrow Z^\flat$ such that the identity map id_X extends to a meromorphic map $\eta : Z \to Z^\flat$. It is not clear that up to equivalence $i^\flat : X \hookrightarrow Z^\flat$ is independent of the choices made in the construction. We proceed in fact to prove that when $i : X \hookrightarrow Z$ is given, up to equivalence $i^\flat : X \hookrightarrow Z^\flat$ is independent of the choice of a dense sequence $(x_i)_{i=0}^\infty$ of points on X and the choice of an integer $m > 0$ such that $\Psi_m : X \hookrightarrow \mathbb{P}^m$ is an embedding, and that furthermore up to equivalence $i^\flat : X \hookrightarrow Z^\flat$ is in fact independent of the choice of a Bergman meromorphic compactification $i : X \hookrightarrow Z$ to start with. The latter will imply that *any* reduction $i^\flat : X \hookrightarrow Z^\flat$ of *any* Bergman meromorphic compactification $i : X \hookrightarrow Z$ gives the minimal Bergman meromorphic compactification (which is unique up to equivalence). In what follows, for a meromorphic map $\alpha : A \to B$ between two Moishezon manifolds, writing $E \subset A$ for the set of indeterminacies of α, by the graph of α, denoted by Graph(α), we will mean the topological closure of Graph$(\alpha|_{A-E})$ in $A \times B$. Graph$(\alpha) \subset A \times B$ is a subvariety. We have the following main result of the current article.

Main Theorem (Theorem 4.1.) *Let (X, ds^2_X) be a canonically embeddable Bergman manifold admitting a Bergman meromorphic compactification $i : X \hookrightarrow Z$. Then, X admits a minimal Bergman meromorphic compactification $i_0 : X \hookrightarrow Z_0$. Furthermore, any two minimal Bergman meromorphic compactifications of X are equivalent in the sense of Definition 3.2.*

Proof. Given two minimal Bergman meromorphic compactifications $i_0 : X \hookrightarrow Z_0$ and $i_0' : X \hookrightarrow Z_0'$, identifying X as an open subset of Z_0, resp. Z_0', the identity map id_X extends meromorphically in both directions, and thus to a bimeromorphic map between Z_0 and Z_0', and it follows that the minimal Bergman meromorphic compactification $i_0 : X \hookrightarrow Z_0$ is uniquely determined up to equivalence. We claim that any minimal Bergman meromorphic compactification $i_0 : X \hookrightarrow Z_0$ is reduced. To see this let $i^\flat : X \hookrightarrow Z_0^\flat$ be a reduction of $i_0 : X \hookrightarrow Z_0$. By construction we have a meromorphic map $\mu : Z_0 \to Z_0^\flat$ which induces the biholomorphism from $i_0(X) \subset Z_0$ and $i_0^\flat(X) \subset Z_0^\flat$ arising from the identity map id_X, i.e., $\mu|_{i_0(X)} = i_0^\flat \circ i_0^{-1}$, where i_0^{-1} denotes the inverse map of the biholomorphism $i_0 : X \xrightarrow{\cong} i_0(X)$. On the other hand, by minimality, there exists a meromorphic map $\gamma : Z_0^\flat \to Z_0$ which induces by restriction the biholomorphism from $i_0^\flat(X) \subset Z_0^\flat$ to $i_0(X) \subset Z_0$ arising from the identity map id_X, i.e.,

$\gamma\big|_{i_0^\flat(X)} = i_0 \circ (i_0^\flat)^{-1}$, where $(i_0^\flat)^{-1}$ denotes the inverse map of the biholomorphism $i_0^\flat : X \xrightarrow{\cong} i_0^\flat(X)$. By definition $\mu\big|_{i_0(X)} : i_0(X) \xrightarrow{\cong} i_0^\flat(X)$ and $\gamma\big|_{i_0^\flat(X)} : i_0^\flat(X) \xrightarrow{\cong} i_0(X)$ are inverses of each other. Hence, Graph(μ) and Graph(γ) are transposes of each other, i.e., $(y, y^\flat) \in \text{Graph}(\mu)$ if and only if $(y^\flat, y) \in \text{Graph}(\gamma)$. Suppose $\mu : Z_0 \to Z_0^\flat$ is not bimeromorphic. Then, for a general point $x^\flat \in Z_0^\flat$, there exists an open neighborhood U of $x^\flat$ and two distinct open subsets $V_1, V_2 \subset Z_0$ such that $\mu\big|_{V_i}$ maps V_i biholomorphically onto U for $i = 1, 2$. Then, for any $y^\flat \in U$, there is $y_i \in V_i$; $i = 1, 2$; such that $\mu(y_i) = y^\flat$. Both $(y_1, y^\flat)$ and $(y_2, y^\flat)$ belong to Graph(μ), and hence both $(y^\flat, y_1)$ and $(y^\flat, y_2)$ must belong to Graph(γ), which is a contradiction since γ is a (meromorphic) map on $Z_0^\flat$. Hence, the claim is proved.

To prove Theorem 4.1 it suffices therefore to prove that *any* reduction $i^\flat : X \hookrightarrow Z^\flat$ of *any* Bergman meromorphic compactification $i : X \hookrightarrow Z$ is the same up to equivalence. Equivalently, we have to show that, given any two reduced Bergman meromorphic compactifications $i_1 : X \hookrightarrow Z_i$; $i = 1, 2$; and identifying X as an open subset of Z_1, resp. Z_2, the identity map id_X extends to a bimeromorphic map $\eta : Z_1 \to Z_2$.

Denote by X_k the image of $i_k : X \hookrightarrow Z_k$; $k = 1, 2$. In place of id_X we will write $f : X_1 \xrightarrow{\cong} X_2$ and consider the problem of extension of Graph(f). By Theorem 2.2.1, Graph(f) extends to an irreducible complex-analytic subvariety $S \subset Z_1 \times Z_2$. We are going to prove that $f : X_1 \xrightarrow{\cong} X_2$ extends to a meromorphic map $F : Z_1 \dashrightarrow Z_2$. Given this, and applying the same statement with X_1 and X_2 interchanged, we will have proved that F is a bimeromorphic map. To prove that f extends meromorphically to Z_1 we are going to argue by contradiction. Supposing that the general fiber of the projection $\pi_1 : S \to Z_1$ consists of $s \geq 2$ points, we obtain by analytic continuation two distinct branches f' and f'' over some nonempty connected open subset $U \subset X_1 \subset Z_1$. Denote by $x_1 \in X_1$, $x_2 \in X_2$ base points such that $f(x_1) = x_2$. Consider $\nu_1 := \mathcal{K}_{X_1, x_1}$ and $\nu_2 := \mathcal{K}_{X_2, x_2}$ chosen such that ν_1, resp. ν_2, is of norm 1 in $H^2(X_1, \omega_{X_1})$, resp. $H^2(X_2, \omega_{X_2})$. Following Mok [Mo2, proof of Theorem 2.2.1, Eqn.(1)], define $K_{X_1}^\flat$ and $K_{X_2}^\flat$ by

$$\mathcal{K}_{X_1}(z, w) = K_{X_1}^\flat(z, w)\big(\epsilon_n \nu_1(z) \wedge \overline{\nu_1(w)}\big) \; ;$$

$$\mathcal{K}_{X_2}(\zeta, \xi) = K_{X_2}^\flat(\zeta, \xi)\big(\epsilon_n \nu_2(\zeta) \wedge \overline{\nu_2(\xi)}\big) \; . \qquad (1)$$

Here ϵ_n stands for some non-zero complex number depending only on n, and the notation $K_{X_1}^\flat$, resp. $K_{X_2}^\flat$, is understood to mean the extension of the function to a meromorphic function in $(z, \overline{w})$, resp. in $(\zeta, \overline{\xi})$, to $Z_1 \times Z_1$, resp. $Z_2 \times Z_2$. From the functional identity given in Mok [Mo2, proof of Theorem 1.2.1, Eqn. (1)] together with the obvious adaptation to the situation of manifolds starting with the functional identity in Mok [Mo2, proof of Theorem 2.2.1] we have the identity

$$K^\flat_{X_2}(f'(z), f(w)) = K^\flat_{X_2}(f''(z), f(w)) = K^\flat_{X_1}(z, w) \tag{2}$$

for any $w \in X_1$ and for any $z \in U$. Thus, for any $\xi \in X_2$ and $z \in U$, writing $K^\flat_{X_2,\xi}(\zeta) = K^\flat_{X_2}(\zeta, \xi)$ we have

$$\frac{K^\flat_{X_2,\xi}(f'(z))}{K^\flat_{X_2,x_2}(f'(z))} = \frac{K^\flat_{X_2,\xi}(f''(z))}{K^\flat_{X_2,x_2}(f''(z))} \,. \tag{3}$$

For the map $\Phi^\sharp_m : Z_2 \dashrightarrow \mathbb{P}^m$ as defined in analogy to the proof of Proposition 3.1 we conclude that

$$\Phi^\sharp_m(f'(z)) = \Phi^\sharp_m(f''(z)) \,, \tag{4}$$

for any $m \geq 1$, contradicting with the assumption that the Bergman meromorphic compactification $i_2 : X_2 \hookrightarrow Z_2$ is reduced. The proof of Theorem 4.1 is complete. □

From the proof of Theorem 4.1 we have

Theorem 4.2. *Let (X, ds^2_X) be a canonically embeddable Bergman manifold admitting a Bergman meromorphic compactification $i : X \hookrightarrow Z$. Then, up to equivalence any reduction $i^\flat : X \to Z^\flat$ is the unique minimal Bergman meromorphic compactification.*

5. Examples of Bergman Meromorphic Compactifications

We provide now examples of Bergman meromorphic compactifications. As mentioned, for a bounded symmetric domain $D \Subset \mathbb{C}^n$ in its Harish-Chandra realization, the Bergman kernel $K_D(z, w)$ is a rational function in $(z, \overline{w})$. It is in fact of the form $K_D(z, w) = \frac{1}{Q_D(z,w)}$, where $Q_D(z, w)$ is a polynomial in $(z, \overline{w})$, as can be found in Faraut–Korányi [FK, pp. 76–77, especially Eqns. (3.4) and (3.9)]. Writing $D \Subset \mathbb{C}^n \subset M$ simultaneously for the Harish-Chandra embedding and the Borel embedding of D into its compact dual manifold M, the Bergman kernel $K_D(z, w)$ extends rationally to M. One can check directly from the explicit forms of the Bergman kernels that $D \subset M$ is a minimal Bergman meromorphic compactification.

More conceptually, the latter is a special case of the following general result concerning minimality of Bergman meromorphic compactifications for complete circular domains.

Theorem 5.1. *Let $G \Subset \mathbb{C}^n$ be a bounded complete circular domain. Suppose the Bergman kernel $K_G(z,w)$ extends to a rational function in $(z,\overline{w})$, S is a compact complex manifold, and $\mathbb{C}^n \subset S$ is a compactification of $\mathbb{C}^n$ birational to the standard compactification $\mathbb{C}^n \subset \mathbb{P}^n$. Then, $G \subset S$ is a minimal Bergman meromorphic compactification. In particular, the Borel embedding $D \subset M$ of a bounded symmetric D into its compact dual manifold M is a minimal Bergman meromorphic compactification.*

Proof. Since $G \Subset \mathbb{C}^n$ is a bounded domain, the Bergman metric is defined on G. For $x \in X$ write $K_{G,x}(z) = K_G(z,x)$, which will be regarded as a rational function on $\mathbb{C}^n$. Write $i : G \hookrightarrow S$ for the inclusion map, which is by the hypothesis a Bergman meromorphic compactification. By Theorem 4.2, to prove Theorem 5.1 it remains to show that $i : G \hookrightarrow S$ is reduced. Suppose otherwise. Then, the reduction $i^\flat : G \hookrightarrow S^\flat$ extends to a meromorphic map $\eta : S \to S^\flat$ which is not bimeromorphic. Hence there exist two disjoint non-empty connected open sets $U_1 \subset G$, $U_2 \subset S - \overline{G}$ such that $i^\flat|_{U_1}$ maps U_1 biholomorphically onto $i^\flat(U_1) := U^\flat$, and $\eta|_{U_2}$ maps U_2 biholomorphically onto $U^\flat$. Thus, there exists a biholomorphism $\varphi : U_1 \to U_2$ such that, writing $z_2 := \varphi(z_1)$, η is holomorphic on both U_1 and U_2, and we have $\eta(z_2) = \eta(z_1) = i^\flat(z_1)$. By Theorem 4.2 the reduction $i^\flat : G \to S^\flat$ is up to equivalence uniquely determined, and can be constructed from any choice of a dense sequence of points $(x_i)_{i=0}^\infty$ on G and any choice of a positive integer m such that, writing $\Psi_m = [K_{G,x_0}, \cdots, K_{G,x_m}]$, $\Psi_m : G \to \mathbb{P}^m$ is a holomorphic embedding on G. We may take $x_0 = 0 \in G$. Since G is a circular domain we have $K_G(e^{i\theta}z, e^{i\theta}w) = K_G(z,w)$ for any $z, w \in G$ and any $\theta \in \mathbb{R}$. Taking $w = 0$ we conclude that $K_G(z,0)$ is a constant. From $\eta(z_2) = \eta(z_1)$ it follows that $K_G(z_2, x_i) = K_G(z_1, x_i)$ for any nonnegative integer i. Since the sequence of points $(x_i)_{i=0}^\infty$ is dense in G, it follows that

$$K_G(z_2, w) = K_G(z_1, w) \tag{1}$$

for any $w \in G$. Expand now $K_G(z,w)$ as a power series at $(0,0) \in G \times G$ and using again the invariance of K_G under the circle group action we have

$$K_G(z,w) = \sum_{|I|=|J|} c_{I\overline{J}} z^I \overline{w^J} = \sum_J h_J(z) \overline{w^J} \ , \ h_J(z) = \sum_{|I|=|J|} c_{I\overline{J}} z^I \ ; \tag{2}$$

where in the summations $I = (i_1, \cdots, i_n)$ and $J = (j_1, \cdots, j_n)$ range over n-tuples of nonnegative integers, and $|I| = i_1 + \cdots + i_n$ is the length of I, etc. For $J = 0$ we have $h_0(z) = K_G(z, 0) = c_{0,0} = K_G(0, 0)$. For J of length 1 it follows that h_J is a linear function. For $1 \leq i, j \leq n$ write $a_{i\bar{j}} = c_{E_i\overline{E_j}}$ where E_k is the unique n-tuple of nonnegative integers of length 1 with the entry 1 at the k-th position. Writing $(g_{i\bar{j}})$ for the matrix expression of the Bergman metric $ds_G^2 = 2\mathrm{Re}(g)$ in terms of Euclidean coordinates, we have by expansion $g_{i\bar{j}}(0) = \frac{a_{i\bar{j}}}{K_G(0,0)}$ and it follows that $(a_{i\bar{j}}) > 0$. Writing $h_j := h_{E_j}$ we conclude that $h_1, \cdots, h_n$ are linearly independent. From (1) and (2) and differentiating against $\overline{w_j}$ at 0 we conclude that

$$h_j(z_2) = h_j(z_1) \quad \text{for} \quad 1 \leq j \leq n\,. \tag{3}$$

From the linear independence of the n linear functions $h_1, \cdots, h_n$ it follows that $z_2 = z_1$, contradicting the definitions of U_1 and U_2. We have thus proven by argument by contradiction that $\eta : S \to S^\flat$ is bimeromorphic, i.e., $i : G \hookrightarrow S$ is reduced and hence a minimal Bergman meromorphic compactification, as desired. □

We examine more generally (not necessarily bounded) domains $G \subset \mathbb{C}^n$ which are biholomorphic to bounded domains, so that the Bergman kernel $K_G(z, w)$ on G is defined, such that $K_G(z, w)$ extends to a rational function in $(z, \overline{w})$. Taking $\mathbb{C}^n \subset \mathbb{P}^n$ to be the standard compactification, $G \subset \mathbb{P}^n$ is then a Bergman meromorphic compactification. Examples of unbounded domains $\mathcal{D} \subset \mathbb{C}^n$ on which the Bergman kernel is rational are given by the unbounded realizations of bounded homogeneous domains of Pyatetskii-Shapiro [Py] as Siegel domains of the first or second kind. Up to affine transformations they are represented as normal Siegel domains $\mathcal{D} := D(V_N, F)$ as given in Xu [Xu, Chapter 3], where the Bergman kernels $K_{\mathcal{D}}$ are completely determined, and they are rational, as given in [Xu, Theorem 3.26, Eqn. (3.131)]. There are also standard realizations of $\mathcal{D}$ as bounded domains with respect to which the Bergman kernels remain to be rational (cf. [Xu, Chapter 4]), especially the canonical bounded realizations D defined by the Bergman mapping. In particular the boundary extension results for holomorphic isometries up to a normalizing constant with respect to the Bergman metric in Mok [Mo2, Theorem 2.1.2] are applicable to the canonical bounded realizations D and to the unbounded realizations $\mathcal{D}$ as Siegel domains $\mathcal{D}$. (To take care of the boundary of $\mathcal{D}$ at infinity one can make use of the more general formulation of extension results in Mok [Mo2, Theorem 2.2.1].) We note that the canonical bounded realizations defined by the Bergman mapping include the Harish-Chandra realizations of bounded symmetric domains, and the unbounded realization as Siegel domains include the Cayley transforms of Korányi–Wolf [KW, 1965] of bounded symmetric domains.

Given a domain $G \subset \mathbb{C}^n$ which is biholomorphic to a bounded domain, and a base point $x_0 \in G$, the Bergman mapping $\sigma : (G; x_0) \to (\mathbb{C}^n; 0)$, $\sigma(z) = (\zeta_1, \cdots, \zeta_n)$ is a germ of biholomorphism at x_0, uniquely defined up to a linear transformation on the target Euclidean space, which may be given by the formula (cf. [Xu, Chapter 4, §1, Eqn. (4.1)])

$$\zeta_k = \frac{\partial}{\partial \overline{w_k}} \log \frac{K_G(z, \overline{w})}{K_G(z_0, \overline{w})}\bigg|_{w=x_0}$$

(Alternatively, as was originally done, the Bergman mapping can be normalized by requiring the Bergman metric to agree with the Euclidean metric at $0 = \sigma(x_0)$, in which case it is uniquely determined up to unitary transformations.) When $G \Subset \mathbb{C}^n$ is a bounded symmetric domain in its Harish-Chandra realization, $x_0 = 0$, the Bergman mapping at 0 is a linear map. In this case we have $\zeta_k = \frac{1}{K_G(0,0)} \frac{\partial}{\partial \overline{w_k}} K_G(z, \overline{w})\big|_{w=0}$. In general if $G \subset \mathbb{C}^n$ is biholomorphic to a bounded domain, $x_0 \in G$, and the Bergman mapping $\sigma : (G; x_0) \to (\mathbb{C}^n; 0)$ extends to a biholomorphism, still to be denoted by $\sigma : G \xrightarrow{\cong} D \subset \mathbb{C}^n$, $\sigma(x_0) = 0$, then the Bergman mapping on D at 0 is a linear map (cf. Xu [Xu, Chapter 4, Theorem 4.2]). From the latter observation and the fact the arguments in the proof of Theorem 5.1 given in Eqns. (1)–(3) there, we have readily the following result giving a sufficient condition for a Bergman meromorphic compactification to be minimal when the compactifying manifold is birational to $\mathbb{P}^n$.

Theorem 5.2. *Let $G \subset \mathbb{C}^n$ be a domain biholomorphic to a bounded domain, so that the Bergman kernel $K_G(z, w)$ and the Bergman metric ds^2_G are defined. Suppose the Bergman kernel $K_G(z, w)$ extends to a rational function in $(z, \overline{w})$, S is a compact complex manifold, and $\mathbb{C}^n \subset S$ is a compactification of $\mathbb{C}^n$ birational to the standard compactification $\mathbb{C}^n \subset \mathbb{P}^n$. Suppose furthermore that there exists some point $x_0 \in G$ such that the Bergman mapping $\sigma : (G; x_0) \to (\mathbb{C}^n; 0)$ extends to a birational map on $\mathbb{C}^n$. Then, $G \subset S$ is a minimal Bergman meromorphic compactification.*

As special cases of Theorem 5.2 we have

Corollary 5.1. *Let $\mathcal{D} \subset \mathbb{C}^n$ be the realization of a bounded homogeneous domain as a Siegel domain of the first or second kind, S be a compact complex manifold, and $\mathbb{C}^n \subset S$ be a compactification birational to the standard compactification $\mathbb{C}^n \subset \mathbb{P}^n$. Then, the inclusion $i : \mathcal{D} \hookrightarrow S$ is a minimal Bergman meromorphic compactification. Furthermore, the canonical bounded realization $D \Subset \mathbb{C}^n$ (uniquely determined up to linear transformations when the base point is fixed) is also a minimal Bergman meromorphic compactification.*

Proof. Fix any base point $x_0 \in \mathcal{D}$, and let $\sigma : (\mathcal{D}; x_0) \to (\mathbb{C}^n; 0)$ be the Bergman mapping as a germ of biholomorphism. By Xu [Xu, Chapter 4, Theorem 4.7], σ extends to a biholomorphism of $\mathcal{D}$ onto its image D, yielding the canonical bounded realization $\sigma : \mathcal{D} \xrightarrow{\cong} D \subset \mathbb{C}^n$. By Theorem 5.2 it suffices to check that σ extends further to a birational map on $\mathbb{C}^n$. This is implicit in [Xu, Chapter 4] as given by the explicit calculations there in terms of a factorization $\sigma = \sigma_3 \circ \sigma_2 \circ \sigma_1$. Here each $\sigma_i, 1 \leq i \leq 3$, is a birational map from [Xu, Theorem 4.3 (for σ_1), Theorem 4.5 (for σ_2), and Theorem 4.7 (for σ_3)]. The birationality of each σ_i comes from the description of σ_i as a matrix of functions in upper triangular form. Typically, if f is a birationl map on $\mathbb{C}^k$, g is a birational map on $\mathbb{C}^\ell$, and $h : \mathbb{C}^k \dashrightarrow \mathbb{C}^\ell$ is a rational map, then $\Phi(z, w) := (f(z), g(w) + h(z))$ is a birational map on $\mathbb{C}^{k+\ell}$. We refer the reader to [Xu, Chapter 4] for details.

The last statement that $D \Subset S$ is a minimal Bergman meromorphic compactification follows since minimality is unchanged under a birational map. (Alternatively, noting that the Bergman mapping of D at 0 is a linear map, minimality of the Bergman meromorphic compactification $D \Subset S$ follows from the proof of Theorem 5.1.) □

Remark For all known examples of Bergman meromorphic compactifications $i : X \hookrightarrow Z$ of canonical embeddable Bergman manifolds, including those given here, the Bergman kernel forms $\mathcal{K}_X(z, w)$ extend meromorphically in $(z, \overline{w})$ from $X \times X$ to $Z \times Z$. However, as mentioned in the Remark in §2, the latter extension property of the Bergman kernel form does not *a priori* hold when one descends to a quotient manifold obtained by a reduction of $i : X \hookrightarrow Z$ and by desingularization, and that is the reason for defining Bergman meromorphic compactifications by imposing slightly weaker extension properties related to the Bergman kernel form as given in Definition 2.1.

Acknowledgements

On the invitation of the research group in Differential Geometry at Fudan University, including Professors Chaohao Gu, Hesheng Hu and Jiaxing Hong, the author has given series of lectures on complex geometry over a number of years starting from 2006, including those on holomorphic isometries on bounded domains with respect to the Bergman metric. The author wishes to thank the research group for their kind invitation, and to the organizers especially Professor Yuxin Dong for their efforts and their interest in the subject. The topic of the current article is an outgrowth of the author's work on extension problems on holomorphic isometries between Bergman manifolds, and it is befitting that the author dedicates this article to the memory of Professor Chaohao Gu as a tribute to his leadership through decades of work in Differential Geometry and related areas.

References

[Ca] Calabi, E., Isometric imbedding of complex manifolds, *Ann. Math.* **58** (1953), 1-23.

[CU] Clozel, Laurent and Ullmo, E., Modular correspondences and invariant measures, *J. Reine Angew. Math.* **558** (2003), 47-83.

[FK] Faraut, J. and Korányi, A., Function spaces and reproducing kernels on bounded symmetric domains, *J. Funct. Anal.* **88** (1990), 64-89.

[KW] Korányi, A. and Wolf, J. A., Realization of hermitian symmetric spaces as generalized half-planes, *Ann. of Math.* **81** (1965), 265-288.

[Mo1] Mok, N., Geometry of holomorphic isometries and related maps between bounded domains, in *Geometry and Analysis*, Vol. II, Advanced Lectures in Mathematics 18, Higher Educational Press, Beijing, 2011, pp. 225-270.

[Mo2] Mok, N., Extension of germs of holomorphic isometries up to normalizing constants with respect to the Bergman metric, *J. Eur. Math. Soc.* **14** (2012), 1617-1656.

[Py] Pyatetskii-Shapiro, I. I., *Automorphic Functions and the Geometry of Classical Domains*, Gordon and Breach Science Publishers, New York-London-Paris, 1969.

[Xu] Xu, Y., *Theory of Complex Homogeneous Bounded Domains*, Science Press, Kluwer Academic, London, 2005.

The Application of Conditional Nonlinear Optimal Perturbation to Targeted Observations for Tropical Cyclone Prediction

Mu Mu

Key Laboratory of Ocean Circulation and Wave, Institute of Oceanology, Chinese Academy of Sciences, Qingdao 266071, China

Feifan Zhou

Laboratory of Cloud-Precipitation Physics and Severe Storms (LACS), Institute of Atmospheric Physics, Chinese Academy of Sciences, Beijing 100029, China

Xiaohao Qin

State Key Laboratory of Numerical Modeling for Atmospheric Sciences and Geophysical Fluid Dynamics (LASG), Institute of Atmospheric Physics, Chinese Academy of Sciences, Beijing 100029, China

Chen Boyu

National Meteorological Center of China Meteorological Administration, Beijing 100081, China

This paper briefly but systematically introduces the application of Conditional Nonlinear Optimal Perturbation (CNOP) to targeted observations for tropical cyclone predictions. CNOP is a natural extension of the singular vector (SV) method into the nonlinear regime, and it has recently been used to identify the sensitive areas in typhoon targeted observations.

First, it was demonstrated that CNOP is different from the internationally used method (SV) for targeted observation and can take nonlinear processes into account. Then, it was revealed that the locations of initial errors have a great effect on their nonlinear growth. For a given norm of initial errors and basic state, the CNOP-type initial errors introduced into their corresponding sensitive areas cause the largest changes to the final verification forecasts. These results suggest that the application of CNOP to targeted observation has a theoretical basis.

Then, the properties of the CNOP-sensitive areas were examined, and the results indicated that the CNOP-sensitive areas have favorable features for targeted observations. Next, observing system simulation experiments (OSSEs) and the observation system experiments (OSEs) were conducted to evaluate the efficiency of the CNOP-sensitive areas. The impacts on typhoon forecasting skills achieved by assimilating simulated or real observation data in the CNOP- and SV-sensitive areas were compared. The results indicate that greater benefits were obtained by assimilating data in CNOP-sensitive areas averagely. Finally, the conditions under which the CNOP-sensitive areas would be more effective were summarized, and the execution of the CNOP method with a more advanced model and more reasonable initial constraint were studied.

All the above results demonstrated that the CNOP is a useful tool for identifying the sensitive areas for targeted observations. The method can take the nonlinear processes of physical issues into account and possesses a solid theoretical foundation.

1. Introduction

Targeted observation (also called adaptive observation) is a new observational methodology by which observational capabilities are intensified in areas where additional observations are expected to significantly improve a forecast. These areas are considered as "sensitive", in the sense that changes to the initial conditions in these areas are expected to have a larger impact on the forecast than changes in other areas. Obviously, the key point in targeted observation is to find the sensitive areas.

Currently, there are several strategies used for identifying the sensitive areas, including singular vectors (SVs, Palmer *et al.*, 1998), adjoint sensitivities (Ancell and Mass, 2006), the adjoint-derived sensitivity steering vector (ADSSV) (Wu *et al.*, 2007), the ensemble transform (ET, Bishop and Toth, 1999), the ensemble Kalman filter (EnKF, Hamill and Snyder, 2002), and the ensemble transform Kalman filter (ETKF, Bishop *et al.*, 2001). These strategies have been tested in field experiments such as the Fronts and Atlantic Storm-Track Experiment (FASTEX; Snyder, 1996; Joly *et al.*, 1997), the North Pacific Experiment (NORPEX; Langland *et al.*, 1999), the Winter Storm Reconnaissance Programs (WSR; Szunyogh *et al.*, 2000; 2002), the Dropwindsonde Observations for Typhoon Surveillance near the Taiwan region (DOTSTAR; Wu *et al.*, 2005), the Observing System Research and Predictability Experiment (THORPEX) Pacific Asian Regional Campaign

(T-PARC; Aberson, 2011), and the Atlantic THORPEX Regional Campaign (ATReC; Rabier *et al.*, 2008). Forecasts are generally improved by assimilation of targeted observations (Gelaro *et al.*, 1999; Langland, 2005; Wu *et al.*, 2005; Buizza *et al.*, 2007; Rabier *et al.*, 2008).

The strategies mentioned above are all constrained by linear approximations. To study the effect of nonlinearity, Mu *et al.* (2003) proposed the novel approach of conditional nonlinear optimal perturbation (CNOP). CNOP is an extension of the SV method into the nonlinear regime, and it has been applied to some research fields such as El Niño-Southern Oscillation (ENSO) predictability (Mu *et al.*, 2007; Duan and Mu, 2009; Duan and Zhang, 2010; Duan and Luo, 2010; Peng *et al.*, 2011), the nonlinear behavior of baroclinic unstable flows (Rivier *et al.*, 2008), ensemble forecasting (Mu and Jiang, 2008), the transitions between multiple equilibria states of the ecosystem (Sun and Mu, 2009), and the North Atlantic Oscillation (Jiang *et al.* 2013).

Recently, Mu *et al.* (2009) suggested that CNOP can be used to identify the sensitive areas for typhoon targeted observations because the forecasts benefit more from reductions of CNOP-type initial errors than from reductions of SV-type initial errors. Then, Chen and Mu (2012) revealed that the initial errors which have the largest impact on typhoon forecasts are located in particular areas and that, generally, the initial errors in the sensitive areas identified by CNOP (CNOP-sensitive areas) have the largest forecast impact on the forecast. These studies demonstrated that sensitive areas do exist and confirmed the theoretical basis of applying CNOP to targeted observations. Zhou and Mu (2011, 2012a,b) further demonstrated the advantages of the CNOP method in identifying the sensitive areas by studying the properties of the CNOP-sensitive areas with respect to variations of the horizontal resolution, the verification area design and the optimization time period. Furthermore, Qin (2010a,b) and Qin and Mu (2011a,b) performed observing system simulation experiments (OSSEs) to assess whether the sensitive areas identified by CNOP can be considered as dropping sites in realtime targeting. Observing system experiments (OSEs) using the DOTSTAR Data have also been conducted by Chen (2011) and Chen *et al.* (2013) to demonstrate the utility of the CNOP method. In addition, Qin *et al.* (2013) summarized the conditions under which the CNOP-sensitive areas would be more beneficial, and Wang *et al.* (2011) combined the CNOP method and the Weather Research Forecasting (WRF)model and used background error as an initial constraint to study targeted observations for typhoon predictions.

This study will summarize the above works on the application of the CNOP method to typhoon targeted observations. The structure of this paper is as follows. Section 2 provides an introduction to CNOP and SV and their impacts on typhoon forecasts. The theoretical basis of applying CNOP to targeted observations is also introduced in this section. Section 3 shows the favorable features of the CNOP sensitive areas in targeted observations. Section 4 estimated the effects of CNOP sensitive-areas by both OSSEs and OSEs. Section 5 summarizes the conditions under which the CNOP-sensitive areas would be more beneficial and shows the execution of the CNOP method with a more advanced model and more reasonable initial constraint. A brief summary and discussion are provided in the final section.

2. Theoretical Basis of CNOP and SV for Targeted Observations

2.1. *CNOP method*

A detailed description of CNOP can be found in Mu *et al.* (2003). Briefly, CNOP is defined as follows. An initial perturbation $\check{S}X_0^*$ of vector X_0 is called a CNOP if and only if

$$J(\check{S}X_0^*) = \max_{\delta X_0^T C_1 \delta X_0 \leq \beta} J(\check{S}X_0) \tag{1}$$

where

$$J(\check{S}X_0) = [PM(X_0 + \check{S}X_0) - PM(X_0)]^T C_2 [PM(X_0 + \check{S}X_0) - PM(X_0)] \tag{2}$$

δX_0,which is the first guess of the initial perturbation, should be adjusted to satisfy the constraint condition $\delta X_0^T C_1 \delta X_0 \leq \beta$. The presumed positive constant β represents the magnitude of the initial perturbation. The symmetric positive definite matrices C_1 and C_2 represent the norms of initial perturbation and the forecast errors at prediction time. M is a nonlinear propagator, and P is a local projection operator and takes value 1(0) within (without) the verification region. J is the cost function. The superscript "T" denotes the transpose of the vectors or matrices. Note that the norms C_1 and C_2 may be the same, depending on the physical problem. It is clear that CNOPs depend on the nonlinear model M, the initial state vector X_0, and the parameters β, P , C_1 and C_2.

2.2. SV method

Define L as the forward tangent propagator corresponding to M. According to Barkmeijer *et al.* (2003), the first singular value σ_1 of L satisfies (with respect to the norms and C_1 and C_2) the following:

$$\sigma_1^2 = \max_{(\delta X_0)^T C_1(\delta X_0)\neq 0} \frac{[L(\delta X_0)]^T C_2[L(\delta X_0)]}{(\delta X_0)^T C_1(\delta X_0)}. \tag{3}$$

In addition, if $\mathbf{v_j}$ is the singular vector of L, then

$$(C_1)^{-1}(L^T C_2 L)\mathbf{v_j} = \sigma_{\mathbf{j}}^{\mathbf{2}}\mathbf{v_j} \tag{4}$$

where superscript "-1" denotes the inverse of the matrix. σ_j is the singular value corresponding to $\mathbf{v_j}$. The first SV (FSV) maximizes the linear development of the initial perturbations, the second SV maximizes the development under the constraint of being orthogonal to the first SV, and the third SV maximizes the development under the constraint of being orthogonal to the first two SVs, and so on (Peng and Reynolds, 2006). A local projection operator P (same meaning as in Eq. (2)) is employed to localize the development of the perturbation in the verification region. Note that, here, the initial perturbation δX_0 should be sufficiently small, and the propagator time interval should be of moderate length.

The FSV can be obtained by solving the following linear optimization problem (Ehrendorfer and Errico, 1995):

$$J(\delta X_0^*) = \max_{\delta X_0^T C_1 \delta X_0 \leq \beta} J(\delta X_0) \tag{5}$$

where

$$J(\delta X_0) = [PL(\delta X_0)]^T C_2[PL(\delta X_0)]. \tag{6}$$

According to the linear characteristics of SV, the FSV defined by Eqs. (5) and (6) equals the FSV defined by Eq. (4) when it has been unit normalized.

In this paper, except when clarified, the model designs are as follows. The fifth-generation Pennsylvania State University-National Center for Atmospheric Research (PSU-NCAR) Mesoscale Model (MM5; Dudhia, 1993) and its corresponding adjoint system (Zou *et al.*, 1997) are used. The National Centers for Environment Predictions (NCEP) FNL (Final) Operational Global Analysis interpolated into MM5 grids serves as the basic states. The model horizontal resolution is 60km, and there are 11 vertical levels. When the FSV is considered, it will be obtained by solving Eqs. (5) and (6) with the optimization algorithm spectral projected gradient 2 (SPG2) (Birgin *et al.* 2001). When other SVs are considered, the SVs are obtained by using the Lanczos algorithm (Ehrendorfer and Errico, 1995). The metric of total dry energy (see Mu *et al.* 2009) has been chosen as the norm for C_1 and C_2.

2.3. *The impacts of CNOP and SV on typhoon forecasts*

From the definition of CNOP (FSV), it is observed that CNOP (FSV) can be taken as a type of initial errors whose nonlinear (linear) evolutions would be the largest. Mu *et al.* (2009) studied the impacts of CNOP-type and FSV-type initial errors on the forecasts of three typhoons, Mindulle (2004), Meari (2004), and Matsa (2005). It was observed that, when the initial constraint value is small, the CNOP-type errors resemble the FSV-type errors. When the initial constraint value is sufficiently large such that the magnitudes of the temperature and wind components of the CNOP are comparable with the counterparts of current analysis errors, the CNOP patterns are much different from that of the FSV (figure 1), and the CNOP reflects the impact of nonlinear processes. In addition, the CNOP-type errors have a larger impact on the forecasts of typhoons than the FSV-type errors.

Moreover, Mu *et al.* (2009) designed a series of sensitivity experiments to determine what benefit can be obtained by reductions of CNOP-type errors versus FSV-type errors. The results indicated that reductions of CNOP-type errors in the initial states provide more benefits than reductions of FSV-type errors (table 1). Because FSV has been widely used to identify sensitive areas for targeted observations, the above results suggest that it is worthwhile to use CNOP as a method to identify the sensitive areas for targeted observation in typhoon prediction.

2.4. *The evolution of the initial errors from CNOP-, FSV- and CSV-, sensitive areas and random areas*

As related in section 2.3, it is worthwhile to use CNOP to identify sensitive areas, so in this section, CNOP is used to identify sensitive areas. Similar to the utility of SV in targeted observations, a vertically integrated energy (VIE) of CNOP is calculated. The CNOP-sensitive areas are defined according to the CNOP-VIE. Similarly, FSV-sensitive areas are defined according to FSV-VIE. For the composite of the SV (short for CSV or SVs), the first several SVs are combined according to their singular values (Chen and Mu, 2012), and the CSV-sensitive areas are defined according to the CSV-VIE.

Then, the random initial errors are introduced into the CNOP-, FSV- and CSV- sensitive areas and the randomly selected areas. Their impacts on tropical cyclone (TC) forecasts are explored (Chen and Mu, 2012). Two tropical cyclones, Longwang (2005) and Sinlaku (2008), are studied.

For simplicity, the CNOP-, CSV-, and FSV-sensitive areas and randomly selected areas are designated as CNOP_Sen, CSV_Sen, FSV_Sen, and Ran_Area, respectively. The random initial errors (Ran_Err) are added to the four areas. Notably, the random initial errors have the same size measured by the dry energy norm. The four experiments noted by RA-CN, RA-CS, RA-FS, and RA-RA, respectively, represent that the initial errors are introduced into the CNOP_Sen, CSV_Sen, FSV_Sen, and Ran_Area.

First, the distributions of CNOP_Sen, CSV_Sen, and FSV_Sen are checked. For the case of TC Longwang (2005), the maximum of CNOP-VIE, CSV-VIE, and FSV-VIE are defined as the centers of CNOP_Sen, CSV_Sen, and FSV_Sen (indicated by the smaller squares in figures 2 and 3, differing from the definition in section 2.3). Considerable difference exists between the locations of FSV_Sen and CSV_Sen (figure 2a,c), whereas the CNOP_Sen and CSV_Sen are slightly different (figure 2b,c). The CNOP_Sen for TC Sinlaku (2008) is defined as a minimized square area that is large enough to contain the two CNOP VIE maxima. FSV_Sen and CSV_Sen are located in the same position, and they have the same size as CNOP_sen (figure 3). In addition, the extents of CNOP_Sen, CSV_Sen, and FSV_Sen for TC Longwang (2005) are defined as a square area with 44 grid points, and the extent for TC Sinlaku (2008) is defined as a square area with 1212 grid points.

Thirty-seven randomly selected areas are considered in the RA-RA experiment for TC Sinlaku (2008) and TC Longwang (2005). The statistical averages of the verification dry energy norms of the 40 evolved Ran_Errs from the 40 local areas (including CNOP_Sen, CSV_Sen and FSV_Sen) are obtained. The geographical distribution of the 40 local areas, including CNOP_Sen, CSV_Sen, FSV_Sen, and 37 Ran_Areas, is shown in figure 4. Notably, the locations of these areas are defined around the initial storm center to the far outside, and the distance between the centers of the two adjacent areas is defined as two-grid spacing for TC Sinlaku (2008) and as one-grid spacing for TC Longwang (2005).

In the four types of experiments (RA-CN, RA-CS, RA-FS, and RA-RA), 40 Ran_Errs are generated for both TC Longwang (2005) and TC Sinlaku (2008). Specifically, the random errors are normally distributed. The u-wind, v-wind, and temperature initial states are perturbed, and the random error vectors are zero mean. Their standard deviations are 0.95 for TC Longwang (2005) and 0.32 for TC Sinlaku (2008).

Table 2 lists the results of the experiments for both cases. Every unit of the tables (except units in the first row and column) corresponds approximately to its location in the geographical distribution of the 40 areas as shown in figure 4. For TC Longwang (2005), the statistical value for CNOP_Sen is the largest, followed by the values for the five Ran_Areas close to CNOP_Sen and CSV_Sen on the western and southern sides of the storm center, whereas the value for CSV_Sen is the seventh largest. In addition, the values for the areas on the outer ring are much lower than those for the areas near CNOP_Sen and CSV_Sen. For TC Sinlaku (2008), the value for CNOP_Sen is still the largest, and in general, the values for the areas distributed on the eastern side of the storm center are higher than those for the areas on the western side, including CSV_Sen/FSV_Sen. Thus, the growth rates of Ran_Errs introduced into sensitive areas, such as CNOP_Sen and CSV_Sen, are higher than those introduced into Ran_Area.

These results suggest that, for a given initial error, basic state, and time interval, the growth rate of the initial error strongly depends on where it is fixed, and the initial errors grow rapidly in the CNOP-sensitive areas.

3. The Favorable Features of CNOP-Sensitive Areas for Targeted Observation

From section 2, we know that using CNOP to identify sensitive areas has a solid theoretical basis, and the initial errors in CNOP-sensitive areas grow rapidly. In this section, we will study the properties of the CNOP-sensitive areas to demonstrate that CNOP-sensitive areas have favorable features for targeted observations.

3.1. *Sensitivity of CNOP-sensitive areas with respect to horizontal resolution*

As related in section 2.1, CNOPs depend on the nonlinear model M, so different models due to different resolution may result in different CNOP-sensitive areas. Zhou and Mu (2012a) studied this issue. In their study, a set of experiments were designed in which all the parameters were held constant except for the horizontal resolution.

Three tropical cyclones, TC Matsa (2005), TC Meari (2004), and TC Mindulle (2004), were investigated. A set of 24-h control forecasts, which served as the basic state, were integrated from 0000 UTC 5 Aug, 2005, to 0000 UTC 6 Aug, 2005 (TC Matsa); from 0000 UTC 26 Sep, 2004, to 0000 UTC 27 Sep, 2004 (TC Meari); and from 0000 UTC 28 Jun, 2004 to 0000 UTC 29 Jun, 2004 (TC Mindulle). For each case, the forecasts were run at 120-km, 60-km, and 30-km horizontal resolutions with 11 vertical levels. For TC Matsa, the model domain covered 28*28, 55*55, and 109*109 (y-direction by x-direction) grids, respectively, for 120-km, 60-km, and 30-km horizontal resolutions. For TC Meari, there were 26*28, 51*55, and 101*109 grids for each horizontal resolution, and for TC Mindulle, the domain sizes with respect to each resolution were 21*26, 41*51, and 81*101. For each case with the chosen grids, the real physical domain was the same at all resolutions; thus, the same verification area could be chosen.

For each case, the CNOP-sensitive areas identified using different resolutions were different from each other (figure 5); however, common areas occurred at the three resolutions, and the sizes of the common areas were different from case to case. In general, the sizes of the common areas are larger between sensitive areas at the lower resolution. This can be deduced from the similarities of the energy distributions between each resolution for the three cases (table 3). For the three cases, the similarities between the lower resolutions (60 km and 120 km) were greater than those between the finer resolutions (30 km and 60 km), which illustrates that more small-scale activity can be resolved at higher resolutions.

From the analysis of the similarities, it can be induced that the sensitive areas identified at lower resolutions are also helpful for improving the forecast at finer resolution. This is a favorable feature of CNOP-sensitive areas because the calculation of CNOP with a high resolution would usually require a long time and a large amount of computer resources. So when computation conditions are limited, we can use low resolution to calculate CNOP-sensitive areas, which would also be useful for typhoon targeted observations.

3.2. *Sensitivity of CNOP-sensitive areas with respect to the verification area design*

As indicated in section 2.1, the CNOPs also depend on the verification area design (different parameter P), and this was studied by Zhou and Mu (2011). Tropical Cyclone Rananim, which occurred in the northwest Pacific Ocean in 2004, was studied.

The design of the verification area was as follows. First, they defined a control design in which the verification area included the real cyclone tracks (the best storm tracks) during the integral time period: scheme B (figure 6a, dashed rectangle). Second, the size of the verification area was kept constant as it is moved to other places (schemes A, C, D, E, and F). Small positional variations were denoted as schemes A and C (figure 6a, solid and dotted rectangles, respectively). Large position variations were denoted as schemes D, E, and F (figure 6b, solid, dashed, and dotted rectangles, respectively). Then, both the size and the position variations of the verification area were considered; the designs are shown in figure 6c. Scheme G is designed with small variations (figure 6c, solid rectangle). Schemes H and I have even larger variations. Scheme H has a larger domain, and scheme I has a smaller domain (figure 6c, dashed rectangle and dotted rectangle, respectively).

Generally, different verification area designs may result in different sensitive areas (figure 7). From the comparisons of schemes A, B, and C, it is observed that a small position change of the verification area has minimal influence on the sensitive areas. In addition, the inclusion of the best final position of the cyclone appears to be more important because its exclusion results in very different sensitive areas. Comparisons of schemes B, D, E, and F reveal that the CNOP is sensitive to the large position changes of the verification area, which results in large differences among the identified sensitive areas. The comparisons of schemes G and B indicate that small variations in both size and position also only minimally affect the CNOP-sensitive areas, but a large variation in size or position results in much different CNOP-sensitive areas (comparing the schemes B, H, and I).

From the above study, it can be observed that when the CNOP method is used to identify the sensitive areas, once the general position of the verification area is determined, a small variation in its size or position has minimal influence on the identification of the sensitive areas. This is a favorable characteristic for targeted observations.

3.3. *Sensitivity of CNOP-sensitive areas with respect to the optimization period*

In this part, we discuss whether the CNOP-sensitive areas identified for improving a special time forecast would also benefit forecasts at other times (Zhou and Mu 2012b). Two tropical cyclones Matsa (2005) and Meari (2004) were studied. The optimization time periods were the same as the forecast time periods. The initial time was fixed and forecasts were generated for 12 h, 24 h, and 36 h later. The initial times were set at 1200 UTC

4 Aug, 2005 for the Matsa case and 1200 UTC 25 Sep, 2004 for the Meari case.

In the Meari case, the sensitive areas identified for a special forecast time are consistent with those identified for other forecast times when the initial time is fixed (figure 8). This result indicates that targeted observations deployed to improve a special time forecast would also favorably affect the forecasts at other times. In the Matsa case; however, although there are obvious differences in the sensitive areas identified for the different forecast times, there are similarities (figure 9). This indicates that although the targeted observations deployed for a special time forecast may not greatly improve the forecast of other times, but they can still be beneficial for the forecast of other times. In addition, for both cases, the closer the forecast times, the higher the similarities of the sensitive areas.

Generally, both cases demonstrated that the CNOP-sensitive areas identified for improving a special time forecast would also benefit forecasts at other times, although the benefits may be different from case to case.

The above studies on the properties of CNOP-sensitive areas revealed that CNOP-sensitive areas have favorable features for targeted observations.

4. Examination of the CNOP-sensitive Areas

In this section, we will evaluate the effects of CNOP-sensitive areas with several cases by conducting observing system simulation experiments (OSSEs), and observing system experiments (OSEs).

4.1. *Observing system simulation experiments (OSSEs)*

To assess whether the sensitive regions calculated by CNOPs should be considered for real-time targeting dropsonde sites, Qin (2010a) and Qin and Mu (2011a) performed observing system simulation experiments (OSSEs).

Generally, three basic components should be included in OSSEs (Hoffman *et al.*, 1990): a four dimensional reference atmosphere, often called the nature run, the purpose of which is considered to be the 'truth'; a procedure to obtain simulated observations by sampling the nature run and adding errors; and a data assimilation system, which comprises a forecast model and the analysis procedure.

Qin (2010a) studied three typhoons, Nock-Ten (2004), Matsa (2005), and Morakot (2009), and conducted nature experiments with the MM5 model but used a higher horizontal resolution of 30 km. The simulated observation data were produced by adding error to the nature run. There were two sets of data at the targeting time. Take Nock-Ten for example, one dataset is the routine observation data that comprised the fixed stations on land and buoys in the ocean (figure 10a), and the other dataset is the additional observation data obtained by dropsondes distributed in the CNOP-sensitive areas (figure 10b) or randomly selected areas (figure 10c).

The effects (vertical mean of the variables) of utilizing additional data for Nock-Ten prediction are shown in Table 4. The TPE (vertical integrated total perturbation energy) improvement produced by the CNOP is the most significant (12.74%), whereas the effects caused by dropping sondes randomly have much less impact (some of them even have negative effects). These results indicate that additional observations in randomly selected areas are useless for the forecast. In contrast, the sensitive regions identified by CNOP are better locations for additional observations. The other two cases produced similar results.

In addition, Qin and Mu (2011a) performed OSSEs to evaluate the influence of additional dropsonde observation data in CNOP- and SVs-sensitive areas on typhoon track forecasts. In that study, the true state was considered to be forecasts from 0 to 72 h using the ERA-Interim reanalysis from ECMWF. The forecast typhoon centers at 6-h intervals were collected to represent the 'true' typhoon tracks. Forecasts during the same period, using the reanalysis from NCEP, were considered the control run. The difference between the typhoon center positions of the control run and the nature run was defined as the error in the typhoon track forecast without dropsonde data. After identifying the sensitive regions for the optimization period (24–48 h), simulated dropsondes were deployed for 24 h over these regions to obtain observational data, which represent the sum of the forecasts of the nature run at that time and randomly produced observation errors with the order of 0.1 times of the analysis. The simulated additional dropsonde data included horizontal wind speed, horizontal wind direction, and temperature at 850, 500 and 200 hPa. The 3D-Var assimilation system of MM5 was used to assimilate the additional dropsonde data to produce an analysis at 24 h, which can be run to predict the locations of the typhoon centers in the following 48 h (from 24 to 72 h). The differences between these typhoon center positions and the nature run were defined as the typhoon track forecast errors with dropsondes. The difference between these errors and the errors without dropsondes were used to indicate the influence of CNOP-sensitive areas on typhoon track forecasts.

Seven typhoon events (with large track forecast errors, table 5) originating in the western North Pacific during the 2009 season were selected for analysis. Generally, the CNOP-sensitive areas form (half) an annulus around the typhoon centers at the targeting time for most of the typhoon cases (five of seven), and, in addition, the SV-sensitive areas presented a maximum at the rear left quadrant with respect to the storm motion, approximately 500 km from the center of the storms in five of seven cases (figure 11, e.g., Typhoon Mirinae). Dropsonde sites were selected, and the distance between the adjacent sites was appropriate (approximately 150 km), and the total number of sites is the same in the CNOP- and SV-sensitive areas (e.g., figure 11). The track errors with and without dropsondes were compared, and a varying degree of improvement in typhoon track forecasts for six of the seven cases, after assimilating simulated dropsonde data obtained for the sensitive regions (figure 12), was observed. Moreover, the improvements were not only obtained for the optimization period used to calculate CNOPs and SVs but also for the subsequent 24 h. During the 24–72 h period, the deployment of dropsondes according to CNOP-sensitive areas could reduce track forecast errors by 13%–46% and by 14%–25% according to SV-sensitive areas.

The deployment of dropsondes in the CNOP-sensitive areas had an overall positive influence on typhoon track forecasts, suggesting that CNOP can be utilized as an adaptive method to determine sensitive regions in adaptive observations.

4.2. *Observing system experiments (OSEs)*

In the above section, we demonstrated the utility of CNOP-sensitive areas in OSSEs by using the ideal observations. In this section, real observations are used to demonstrate the utility of CNOP-sensitive areas (Chen, 2011; Chen *et al.* 2013).

Twenty typhoons that occurred between 2004 and 2009 in the typhoon surveillance program Dropwindsonde Observations for Typhoon Surveillance near the Taiwan Region (DOTSTAR) were studied. The dropsonde observations were collected under this program. The observational data provided by the Joint Typhoon Warning Center (JTWC) were deemed the truth. The sonde data those were close to the CNOP (FSV) sensitive areas were used as dropsonde observations in CNOP (FSV) sensitive areas. In this study, the CNOP-sensitive areas and FSV-sensitive areas were defined by using the MM5 model, whereas both the MM5 and the weather research and forecasting (WRF) model were used to assimilate the dropsonde observations.

Five types of experiments were designed and conducted: (1) no observations are assimilated; (2) all observations are assimilated; (3) only the CNOP targeted observations (observations in the CNOP sensitive area) are assimilated; (4) only the FSV targeted observations (observations in the FSV sensitive area) are assimilated; and (5) randomly targeted observations (observations within a randomly selected area) are assimilated.

Figure 13 shows the ratios of track forecast errors from EXP 2-5 to those from EXP 1 at 24 and 36 h using MM5 for all 20 typhoons. Most of the ratios are above zero, indicating that the adaptive observational data obtained from dropwindsondes can improve typhoon track forecasts. The average MM5 results indicate that track forecast errors were reduced by 4.8% (4.3%) at 24 (36) h if all dropwindsonde data were used, which was the most significant improvement among all four experiments (EXP 2-5), except for CNOP at 36 h. Assimilating observational data from several sondes deployed in sensitive areas identified by CNOP reduced the typhoon track forecast errors by 4.3% (4.9%) at 24 (36) h. The corresponding reduction in errors based on the FSV-sensitive areas (EXP 4) and the random deployment of sondes (EXP 5) was 3.8% (1.7%) and 2.5% (1.2%), respectively, at 24 h (36 h). Consequently, dropping sondes at different sites leads to different degrees of improvement, and, on average, the impact of dropwindsonde data on 36 h forecasts is less than that for 24 h forecasts. Although using all of the dropwindsonde data led to the greatest improvement in the typhoon track forecasts in most cases, using data from only three or four dropwindsondes deployed according to the CNOP-sensitive areas gave a comparable improvement on the accuracy of the forecasts, and if this result is more widely applicable, it could provide considerable economic and human benefits. These findings also imply that CNOP represents a useful approach to determining sensitive areas for targeted observations. Overall, similar results were gained from the WRF model (figures not shown), except that the FSV results were better than the MM5 results. These findings indicate that the sensitive areas identified by MM5 can also be utilized in the WRF model.

Table 6 lists the cases where the reduction in the typhoon track forecast error was more than 10% for EXP 2-5. For MM5, track forecast errors were reduced by more than 10% in half of the 20 cases when all of the available dropwindsonde data were utilized, which was more than that achieved by the use of CNOP/FSV sensitive areas or without sensitivity. In cases where the forecast time was extended to 36 h, improvements of more than 10% occurred a similar number of times in each of the four experiments, with EXP 3 (CNOP) being the most successful. Based on these results, dropping sondes according to CNOP-sensitive areas appears to offer the greatest potential for improving typhoon track forecasting.

All the results suggest that CNOP provides a suitable approach to determining sensitive areas during the targeted observation of typhoons.

5. The Ability of Operational Application of the CNOP Method

The above sections have demonstrated that CNOP is a valuable tool in targeted observations to identify sensitive areas. We further studied the conditions under which CNOP-sensitive areas are valid for tropical cyclone targeted observations and the execution of CNOP method on the WRF model. These will provide theoretical basis for the future operational applications of CNOP.

5.1. *Conditions leading to significant forecast skill improvements*

Observing system simulated experiments (OSSEs) were conducted on 20 tropical cyclones that developed over the western North Pacific during 2010 using Mesoscale Model 5 and its 3D Var assimilation system. Of the 20 cases, 13 displayed neutral or improved track forecasts with improvements between 0% and 51.2% (figure 14). Eliminating the initial errors within the CNOP pattern, which are related to either the storm directly or the surrounding regimes indirectly, reduced the subsequent track forecast errors. The remaining seven TCs displayed deterioration in the accuracy of the track forecasts over the 48-h forecast period. Accurate forecasts made without targeted observations, a low sensitivity of forecast errors to initial errors, or major forecast errors associated with regimes other than the TC can lead to a decline in the accuracy of TC track forecasts.

Following an analysis of the potential causes of inaccuracy in track forecasts, Qin *et al.* (2013) identified that TC cases with profound positive effects on track forecasts often satisfy following four conditions: i) an inaccurate initial forecast without additional observation data, ii) proper sensitivity of the forecast errors to the initial errors, iii) a large proportion of the forecast errors fall within the verification region, and iv) the TC system is the dominant regime in the verification region at verification time. Seven TCs satisfied these four conditions and displayed a mean reduction of 28.75% in track forecast errors over periods of 12 to 48 h (figure 15). These results suggest that the TC cases satisfying these four conditions often display profound improvements in track forecast errors with the use of dropwindsondes guided by CNOP sensitivity.

5.2. *The execution of the CNOP method using the WRF model*

Wang *et al.* (2011) studied the application of CNOP to typhoon targeted observations using the WRF model. In their study, the norm in the initial constraint is the background error, and the norm in the cost function is the total dry energy. Two typhoons, the fast straight-moving Typhoon Matsa (2005) and the slow-moving recurving Typhoon Shan shan (2006) were studied.

The spatial structures of CNOPs, their energies, nonlinear evolutions and impacts on track simulations were compared with those of the first singular vectors (FSVs). The results indicate that both the CNOPs and the FSVs are localized and evolve into the verification area at the final time with the upscale growth of perturbations. However, the CNOPs are different from the FSVs in spatial patterns, wind maximum distribution, growth rate of energy and impact on track simulation. Compared with FSV, the CNOP and local CNOP have greater impact on the forecast in the verification region at the final time in terms of total energy (figure 16) and have larger, or at least similar, impact on track simulation (figure 17). These results indicate that the CNOP method, with a background error constraint norm at the initial time and total energy norm at the final time, is a reasonable candidatein typhoon targeted observation. Therefore, both CNOP and local CNOP are suggested to be considered in typhoon targeted observation.

The study with the WRF model confirms the applicability of CNOP to the identifying sensitive areas for typhoon targeted observations.

6. Summary and Discussion

In this paper, the recent progresses in targeted observations for tropical cyclone prediction based on the Conditional Nonlinear Optimal Perturbation (CNOP) method were reviewed. The CNOP method was used to identify the sensitive areas for tropical cyclone predictions. The first singular vector

(FSV), as well as the composite singular vectors (CSV or SVs), were also used to identify the sensitive areas for comparison.

First, it was demonstrated that CNOP is different from the internationally used SV method in targeted observation and can take nonlinear processes into account. Then, by studying the influence of the initial errors in various areas on the targeted forecast, it was revealed that the initial errors that have the large impact on the typhoon forecasts are located in particular areas and that generally, the initial errors in the sensitive areas identified by CNOP (CNOP-sensitive areas) have the largest impact on the forecast. And this gives sufficient confidence for the following researches.

Then, the properties of CNOP-sensitive areas were studied, including the variations of the CNOP-sensitive areas with respect to the changes of the horizontal resolution, the verification area design and the optimization time period. It was observed that when the general position of the verification area is designed, small variations have minimal influence on CNOP-sensitive areas. Furthermore, the CNOP-sensitive areas identified at lower resolutions can be helpful for improving the forecast at finer resolution. In addition, the CNOP-sensitive areas identified for improving a special time forecast can also benefit forecasts at other times, although the benefits may be different from case to case. All the above results revealed that the CNOP-sensitive areas have favorable features for targeted observations.

Next, the effects of the CNOP-sensitive areas was examined with many cases. Observing system simulation experiments (OSSEs) were performed to assess whether the CNOP-sensitive areas can be considered as dropsonde deploying sites in real-time targeting. It is demonstrated that the energy of prediction error could be reduced by assimilating the simulated observation data in the CNOP-sensitive areas for three typhoon cases. Another study of seven typhoon events originating in the western North Pacific during the 2009 season revealed that assimilating the ideal observations in the CNOP-sensitive areas resulted in improvements of 13%–46% in typhoon track forecasts, whereas improvements of 14%–25% were obtained by assimilating the ideal observations in the SV-sensitive areas. In addition, the improvements could be achieved for longer forecast times. Observation system experiments (OSEs) using the DOTSTAR Data were conducted to further examine the effects of the CNOP-sensitive areas. The results indicated that assimilating the DOTSTAR data in the CNOP-sensitive areas had a more positive impact on the typhoon track forecast than that in the

FSV-sensitive areas, even better than assimilating all available dropsonde data on some occasions.

Finally, the conditions under which conducting targeted observations in the CNOP-sensitive areas would obtain profound forecast skill improvements, and the extension of the CNOP method by using a more advanced model and more reasonable initial constraints was studied. All the above results demonstrate that the CNOP method is a useful tool in targeted observations to identify sensitive areas, and the CNOP can be successfully employed for operational applications.

Nevertheless, further researches are needed. For example, finer resolution is necessary for more accurate simulations and forecasts. Furthermore, because in the current formulation, the cost function has been designed as the total dry energy in the verification area, it is expected that a new cost function would be better designed and thus could directly relate to our intentions, such as the track or intensity forecasts of the typhoons, and could subsequently guarantee the improvement of typhoon forecasts, and so on.

References

1. Aberson S.D. (2011), The Impact of Dropwindsonde Data from the THORPEX Pacific Area RegionalCampaign and the NOAA Hurricane Field Program on Tropical CycloneForecasts in the Global Forecast System, Mon. Wea. Rev., 139:2689-2703.
2. Ancell B.C. and C.F. Mass (2006), Structure, growth rates, and tangent linear accuracy of adjoint sensitivities with respect to horizontal and vertical resolution, Mon. Wea. Rev., 134:2971-2988.
3. Barkmeijer, J., T. Iversen, and T.N. Palmer (2003), Forcing singular vectors and other sensitivemodel structures, Quart. J. Roy. Meteor. Soc., 129 (592): 2401-2423.
4. Birgin, E.G., J.E. Martinez, and R. Marcos (2001), Algorithm 813: SPG-software for convex-constrained optimization, ACM Transactions on Mathematical Software, 27:340 349.
5. Bishop C.H. and Z. Toth (1999), Ensemble transformation and adaptive observations, J. Atmos. Sci., 56:1748-1765.
6. Buizza, R., C. Cardinali, G. Kelly and J. Thpaut (2007), The value of targeted observations Part II: The value of observations taken in singular vectors-based target areas, Quart. J. Roy. Meteor. Soc., 133:1817-1832.
7. Chen, B.-Y. (2011), Observation system experiments for typhoon Nida (2004) using the CNOP method and DOTSTAR data, Atmos. Oceanic Sci. Lett., 4:118-123.
8. Chen B.-Y., and M. Mu (2012), The Roles of Spatial Locations and Patterns

of Initial Errors in the Uncertainties of Tropical Cyclone Forecasts, Adv. Atmos. Sci., 29:63-78.

9. Chen, B. -Y., M. Mu and X.H. Qin (2013), The impact of assimilating dropwindsonde data deployed at different sites on typhoon track forecasts, Mon. Wea. Rev., in press.
10. Dudhia J. (1993), A nonhydrostatic version of the Penn State/NCAR Mesoscale Model: Validation tests and simulation of an Atlantic cyclone and cold front, Mon. Wea. Rev., 121:1493 1513.
11. Duan, W. S. and Mu Mu (2009), Conditional nonlinear optimal perturbation: applications to stability, sensitivity, and predictability, Science in China (D), 884-906.
12. Duan, W. S. and R. Zhang (2010), Is model parameter error related to a significant spring predictability barrier for El Ni no events? Results from a theoretical model. Adv. Atmos. Sci., 27(5), 10031013, doi:10.1007/s00376-009-9166-4.
13. Duan, W. S. and H. Luo (2010), A new strategy for solving a class of nonlinear optimization problems related to weather and climate predictability, Adv. Atmos. Sci., 27:741-749.
14. Ehrendorfer, M. and R. M. Errico (1995), Mesoscale predictability and the spectrum of optimal perturbations, J. Atmos. Sci., 52:3475-3500.
15. Gelaro, R., R.H. Langland, G.D. Rohaly and T.E. Rosmond (1999), An assessment of the singular-vector approach to targeted observations using the FASTEX dataset, Q. J. R. Meteorol. Soc., 125:3299-3327.
16. Hamill T.M. and Snyder C.(2002), Using improved background-error covariance from an ensemble kalman filter for adaptive observations, Mon. Wea. Rev., 130:1552-1572.
17. Hoffman R.N., C. Grassotti, R.G. Isaacs, J.F. Louis, T. Nehrkorn (1990), Assessment of the impact of simulated satellite lidar wind and retrieved 183 GHz water vapor observations on a global data assimilation system, Mon. Weather Rev. 118: 2513-2542.
18. Joly, A., D. Jorgensen, M.A. Shapiro, A. Thorpe, P. Bessemoulin, K.A. Browning, J.P. Cammas, J.P. Chalon, S.A. Clough, K.A. Emanuel, L. Eymard, R. Gall, P.H. Hildebrand, R.H. Langland, Y. Lemaitre, P. Lynch, J.A. Moore, P.O.G. Persson, C. Snyder, and R.M. Wakimoto (1997), The Fronts and Atlantic Storm-Track Experiments(FASTEX): Scientific objectives and experimental design, Bull. Amer. Meteor. Soc., 78:1917-1940.
19. Langland, R.H., Z. Toth, R. Gelaro, I. Szunyogh, M.A. Shapiro, S.J. Majumdar, R.E. Morss, G.D. Rohaly, C. Velden, N. Bond, and C.H. Bishop (1999a), The North Pacific Experiment (NORPEX-98): Targeted observations for improved North American weather forecasts, Bull. Amer. Meteor. Soc., 80:1363-1384.
20. Langland, R.H. (2005), Issues in targeted observing, Quart. J. Roy. Meteor. Soc., 131:3409-3425.
21. Mu M., W.S. Duan and B. Wang (2003), Conditional nonlinear optimal perturbation and its applications, Nonlinear Processes in Geophysics, 10:493-501.

22. Mu M., W.S. Duan and B. Wang (2007), Season-dependent dynamics of nonlinear optimal error growth and ENSO predictability in a theoretical model, J. Geophys. Res. 112: D10113.
23. Mu, M., and Z.N. Jiang (2008), A new method to generate the initial perturbations in ensemble forecast: conditional nonlinear optimal perturbations, Chinese Science Bulletin, 53:2062-2068S.
24. Mu M., Zhou F.F. and Wang H.L. (2009), A method to identify the sensitive areas in targeting for tropical cyclone prediction: conditional nonlinear optimal perturbation, Mon. Wea. Rev. 137:1623-1639.
25. Palmer T.N., Gelaro R., Barkmeijer J. and Buizza R., (1998), Singular vectors, metrics, and adaptive observations, J. Atmos. Sci., 55:633-653.
26. Peng, Y.H., W.S. Duan and J. Xiang (2011), Effect of stochastic MJO forcing on ENSOpredictability, Adv. Atmos. Sci., 28(6), 12791290, doi: 10.1007/s00376-011-0126-4.
27. Peng, M.S. and C.A. Reynolds, (2006), Sensitivity of tropical cyclone forecasts asrevealed by singular vectors, J. Atmos. Sci., 63:2508-2528.
28. Qin, X.-H. (2010a), The sensitive regions identifiedby the CNOPs of three typhoon events, Atmos. Oceanic Sci. Lett., 3:170-175.
29. Qin, X.-H. (2010b), A comparison study of thecontributions of additional observations in the sensitive regions identified by CNOP and FSV to reducing forecast error variance for the Typhoon Morakot, Atmos. Oceanic Sci. Lett., 3:258-262.
30. Qin X.-H. and M. Mu (2011a), Influence of conditional nonlinear optimal perturbations sensitivity on typhoon track forecasts, Q. J. R. Meteorol. Soc., 138:185-197.
31. Qin X.-H. and M. Mu (2011b), A study on the reduction of forecast error variance by three adaptive observation approaches for tropical cyclone prediction, Mon. Wea. Rev., 139:2218-2232.
32. Qin X.-H., W.S., Duan and M. Mu (2013), Conditions under which CNOP sensitivity is valid for tropical cyclone adaptive observations, Quart. J. Roy. Meteor. Soc., in press.
33. Rabier F., P. Gauthier, C. Cardinali, R. Langland, M. Tsyrulnikov, A. Lorenc, P. Steinle, R. Gelaro, K. Koizumi. (2008), An update on THORPEX-related research in data assimilation and observing strategies, Nonlin. Processes. Geophys, 15:81-94.
34. Riviere, O. G. Lapeyre and O. Talagrand (2008), Nonlinear generalization of singular vectors: behavior in a baroclinic unstable flow, J. Atmos. Sci., 65:1896-1911.
35. Sun, G. D., and M. Mu (2009), Nonlinear feature of the abrupt transitions between multipleequilibria states of an ecosystem model, Adv. Atmos. Sci., 26(2):293-304, doi: 10.1007/s00376-009-0293-8.
36. Szunyogh, I., Z. Toth, S. Majumdar, R. Morss, B. Etherton and C. Bishop (2000), The effect of targetedobservations during the 1999 Winter Storm Reconnaissance program, Mon. Wea. Rev., 128:3520-3537.
37. Szunyogh, I., Z. Toth, A.V. Zimin, S.J. Majumdar and A. Persson (2002),

Propagation of the effect of targeted observations: The 2000 Winter Storm Reconnaissance Program, Mon. Wea. Rev., 130:1144-1165.

38. Snyder C. (1996), Summary of an informal workshop on adaptive observations and FASTEX, Bull. Amer. Meteor. Soc., 77: 953-961.
39. Wang, H.L., M. Mu, and X.Y. Huang (2011), Application of conditional nonlinear optimal perturbations to tropical cyclone adaptive observation using the Weather Research Forecasting (WRF) model, Tellus, 63A:939-957.
40. Wu, C.C., P.H. Lin, S. Aberson, T.C. Yeh, W.P. Huang, K.H. Chou, J.S. Hong, G.C. Lu, C.T. Fong, K.C. Hsu, I.I. Lin, P.L. Lin, and C.H. Liu (2005), Dropwindsonde Observations for Typhoon Surveillance near the Taiwan Region(DOSTAR): An overview, Bull. Amer. Meteor. Soc., 86:787-790.
41. Wu C.C., Chen J.H., Lin P.H. and Chou K.H. (2007), Targeted observations of tropical cyclone movement based on the Adjoint-Derived Sensitivity Steering Vector. J. Atmos. Sci., 64:2611-2626. Zhang F. Q. et al. (2002) Mesoscale predictability of the "surprise" snowstorm of 24–25 January 2000, Mon. Wea. Rev. 130:1617-1632.
42. Zhou, F.F., and M. Mu (2011), The impact of verification area design on tropical cyclone targeted observations based on the CNOP method, Adv. Atmos. Sci., 28(5):997-1010, doi: 10.1007/s00376-011-0120-x.
43. Zhou, F.F., and M. Mu (2012a), The Impact of Horizontal Resolution on the CNOP and on Its Identified Sensitive Areas for Tropical Cyclone Predictions, Adv. Atmos. Sci., 29:36-46, doi: 10.1007/s00376-011-1003-x.
44. Zhou, F.F., and M. Mu (2012b), The time and regime dependences of sensitive areas for tropical cyclone prediction using the CNOP method, Adv. Atmos. Sci., 29:705-716, doi: 10.1007/s00376-012-1174-0.
45. Zou, X., F. Vandenberghe, M. Pondeca and Y.-H. Kuo (1997), Introduction to adjoint techniques and the MM5 adjointmodeling system, NCAR Tech. Note, NCAR/TN-435_STR.

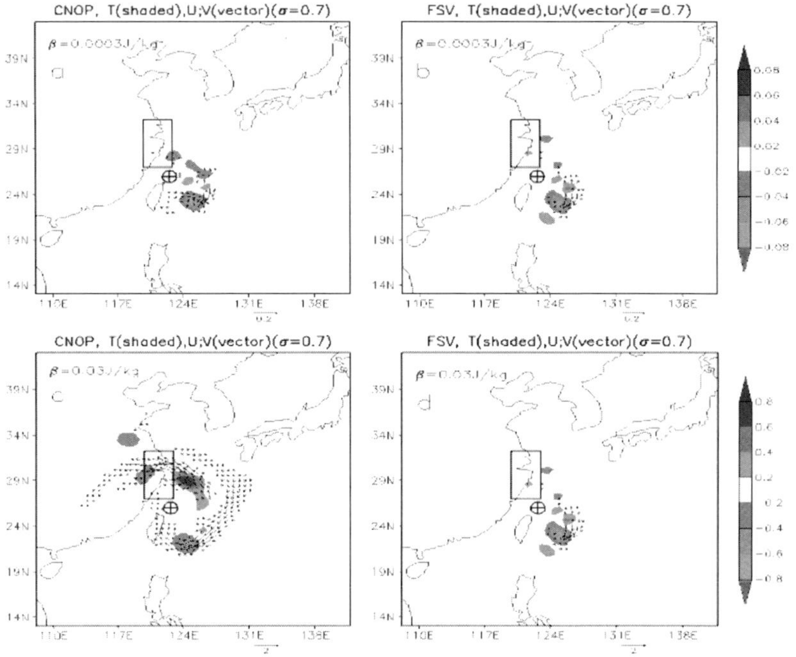

Figure 1. Matsa case. Temperature (shaded, unit: K) and wind (vector, unit: ms^{-1}.) components of CNOP and FSV at $\sigma = 0.7$ over a 24-h optimization time interval initialized at 0000 UTC 5 Aug 2005. (a),(b): CNOP and FSV with initial constraint value $\beta = 0.0003J/kg$; (c),(d): CNOP and FSV with $\beta = 0.0003J/kg$. The boxes indicate the verification areas. The "$\oplus$" symbol indicates the current position of the cyclone.

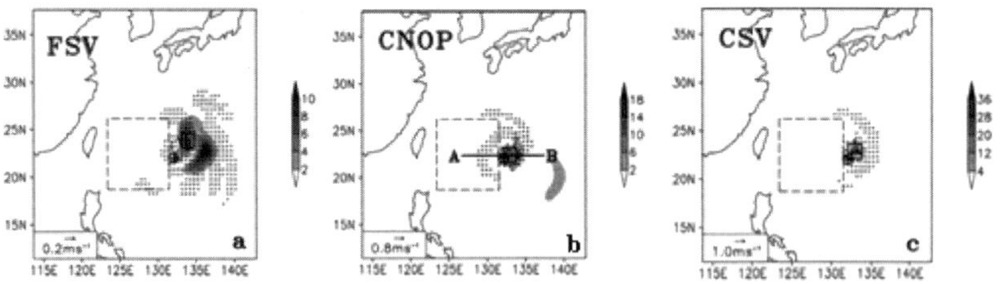

Figure 2. TC Longwang (2005). VIE (shaded; $10Jkg^{-1}$) and wind (vector; ms^{-1}) component of (a) FSV, (b) CNOP, (c) CSV over a24-h optimization time interval initialized at 0000 UTC 30 Sep 2005. The big boxes (dashed) indicate the verification areas; the box (solid) indicates the

sensitive area determined by (a) FSV, (b) CNOP, (c) CSV; the symbol ⊕ indicates the initial cyclone center.

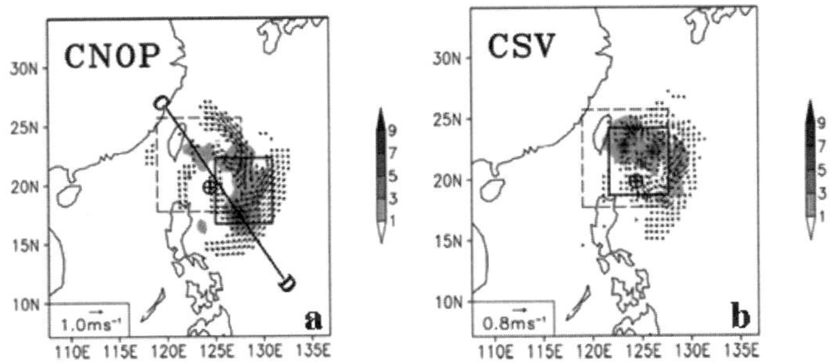

Figure 3. Same as figure 2, but for TC Sinlaku (2008).

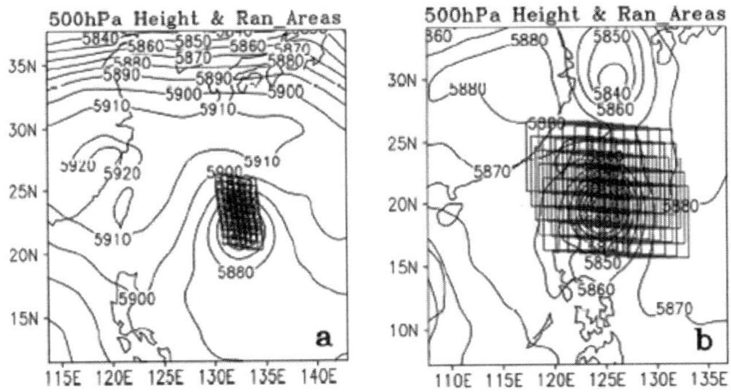

Figure 4. Geopotential height on $500 - hPa$ level (contour; gpm) and geographical distribution of the 40 local areas used in extended experiments (denoted by black boxes) for (a) TC Longwang (2005) and (b) TC Sinlaku (2008).

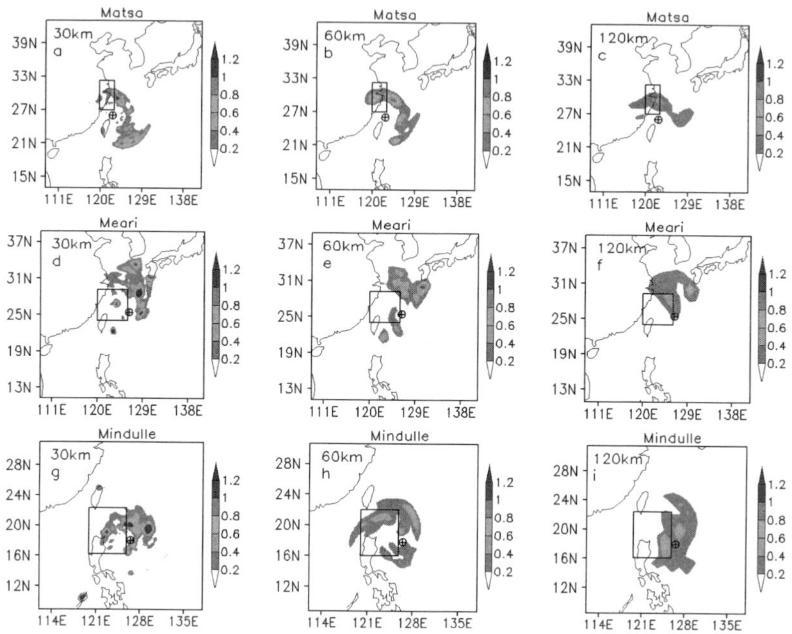

Figure 5. The vertically-integrated energies of CNOP: (a-c) TC Matsa; (d-f) TC Meari; (g-i) TC Mindulle. (a, d, g) at a resolution of $30km$, (b, e, h) at a resolution of $60km$, and (c, f, i) at a resolution of $120km$. The boxes indicate the verification areas. The "⊕" symbol indicates the initial position of the cyclone. (from Zhou and Mu, 2012)

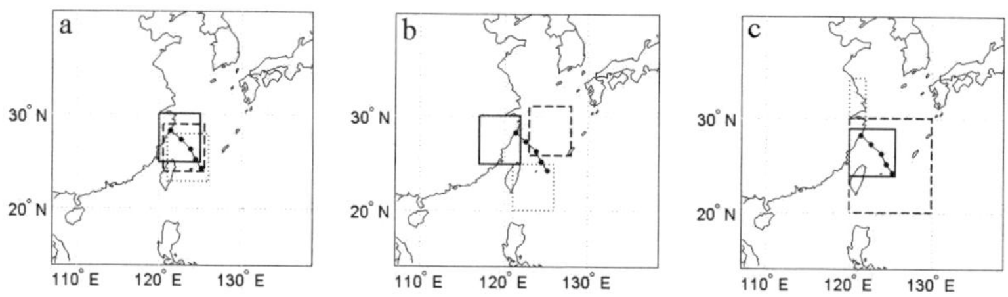

Figure 6. Verification areas for different designs. The solid rectangles in panel (a), (b), and (c) are the verification areas for schemes A, D, and G respectively, while the dashed rectangles are for schemes B, E, and H respectively, the dotted rectangles are for C, F, and I respectively. The observation tracks of the cyclone are also shown in the center of the domain.

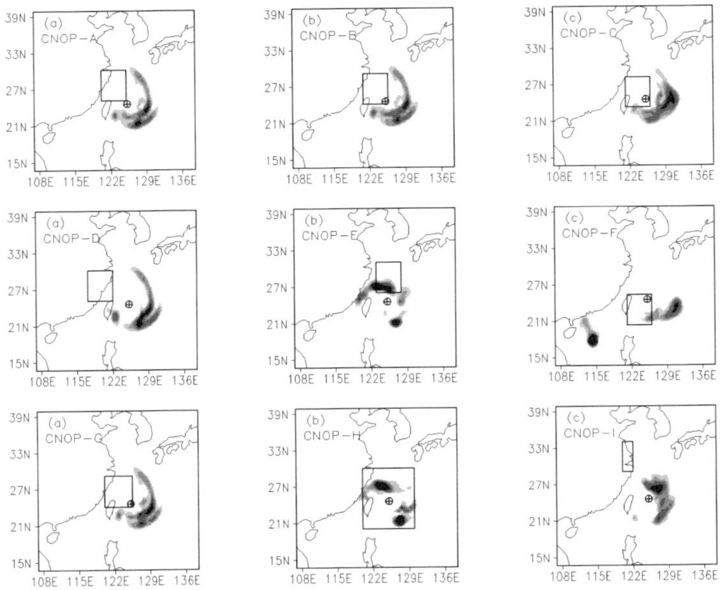

Figure 7. The sensitive areas denoted by the vertically-integrated energies of CNOPs (top panels) and FSVs (bottom panels) with schemes A, B, and C respectively. The rectangles are the verification areas. Both the ring and the plus symbols indicate the initial best position of the cyclone.

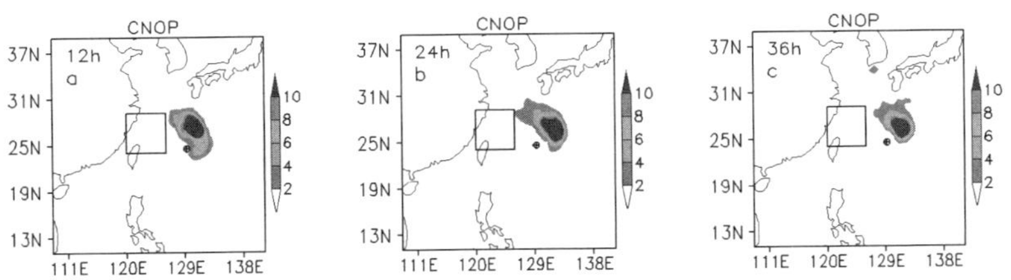

Figure 8. Typhoon Meari. Sensitive areas (shown by the vertically integrated energies (shaded, units: J/kg)) for (a) 12h, (b) 24h, and (c) 36h forecasts respectively. The boxes indicate the verification areas. The indicates the position of the cyclone at 1200 UTC 25 Sep 2004.

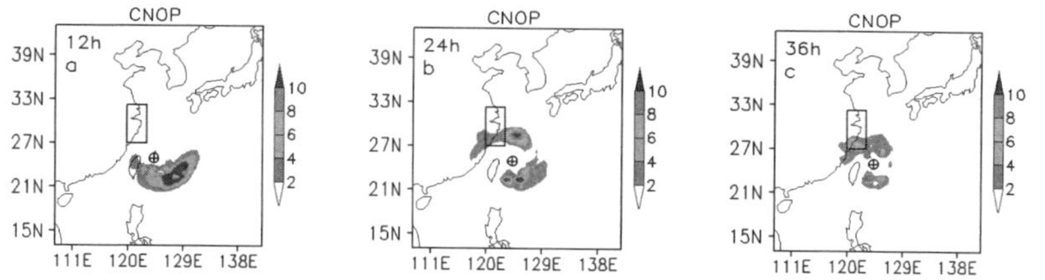

Figure 9. Same as figure 8, for typhoon Matsa at 1200 UTC 04 Aug 2005.

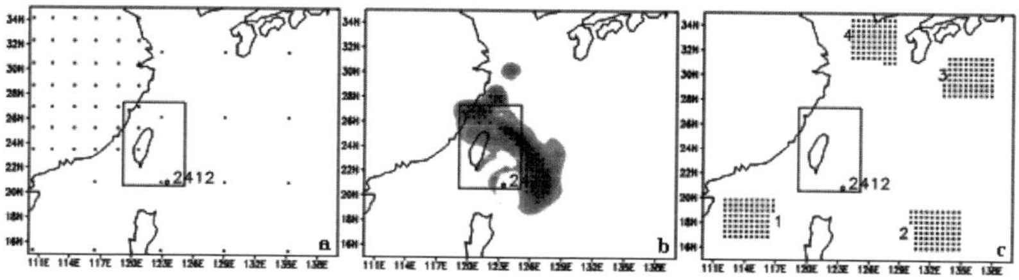

Figure 10. (a) The simulated observation net, (b) the CNOP sensitive areas and (c) the random selected areas denoted by number 1, 2, 3, 4 respectively.

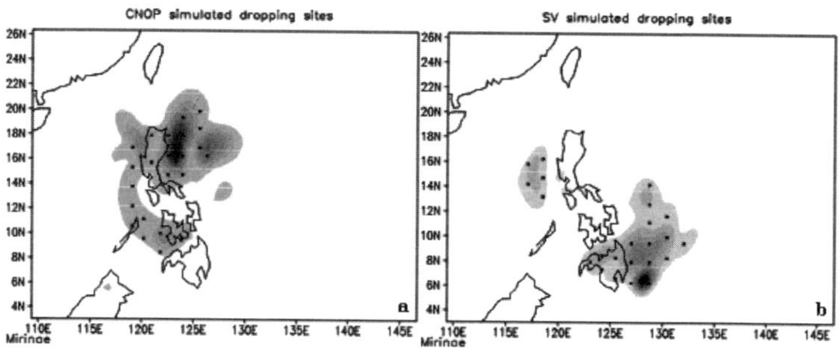

Figure 11. Simulated dropping sites in sensitive regions calculated by CNOPs (a) and SVs (b) for typhoon Mirinae. Shaded regions are the same as those in Figures 1a and b, squares represent the sites for dropping sondes.

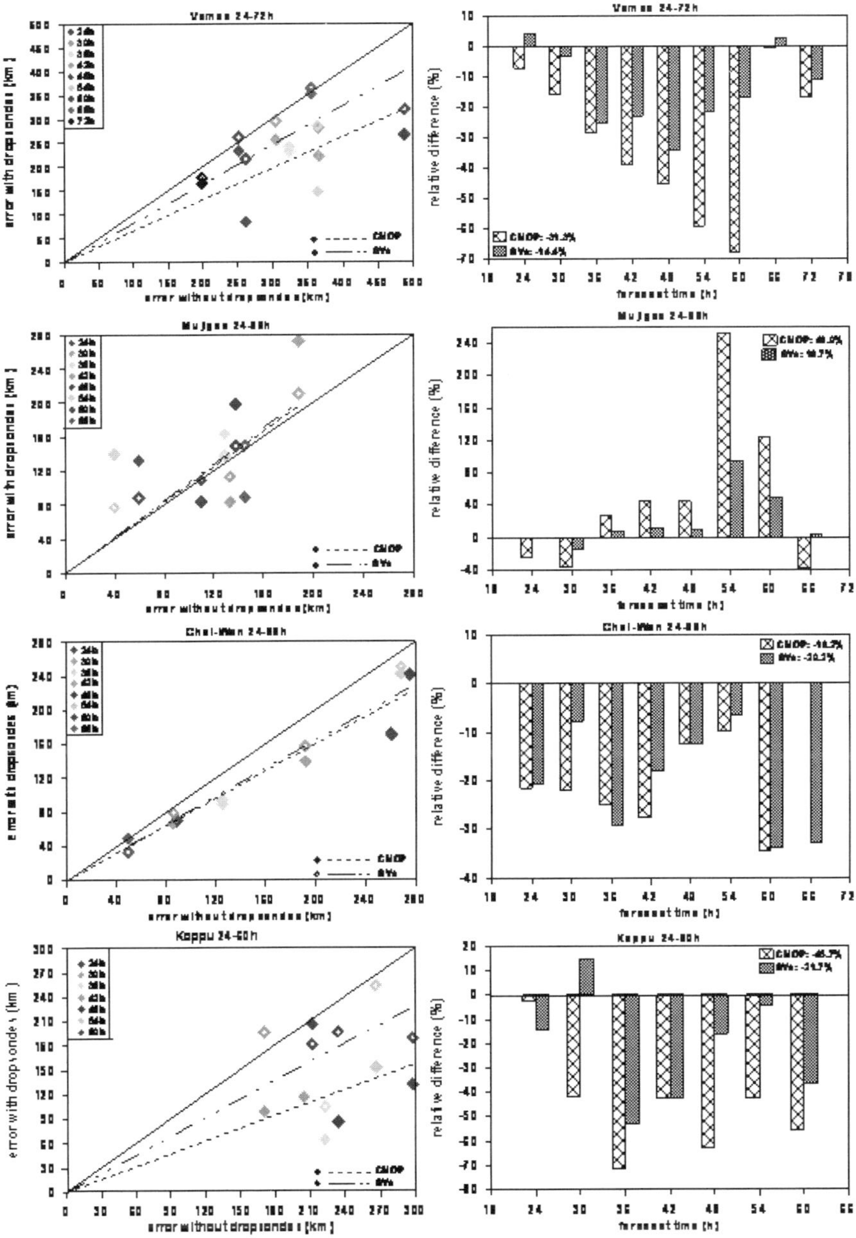
Koppu 24-60h
error with dropsondes (km)
error without dropsondes (km)
relative difference (%)
forecast time (h)
CNOP
SVs

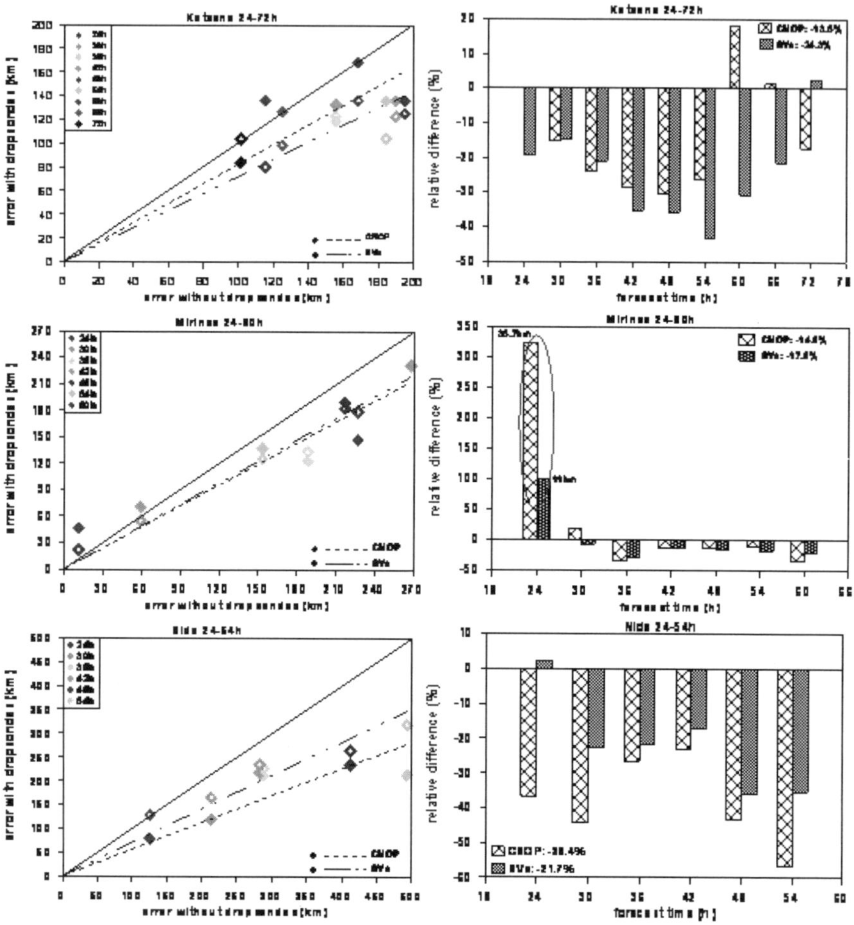

Figure 12. Scatter diagrams of all track forecast errors for 7 typhoon cases (left). The Y-axis represents the track forecast errors with dropsondes, and the X-axis represents those without dropsondes. Filled and empty diamonds denote the results of CNOPs and SVs, respectively. The colour of each diamond indicates the forecast time. Histograms on the right are relative differences corresponding to each case.

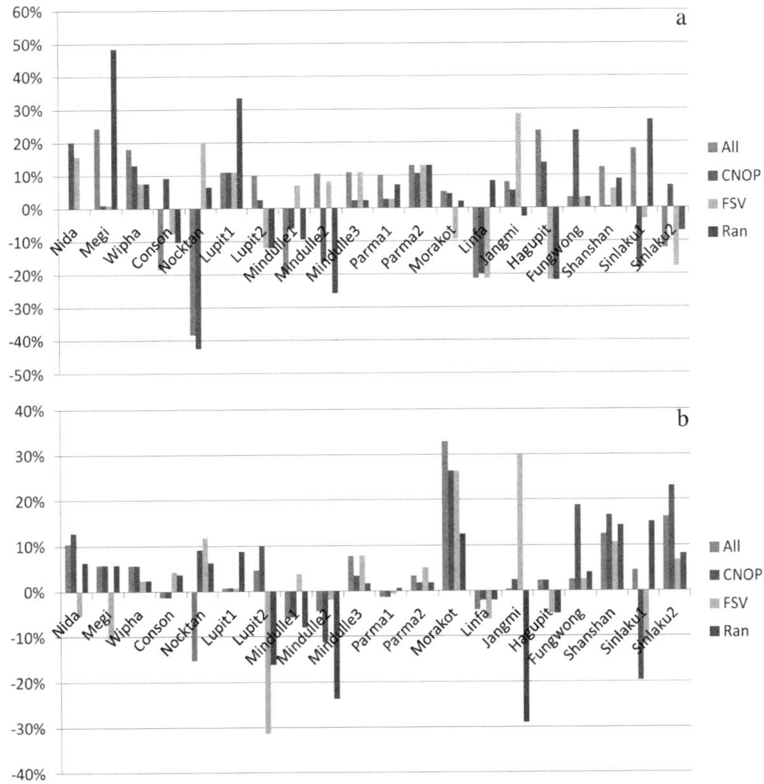

Figure 13. Ratios of track forecast errors of EXP 2 to 5 (All, CNOP, FSV, and Ran) to EXP 1 at 24-(a) and 36-(b)h using MM5 for twenty cases. Positive number means improvement, and negative number means deterioration.

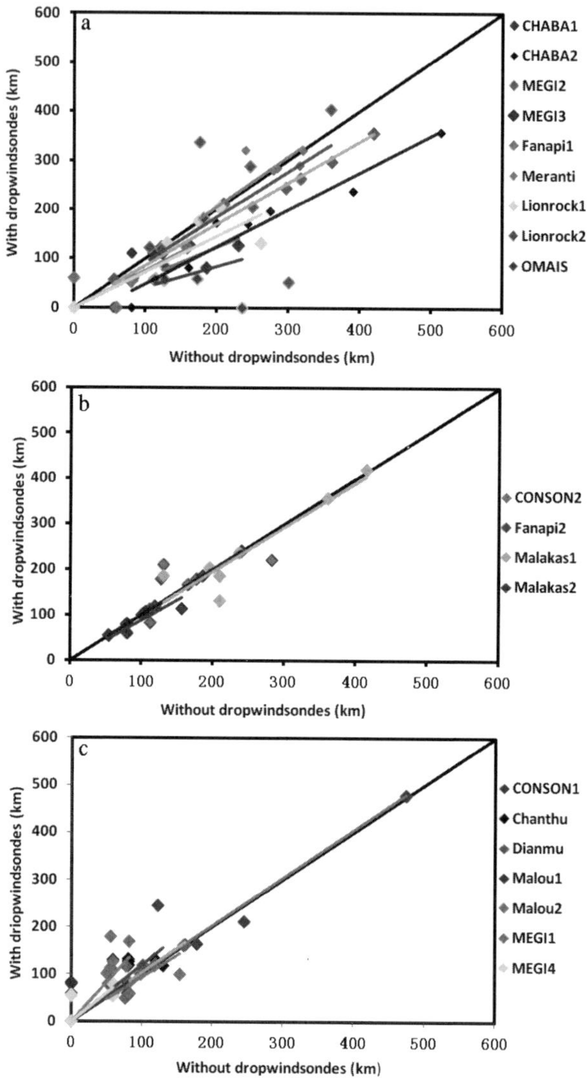

Figure 14. Scatter plots of track forecast errors for each TC. The y axis represents the errors associated with the track forecast made using additional dropwindsonde data, and the X-axis represents those made with no additional data. (a), (b), and (c) show those cases with profound improvements (above 10%), neutral or minor improvements, and a decline in the forecast accuracy, respectively.

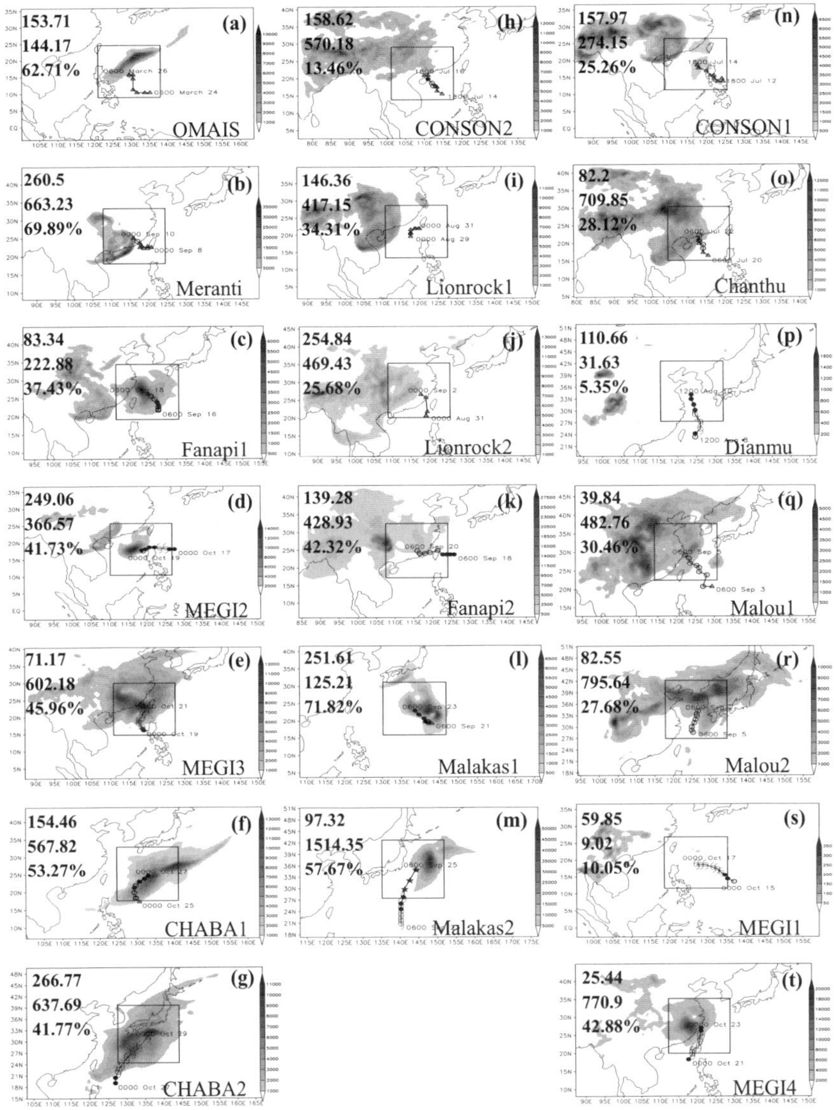

Figure 15. Forecast errors (shaded: $J{\cdot}kg^{-1}$) for those cases that satisfied all four conditions (a)–(g), those that did not meet all four conditions but showed neutral or moderate track forecast improvements (h)–(m), and those

that showed a decline in forecast accuracy (n)–(t). Numbers in the upper left corner of each plot denote the mean track forecast errors (without dropwindsonde data: km) from 12 to 48 h, the scale of the increase in forecast errors between the initial and terminal times, and the proportion of the forecast errors within the verification region. Red numbers indicate that the corresponding condition was not satisfied.

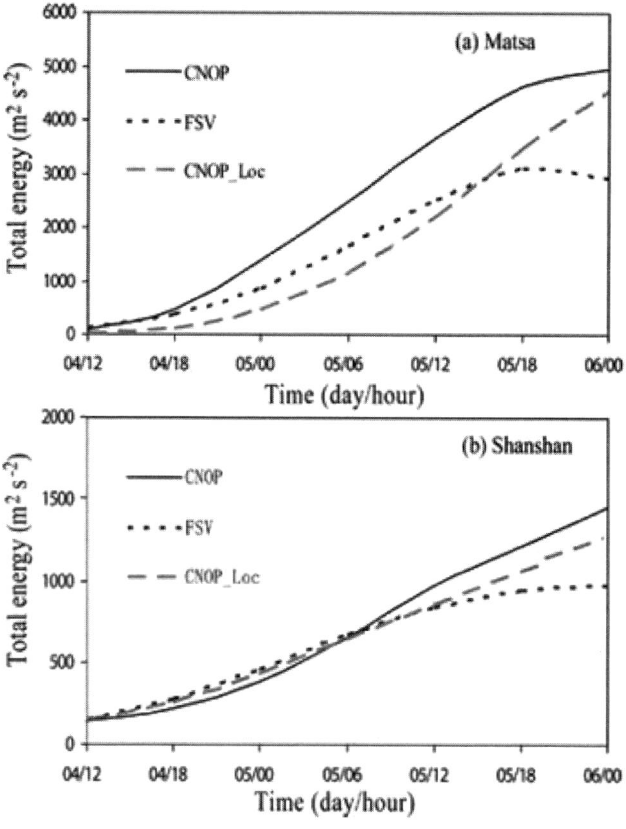

Figure 16. The total energy development of CNOP, FSV and local CNOP during the optimization time period: (a) Matsa and (b) Shanshan.

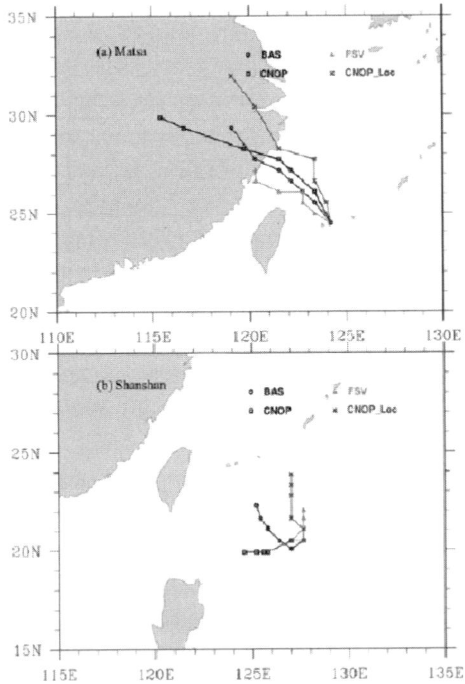

Figure 17. The simulated tracks in the basic trajectories (BAS) and simulated tracks with CNOP, FSV and local CNOP (CNOP_Loc) imposed on the initial condition of BAS. (a) Matsa and (b) Shanshan.

Table 1. The benefits (in percent, %) obtained from the reductions of CNOP or FSV for Matsa, Meari and Mindulle.

	Matsa		Meari		Mindulle	
Extent of reduction reduction	CNOP	FSV	CNOP	FSV	CNOP	FSV
0.75	86.4%	46.5%	94.8%	92.1%	84.8%	25.1%
0.50	69.9%	26.3%	75.0%	67.3%	53.8%	7.5%
0.25	49.4%	15.3%	42.0%	35.7%	24.1%	-5.3%

Table 2. Statistical averages of the verification dry energy norms of the 40 evolved Ran_Errs from the 40 local areas, respectively, for (a) TC Longwang (2005) and (b) TC Sinlaku (2008) (Jkg^{-1}).

(a)

i \ j	**1**	**2**	**3**	**4**	**5**
1	4.05	7.32	10.70	12.39	12.27
2	3.37	6.24	9.94	12.29	13.22***
3	2.83	8.44	14.64	15.10	12.95
4	4.27	27.07	57.20**	29.59	14.86
5	6.19	52.28	92.43*	51.00	15.95
6	4.33	27.51	67.18	42.77	14.31
7	2.38	8.24	13.51	11.52	8.46
8	1.48	3.17	5.28	6.01	6.53

(b)

i \ j	**1**	**2**	**3**	**4**	**5**	**6**	**7**	**8**
1	12.19	20.65	27.80	43.48	28.64	34.45	64.10	43.55
2	21.89	36.50	49.13**	51.02	92.31	66.53	97.53	36.52
3	29.54	40.99	57.98	44.53	129.94	165.45	72.59	26.95
4	22.45	29.25	50.81	74.48	156.69	276.13*	50.16	18.80
5	28.92	33.34	40.31	81.45	116.04	100.57	27.83	9.44

* indicates the CNOP_Sen. ** indicates the CSV_Sen. *** indicates FSV_Sen. i represents the latitudinal direction, and j indicates longitudinal direction.

Table 3. The similarities among the energy distributions obtained at 30-km, 60-km, and 120-km resolutions for TC Matsa (2005), TC Meari (2004), and TC Mindulle (2004).

	30km & 60km	60km & 120km
TC Matsa	0.70	0.78
TC Meari	0.55	0.75
TC Mindulle	0.49	0.72

Table 4. The RMSE of both routine and additional observation relative to only routine observation for Nock-Ten. The first column represents the kind of additional data that is utilized. The variables from the second to the eighth column stands for the vertical mean zonal wind, meridional wind, temperature, surface pressure, specific humidity, vertical wind, and vertical integrated total perturbation energy, respectively. Negative values represent that the RMSEs produced by the corresponding additional data were reduced.

	U	V	T	PP	QV	W	TPE
CNOP	**-2.58%**	**-9.86%**	**-4.01%**	0.62%	0.06%	**-3.60%**	**-12.74%**
random1	**-2.16%**	**-0.96%**	**-1.63%**	**-0.10%**	**-2.41%**	**-5.23%**	**-0.90%**
random2	**-0.28%**	**-1.57%**	0.87%	**-4.78%**	0.00%	1.18%	**-1.56%**
random3	1.60%	3.93%	**-0.54%**	**-5.27%**	0.56%	**-0.59%**	5.74%
random4	**-1.03%**	1.85%	2.62%	**-4.02%**	0.35%	3.43%	2.86%

Table 5. Track forecast errors (km) without dropsondes from 24–72 h for all 7 cases. The second row represents the initial time for each typhoon case (all in 2009, e.g., '082012' represents 1200UTC 20 August). The last row indicates the moving direction during this period.

	Vamco	Mujigae	Koppu	Choi-Wan	Ketsana	Mirinae	Nida
initial time	082012	091000	091306	091700	092618	102918	112806
24 h	251.3	110	211.6	88.7	168.3	11	125.4
30 h	305.0	132.9	171.1	85.9	155.6	59.2	213.6
36 h	323.5	128.8	222.7	125.4	155.6	188.3	287.9
42 h	365.7	188.3	204.6	192.1	189.9	267.9	282.4
48 h	487.2	157.5	233.9	275.7	194.9	216.7	412.3
54 h	364.2	39.7	266.1	268.5	184.4	153.2	493.0
60 h	261.5	59.2	297.2	260.8	165.4	226.8	376.3
66 h	354.7	144.7	537.7	49.2	125.1		
72 h	199.2				101.4		
moving direction	north-ward	west-ward	north-westward	recurved	west-ward	west-ward	stag-nant

Table 6. The number of track forecast error reduction that is above 10% for EXP 2 to 5 of twenty cases.

	ALL	**CNOP**	**FSV**	**Ran**
MM5-24h	10	6	5	4
MM5-36h	4	5	4	3
WRF-24h	13	11	8	8
WRF-36h	10	9	7	5

Isometric Immersions in Minkowski Spaces

Yi-Bing Shen*

Center of Math. Sci., Zhejiang University, Hangzhou 310027, China
yibingshen@zju.edu.cn

In Finsler geometry, the Minkowski space is a generalization of the Euclidean space. We will brief isometric immersions of Finsler manifolds in Minkowski spaces, in particular, minimal immersions. The volume variation and the mean curvature of Finsler submanifolds in a Minkowski space are given. Some Bernstein type theorems and non-trivial examples for minimal surfaces in Minkowski spaces are shown.

1. Introduction

Finsler manifolds are differentiable manifolds endowed with Finsler metrics, which are just Riemannian metrics without quadratic restriction. Riemannian submanifold geometry has been well developed by many geometers, and, however, Finsler submanifolds have not been studied at the same time. There are some fundamental contributions to Finsler submanifold geometry from various points of view ([1], [2], [4], [9], [11], [12], [17], [18], [20], [23], [24], [28]). A few new development of this subject has been made, e.g., see [7], [8], [13], [14], [16], [25], [26] and [27].

Minkowski spaces are vector spaces endowed with Minkowski metrics which are Euclidean metrics without quadratic restriction. The purpose of this paper is to give a brief report on isometric immersions of Finsler manifolds in Minkowski spaces, in particular, minimal immersions.

As is well known, by Nash–Gromov theorems ([10], [19]), a Riemannian manifold may be isometrically immersed into a Euclidean space with sufficiently large dimension. However, the similar conclusion does not hold for Finsler manifolds, although there were some results on isometric immersions of Finsler manifolds into Minkowski spaces ([4], [11], [12], [17]).

*Partially supported by NNSFC (No.11171297).

It is another difference from Riemannian submanifolds that there is no universal volume measure in Finsler geometry. Usually, the Busemann–Hausdorff volume measure and the Holmes–Thompson volume measure are used ([22]). It seems to be not necessary to deduce the so-called Gauss–Codazzi equations for Finsler submanifolds.

A natural idea of studying Finsler submanifolds follows from the volume variation via a certain volume measure. In fact, Finsler geometry is closely related to the calculus of variations. At the 1900 International Congress in Paris D. Hilbert devoted the problem of the geometry of simple integral, namely, Problem 23, to the subject ([24]). So, it is important to investigate minimal Finsler submanifolds.

The contents of the present paper are arranged as follows. A brief for foundations of Finsler geometry is given in Section 2, where we introduce the Finsler metric, the Chern connection and the S-curvature, etc. In Section 3, a non-existence theorem on isometric immersions of Finsler manifolds into a Minkowski space is shown. We then define the mean curvature of the Finsler submanifolds by using the volume variation. Finsler submanifolds with vanishing mean curvature are called minimal submanifolds. In Section 4, it is shown that there is no any compact minimal submanifold in the Minkowski space. A Bernstein theorem on Finsler minimal graphs in the Minkowski 3-space with respect to the Holmes–Thompson volume measure is given. Moreover, The second variation of the volume for minimal hypersurfaces in the Minkowski space is considered. In the last Section 5, we discuss minimal hypersurfaces in the (α, β)-Minkowski space. Some Bernstein type theorems on complete minimal hypersurfaces in the (α, β)-Minkowski space are shown. Moreover, some non-trivial examples of minimal surfaces in the (α, β)-Minkowski space are given.

2. A Brief for Finsler Geometry

2.1. *Finsler metrics*

Let M be an n-dimensional differentiable manifold, and TM its tangent bundle with the local coordinates (x, y) where $x = (x^1, \cdots, x^n) \in M$ is the point and $y = (y^1, \cdots, y^n) \in T_xM$ is the tangent vector. A *Finsler metric* in M is a function $F(x, y) : TM \to [0, +\infty)$ satisfying the following properties: (i) F is smooth on $TM\backslash 0$; (ii) $F(x, \lambda y) = \lambda F(x, y)$ for all $\lambda > 0$; (iii) the induced quadratic form g is positive-definite, where

$$g := g_{ij}(x, y)dx^i \otimes dx^j, \qquad g_{ij} = \frac{1}{2}[F^2]_{y^i y^j}. \tag{1}$$

The pair (M, F) is called a Finsler manifold. Here and from now on, we will use the following convention of index ranges unless other stated:

$$1 \leq i, j \cdots \leq n; \qquad 1 \leq \alpha, \beta \cdots \leq n + p.$$

The second property of F implies, by the Euler theorem,

$$y^i F_{y^i} = F, \qquad y^j F_{y^i y^j} = 0, \qquad F^2 = g_{ij} y^i y^j.$$

Clearly, F is Riemannian if g_{ij} are independent of y. A *Minkowski metric* is a Finsler metric F which is independent of x. A vector space with a Minkowski metric is called a *Minkowski space*, which is a generalization to the Euclidean space.

The *Cartan torsion* $\mathbf{C}$ of (M, F) is a symmetric 3-tensor field defined on the pulled-back bundle $\pi^* TM$

$$\mathbf{C} = C_{ijk}(x, y) dx^i \otimes dx^j \otimes dx^k, \qquad C_{ijk} := \frac{1}{4}[F^2]_{y^i y^j y^k} = \frac{1}{2}\frac{\partial g_{ij}}{\partial y^k} \tag{2}$$

whose average $\mathbf{I}$ is called the *mean Cartan torsion*:

$$\mathbf{I} = \mathbf{I}_k dx^k, \qquad \mathbf{I}_k = g^{ij} C_{ijk}, \qquad (g^{ij}) = (g_{ij})^{-1}. \tag{3}$$

$\mathbf{C}$ and $\mathbf{I}$ are also called the *Cartan tensor* and the *Cartan form*, respectively.

The well known Deicke theorem ([3]) implies that

$$F \text{ is Riemannian } \Longleftrightarrow \mathbf{C} \equiv 0 \Longleftrightarrow \mathbf{I} \equiv 0.$$

2.2. *Chern connection and curvature*

The *Chern connection* is defined on $\pi^* TM$ and its connection forms ω^i_j are determined uniquely by the following structure equations

$$\begin{aligned} &dx^j \wedge \omega^i_j = 0, \\ &dg_{ij} - g_{kj}\omega^k_i - g_{ik}\omega^k_j = 2C_{ijk}(dy^k + \omega^k_l y^l). \end{aligned} \tag{4}$$

The curvature form of the Chern connection is defined as

$$\begin{aligned} \Omega^i_j &:= d\omega^i_j - \omega^k_j \wedge \omega^i_k \\ &= \frac{1}{2} R_j{}^i{}_{kl} dx^k \wedge dx^l + P^i_{j\,kl} dx^k \wedge (dy^l + \omega^l_s y^s). \end{aligned} \tag{5}$$

The *flag curvature* tensor of F is defined by

$$\mathfrak{R} = R^i{}_k dx^k \otimes \frac{\partial}{\partial x^i}, \quad R^i{}_k := y^j R_j{}^i{}_{kl} y^l, \tag{6}$$

and the *flag curvature* of a 2-plane $\Pi_y = \mathrm{span}\{y, v\} \subset T_xM$ at a point $x \in M$ is defined by

$$K(\Pi_y)(v) := g_{ij} R^i{}_k v^j v^k / F^2,$$

where v is a unit vector orthogonal to y.

Let $\{e_1, \cdots, e_{n-1}, e_n = \ell\}$ be an orthonormal basis in T_xM with respect to g_y, where $\ell = y/F$ is the vector dual to the Hilbert form $\omega := F_{y^i} dx^i$. The *Ricci curvature* of F at $x \in M$ is defined by

$$Ric(y) := \sum_{k=1}^{n-1} K(\Pi_y)(e_k) = R^i{}_i. \tag{7}$$

2.3. *Volume measures and* S*-curvature*

For a Finsler manifold (M, F), there are several volume measures, which are reduced to the usual Riemannian volume measure when F is Riemannian.

The *Busemann-Hausdorff volume form* at $x \in M$ is defined by

$$\begin{aligned} dV_F &:= \sigma_{BH}(x) dx^1 \wedge \cdots \wedge dx^n, \\ \sigma_{BH} &:= \frac{\mathrm{Vol}(B^n(1))}{\mathrm{Vol}\{y \in T_xM | F(x,y) < 1\}}, \end{aligned} \tag{8}$$

where Vol denotes the Euclidean volume, $B^n(1)$ is the unit ball in R^n.

Let $S_xM = \{y \in T_xM | F(x,y) = 1\}$ be the indicatrix at $x \in M$. The *Holmes–Thompson volume form* at $x \in M$ is defined by

$$\begin{aligned} dV_F &:= \sigma_{HT}(x) dx^1 \wedge \cdots \wedge dx^n, \\ \sigma_{HT} &:= \frac{1}{a_{n-1}} \int_{S_xM} \det(g_{ij}) d\eta \\ d\eta &:= \sum_{i=1}^{n} (-1)^{i-1} y^i dy^1 \wedge \cdots \wedge \widehat{dy^i} \wedge \cdots \wedge dy^n, \end{aligned} \tag{9}$$

where "$\widehat{\ }$" means the corresponding term is cancelled, a_{n-1} denotes the volume of the $(n-1)$-sphere in R^n. This form is also called the volume form induced from the projective sphere bundle.

For a general volume form $dV_F = \sigma_F(x) dx^1 \wedge \cdots \wedge dx^n$, the distortion of F is defined by

$$\tau(x,y) := \ln \frac{\sqrt{\det(g_{ij})}}{\sigma_F}.$$

For a vector $y \in T_xM \setminus \{0\}$, let $\gamma(t)$ be the geodesic with $\gamma(0) = x$ and $\dot{\gamma}(0) = y$. The **S**-*curvature* of (M, F, dV_F) is defined by

$$\mathbf{S}(x,y) := \frac{d}{dt}\left[\tau(\gamma(t), \dot{\gamma}(t)\right] |_{t=0}. \tag{10}$$

Clearly, $\mathbf{S}(x, \lambda y) = \lambda \mathbf{S}(x, y)$ for $\lambda > 0$. It vanishes identically if F is Riemannian.

3. Isometric Immersions

3.1. *Isometric immersions*

Let (M, F) and $(\widetilde{M}, \widetilde{F})$ be Finsler manifolds, $f : M \to \widetilde{M}$ a smooth map, whose tangent map is $df \equiv f_* : TM \to T\widetilde{M}$. If $\operatorname{rank} df = \dim M$ and $F = f^*\tilde{F}$, then f is called an *isometric immersion* (resp. an *isometric imbedding* when $f(M) \subset \widetilde{F}$ has no self-intersection point). For simplicity, we call (M, F) an immersed (resp. imbedded) *submanifold* in $(\widetilde{M}, \widetilde{F})$. In the standard local coordinates (x, y) and $(\tilde{x}, \tilde{y})$ we have

$$y^i f_i^\alpha = \tilde{y}^\alpha, \quad [F]_{y^i} = [\widetilde{F}]_{\tilde{y}^\alpha} f_i^\alpha, \quad f_i^\alpha = \frac{\partial f^\alpha}{\partial x^i}$$

and

$$g_{ij}(x,y) = \tilde{g}_{\alpha,\beta}(\tilde{x}, \tilde{y}) f_i^\alpha f_j^\beta, \qquad \tilde{x} = f(x), \quad \tilde{y} = f_*(y). \tag{11}$$

Let G^i and $\widetilde{G}^\alpha$ be geodesic coefficients of F and $\widetilde{F}$, respectively.[24] It follows that

$$G^k = g^{kl} f_l^\beta \tilde{g}_{\beta\alpha}\left(\widetilde{G}^\alpha + f_{ij}^\alpha y^i y^j\right), \quad f_{ij}^\alpha = \frac{\partial f_i^\alpha}{\partial x^j}. \tag{12}$$

We see from (11) that F is Riemannian if $\widetilde{F}$ is Riemannian, so that (M, F) is a Riemannian submanifold. As is well known, by the Nash theorem [19]), any Riemannian manifold can be isometrically imbedded into a Euclidean space with sufficiently higher dimension. However, for general Finsler manifolds the problem becomes very difficult. ([11], [12], [17], etc.)

Define the norm $||\mathbf{C}||_x$ of the Cartan tensor at $x \in M$ by

$$||\mathbf{C}||_x := \sup_{y,v \in S_xM} \frac{|\mathbf{C}_{(x,y)}(v,v,v)|}{|g_{(x,y)}(v,v)|^{3/2}},$$

where S_xM is the indicatrix of (M, F) at x. We then have the following

Theorem 1. ([23]) *If the norm of the Cartan tensor of (M, F) is infinity, then (M, F) can not be isometrically immersed into any Minkowski space.*

Proof. We use reductio ad absurdum. Assume that $f : (M, F) \to (\widetilde{V}, \widetilde{F})$ is an isometric immersion, where $(\widetilde{V}, \widetilde{F})$ is a Minkowski space with the Minkowski metric $\widetilde{F} = \widetilde{F}(\tilde{y})$. Let $\widetilde{C}$ be the Cartan tensor of $\widetilde{F}$. From (11) we see easily

$$F(y) = \widetilde{F}(f_* y), \qquad g_y(u, v) = \widetilde{g}_{f_* y}(f_* u, f_* v),$$
$$\mathbf{C}_y(u, v, w) = \widetilde{\mathbf{C}}_{f_* y}(f_* u, f_* v, f_* w), \qquad \forall y, u, v, w \in TM.$$

It follows that

$$\begin{aligned} ||\mathbf{C}||_x &= \sup_{y,v \in T_x M} \frac{|\mathbf{C}_{(x,y)}(v, v, v)|}{|g_{(x,y)}(v, v)|^{3/2}} \\ &= \sup_{y,v \in T_x M} \frac{|\widetilde{\mathbf{C}}_{(f(x), f_* y)}(f_* v, f_* v, f_* v)|}{\tilde{g}_{(f(x), f_* y)}(f_* v, f_* v)|^{3/2}} \\ &\leq \sup_{\tilde{y}, \tilde{v} \in T_{f(x)} \widetilde{M}} \frac{|\widetilde{\mathbf{C}}_{(f(x), \tilde{y})}(\tilde{v}, \tilde{v}, \tilde{v})|}{\tilde{g}_{(f(x), \tilde{y})}(\tilde{v}, \tilde{v})|^{3/2}} = ||\tilde{\mathbf{C}}||_{f(x)} \\ &= ||\widetilde{\mathbf{C}}||_{\tilde{x}_0} < +\infty. \end{aligned}$$

Here $\tilde{x}_0 \in \widetilde{V}$ is an arbitrary point since $\widetilde{\mathbf{C}}$ is independent of the point of $\widetilde{V}$. It contradicts to that $||\mathbf{C}||$ is infinity.

□

This theorem means that the norm of the Cartan tensor of any isometrically immersed submanifold in a Minkowski space must be finite.

Example 1. Consider the following Finsler metric F defined on the unit n-sphere $\mathbf{S}^n \subset \mathbf{R}^{n+1}$.

$$F(x, y) = \frac{(\sqrt{|y|^2 - (|x|^2|y|^2 - < x, y >^2)} + < x, y >)^2}{(1 - |x|^2)^2 \sqrt{|y|^2 - (|x|^2|y|^2 - < x, y >^2)}},$$

where $<,>$ denotes the Euclidean inner product. It is a forward complete Finsler metric with zero flag curvature, of which the norm of the Cartan tensor is infinity.[23]

3.2. *Mean curvature*

Let $f : (M, F) \to (\widetilde{V}, \widetilde{F})$ be an isometric immersion in an $(n + p)$-dimensional Minkowski space $(\widetilde{V}, \widetilde{F})$. Fixed a point $x \in M$, $T_x M$ is a linear n-subspace of $\widetilde{V}$. Letting $e_i = \frac{\partial}{\partial x^i}|_x$, by (11), we see that $F(e_i) = \widetilde{F}(f_i^\alpha \tilde{e}_\alpha)$

at x and $\tilde{x} = f(x)$, where $\{\tilde{e}_\alpha\}$ is an oriented basis in $\widetilde{V}$. Clearly, the matrix $z \equiv (f_i^\alpha) \in GL(n, n+p)$. Suppose

$$dV_F = \sigma_F(F, e)dx^1 \wedge \cdots \wedge dx^n$$

is a general volume form of (M, F). Define a function $\mathcal{F}_{\tilde{e}}^\sigma : GL(n, n+p) \to \mathbf{R}$ by

$$\mathcal{F}_{\tilde{e}}^\sigma := \sigma_F(F, e), \tag{13}$$

which satisfies the following [22]

$$\mathcal{F}_{T\tilde{e}}^\sigma(z) = \mathcal{F}_{\tilde{e}}^\sigma(zT), \quad \forall\, T \in GL(n+p, n+p),$$
$$\mathcal{F}_{\tilde{e}}^\sigma(Qz) = \det(Q)\mathcal{F}_{\tilde{e}}^\sigma(z), \quad \forall\, Q \in GL(n, n).$$

Let $\mathfrak{D} \subset M$ be a compact domain with smooth boundary $\partial\mathfrak{D}$. Consider a variation $f_t : M \to (\widetilde{M}, \widetilde{F})$, $t \in (-\varepsilon, \varepsilon)$, such that $f_0 = f$ and $f_t = f_0$ on $\partial\mathfrak{D} \cup (M \setminus \{\mathfrak{D}\})$. Then $\{f_t\}$ induce a family of Finsler metrics $F_t = f_t^*\widetilde{F}$ on $\mathfrak{D}$. The variation vector field along f is

$$\widetilde{X} := \frac{\partial f_t}{\partial t}\bigg|_{t=0} = \widetilde{X}^\alpha \tilde{e}_\alpha, \qquad \widetilde{X}\bigg|_{\partial\mathfrak{D}} = 0.$$

By (13), the volume form of (M, F_t) is

$$dV_{F_t} = \sigma_t(x)dx^1 \wedge \cdots \wedge dx^n = \mathcal{F}_{\tilde{e}}^\sigma(f_t(x), df_t(x))dx^1 \wedge \cdots \wedge dx^n,$$

where $df_t(x) := ((f_t)_i^\alpha(x))$ is determined by $(f_t)_i^\alpha \tilde{e}_\alpha = e_i$. It follows from the property of $\mathcal{F}_{\tilde{e}}^\sigma$ that $\sigma_t(x)$ above is independent of the choose of the basis $\{\tilde{e}_\alpha\}$ in $\widetilde{V}$. So, we have the volume functional

$$V(t) := \mathrm{Vol}_{F_t}^\sigma(\mathfrak{D}) = \int_{\mathfrak{D}} dV_{F_t} = \int_{\mathfrak{D}} \sigma_t(x)dx^1 \wedge \cdots \wedge dx^n.$$

It follows that[23]

$$\begin{aligned} V'(0) &= \frac{dV(t)}{dt}\bigg|_{t=0} = \int_{\mathfrak{D}} \frac{\partial \sigma_t}{\partial t}\bigg|_{t=0} \theta^1 \wedge \cdots \wedge \theta^n \\ &= \int_{\mathfrak{D}} \frac{d}{dt}[\log \sigma_t]\bigg|_{t=0} dV_F := \int_{\mathfrak{D}} B_x^\sigma(\widetilde{X})dV_F, \end{aligned} \tag{14}$$

where

$$B_x^\sigma(\widetilde{X}) := \frac{d}{dt}[\log \sigma_t(x)]\bigg|_{t=0} = \frac{d}{dt}[\log \mathcal{F}_{\tilde{e}}^\sigma(f_t(x), df_t(x))]\bigg|_{t=0}.$$

On the other hand, at each point $x \in M$, there is a canonical projection $P_{f_*} : \widetilde{V} \to T_xM$ so that $P_{f_*}(\widetilde{X})$ is a smooth vector field on M.[23] The *mean curvature* of the isometric immersion f is defined by

$$\mathcal{H}_f^\sigma(\widetilde{X})\bigg|_x := B_x^\sigma(\widetilde{X}) - \mathrm{div}_F\,(P_{f_*}^\sigma(\widetilde{X}))\bigg|_x, \tag{15}$$

where the divergence div_F is determined by the volume form dV_F, i.e.,

$$(\mathrm{div}_F X)dV_F := d(X \rfloor dV_F) \qquad \forall \text{ any vector field } X \text{ on } M.$$

Thus, by the divergence theorem, (14) can be rewritten as

$$V'(0) = \int_{\mathfrak{D}} \mathcal{H}_f^\sigma(\widetilde{X})dV_F, \tag{16}$$

where $\widetilde{X}$ is the variation vector field along f. It can be shown that[23] $\mathcal{H}_f^\sigma(\widetilde{X})$ linearly depends only on $\widetilde{X}_{f(x)}$ at each point $x \in M$. Moreover, it vanishes on $f_*(T_xM)$. Hence, $\mathcal{H}_f^\sigma$ is an 1-form along f in $\widetilde{V}$, which is called the *mean curvature form.* ([13], [28], etc.)

4. Minimal Immersions

4.1. *Minimal submanifolds*

Let (M, F, dV_F) be a submanifold in a Minkowski space $(\widetilde{V}, \widetilde{F})$. If $V'(0) = 0$ for any compact domain $\mathfrak{D} \subset M$ and any variation vector field $\widetilde{X}$, then (M, F) is called a *minimal submanifold* in $(\widetilde{V}, \widetilde{F})$ with respect to the volume form $dV_F = \sigma_F dx^1 \wedge \cdots \wedge dx^n$, or simply a *$\sigma_F$-minimal submanifold.* In particular, for σ_{BH} and σ_{HT} defined in (8) and (9) it is called the *BH-minimal submanifold* and the *HT-minimal submanifold*, respectively.

Form (16) we see that (M, F) is minimal if and only if $\mathcal{H}_f^\sigma = 0$ identically.

Theorem 2. ([22]) *There is no any compact minimal submanifold in a Minkowski space $(\widetilde{V}, \widetilde{F})$.*

Proof. Let $f : (M, F) \to (\widetilde{V}, \widetilde{F})$ be an isometric immersion. It follows from (13) that the volume form of (M, F) is

$$dV_F^\sigma = \mathcal{F}_f^\sigma(z)dx^1 \wedge \cdots \wedge dx^n, \qquad z = (f_i^\alpha)\,.$$

Since $(\widetilde{V}, \widetilde{F})$ is a Minkowski space, we have[22] $B_x^\sigma(f) = n$ in (15), i.e.,

$$\mathcal{H}_f^\sigma = n - \mathrm{div}_F\,(P_{f_*}^\sigma(\widetilde{X})).$$

We then have

$$\int_M \mathcal{H}_f^\sigma(f)dV_F^\sigma = n\mathrm{Vol}_F^\sigma(M).$$

This completes the proof of Theorem 2. □

4.2. *HT-minimal immersions*

For a Minkowski $(n+p)$-space $(\widetilde{V}, \widetilde{F})$, consider the standard imbedding $(n+p-1)$-sphere $\mathcal{I} : \mathbf{S}^{n+p-1} \to \widetilde{V}$ under the Euclidean topology. We endow $\mathbf{S}^{n+p-1}$ with the Finsler metric $\mathcal{I}^*\widetilde{F}$, so that $(\mathcal{I} : (\mathbf{S}^{n+p-1}, \mathcal{I}^*\widetilde{F}) \to (\widetilde{V}, \widetilde{F})$ is an isometric immersion. Let $f : (M, F) \to (\mathbf{S}^{n+p-1}, \mathcal{I}^*\widetilde{F})$ be an isometric immersion. Then the composition $\phi \equiv \mathcal{I} \circ f : (M, F) \to (\widetilde{V}, \widetilde{F})$ is also an isometric immersion. On the other hand, by the Holmes–Thompson volume form dV_{HT} defined by (9), it is well known that f is HT-minimal if and only if f, as a map, is harmonic.[15] The following result is a generalization to Takahashi's theorem on minimal submanifolds in the Euclidean sphere.

Proposition 1. ([15]) *The isometric immersion $f : (M, F) \to (\mathbf{S}^{n+p-1}, \mathcal{I}^*\widetilde{F})$ is HT-minimal if and only if there is a function $\lambda(x)$ on M such that*

$$\mathcal{H}_\phi^{\sigma_{HT}} = \lambda \phi,$$

where $\mathcal{H}_\phi^{\sigma_{HT}}$ is the mean curvature of the composition $\phi \equiv \mathcal{I} \circ f$.

For the isometric immersion $f : (M, F) \to (\widetilde{V}, \widetilde{F})$, the mean curvature of f with respect to dV_{HT} can be written explicitly as[13]

$$\begin{aligned}
&\mathcal{H}_f^{HT} = \frac{1}{c_{n-1}} \left(\int_{S_x M} \frac{\widetilde{g}_{\alpha\beta} h^\beta}{F^2} \sqrt{\det(g)} d\nu \right) d\tilde{x}^\alpha, \\
&h^\alpha := f_{ij}^\alpha y^i y^j - f_k^\alpha G^k, \qquad f_i^\alpha := \frac{\partial f^\alpha}{\partial x^i}, \quad f_{ij}^\alpha := \frac{\partial^2 f^\alpha}{\partial x^i \partial x^j}, \\
&G^k = g^{kl} f_l^\beta \tilde{g}_{\beta\alpha} \left(f_{ij}^\alpha y^i y^j \right),
\end{aligned} \tag{17}$$

where G^k is computed by (12) since $\widetilde{G}^\alpha = 0$ for the Minkowski space. Set

$$h_\alpha := \widetilde{g}_{\alpha\beta} h^\beta = T_{\alpha\beta} f_{ij}^\beta y^i y^j, \qquad T_{\alpha\beta} := \widetilde{g}_{\alpha\beta} - \widetilde{g}_{\alpha\gamma} \widetilde{g}_{\beta\delta} f_i^\gamma f_j^\delta g^{ij}. \tag{18}$$

From (17) and (18) we see that f is HT-minimal if and only if

$$\int_{S_x M} \frac{1}{F^2} T_{\alpha\beta} f_{ij}^\beta y^i y^j \sqrt{\det(g)} d\nu = 0 \qquad \forall \alpha. \tag{19}$$

We now assume $p = 1$, i.e., $M \to \widetilde{V}^{n+1}$. On putting

$$S_x := \{ y \in \mathbf{R}^n | a_{ij} y^i y^j = 1 \}, \quad \Phi := \frac{\det(\tilde{g}_{\alpha\beta})}{F^{n+2}}, \quad a := \det \left(\sum_\alpha f_i^\alpha f_j^\alpha \right),$$

one can claim ([14]) that $f : (M, F) \to (\widetilde{V}^{n+1}, \widetilde{F})$ is HT-minimal if and only if

$$\sum_{\alpha} B^{ij} f^{\alpha}_{ij} n^{\alpha} = 0, \tag{20}$$

where

$$B^{ij} := \int_{S_x} \Phi y^i y^j \sqrt{a} d\eta = \int_{S_x} \Phi y^i y^j dV_{S_x},$$

and $\{n^{\alpha}\}$ are determined by the following system

$$\sum_{\alpha} f^{\alpha}_i n^{\alpha} = 0, \qquad \sum_{\alpha} (n^{\alpha})^2 = 1.$$

Now assume that M is a graph of $\widetilde{V}^{n+1}$ defined by

$$f(x_1, \cdots, x_n) = (x_1, \cdots, x_n, u(x_1, \cdots, x_n)) \tag{21}$$

for $x = (x_1, \cdots, x_n) \in U \subseteq R^n$. Thus, we have

$$a = 1 + |\nabla u|^2, \qquad (n^1, \cdots, n^n, n^{n+1}) = a^{-1/2}(-u_1, \cdots - u_n, 1),$$

where $\nabla u = (u_1, \cdots, u_n)$ and $u_i = \partial u / \partial x^i$. Thus, the graph (21) is HT-minimal if and only if

$$B^{ij}(x, u, \nabla u) u_{ij} = 0, \qquad u_{ij} := \frac{\partial^2 u}{\partial x^i \partial x^j}. \tag{22}$$

Since $\Phi = (\det(\tilde{g}_{\alpha\beta})/F^{n+2}) > 0$, then

$$B^{ij}\xi_i\xi_j = \int_{S_x} \Phi y^i y^j \xi_i \xi_j dV_{S_x} = \int_{S_x} \Phi (y^i \xi_i)^2 dV_{S_x} \geq 0$$

for $\xi \in \mathbf{R}^n$, where the equality holds if and only if $y^i \xi_i = 0$ for $y \in S_x$, i.e., $\xi = 0$. It implies that the equation (22) is elliptic.

On the other hand, for $y \in S_x$, we see that $\tilde{y} = df(y) \in S^n \subset \widetilde{V}^{n+1}$. So, we have

$$\min\{\Phi(x, y) : y \in S_x\} \geq \min\{\frac{\det(\tilde{g}_{\alpha\beta})}{\widetilde{F}^{n+2}} : \tilde{y} \in S^n\} = \kappa_1 > 0,$$

$$\max\{\Phi(x, y) : y \in S_x\} \leq \max\{\frac{\det(\tilde{g}_{\alpha\beta})}{\widetilde{F}^{n+2}} : \tilde{y} \in S^n\} = \kappa_2 > 0,$$

where κ_1 and κ_2 are constants. Hence, we have

$$\kappa_1 a^{ij}\xi_i\xi_j \leq B^{ij}\xi_i\xi_j \leq \kappa_2 a^{ij}\xi_i\xi_j \qquad \forall\, \xi \in \mathbf{R}^n,$$

which means that (22) is of mean curvature type. Since the equation of mean curvature type in 2 independent variables has only trivial solution[21], then we have proved the following Berstein theorem.

Theorem 3. ([14]) *Any complete HT-minimal graphs in the 3-dimensional Minkowski space $(\widetilde{V}^3, \widetilde{F})$ are planes.*

The above theorem is a generalization to the classical Bernstein theorem that any complete minimal graphs in the 3-dimensional Euclidean space are planes.

4.3. *Stability of minimal immersions*

Let (M, F, dV_{HT}) be a HT-minimal n-hypersurface in the Minkowski $(n+1)$-space $(\widetilde{V}, \widetilde{F})$. Let $\mathfrak{D} \subset M$ be a compact domain with smooth boundary $\partial\mathfrak{D}$. Recall the first variation (16) of the volume $V(\mathfrak{D})$, where $\mathcal{H}_f^\sigma = \mathcal{H}_f^{HT}$ is defined by (17). It was been proved in [16] that the second variation of the volume is

$$V''(0) = \frac{n}{c_{n-1}} \int_{SM} \{||W^\perp||^2 + h^*(\tilde{\nabla}_{\tilde{X}^T}\tilde{X}) - h^*(\tilde{X})\mathrm{tr}\tilde{g}(\tilde{\nabla}\tilde{X}, df) \\ - \tilde{g}(\tilde{h}(\ell, \tilde{X}^T), W) + \tilde{g}(\tilde{h}(\ell, W^T), \tilde{X})\} dV_{SM}, \tag{23}$$

where

$$W = \tilde{\nabla}_\ell \tilde{X}, \quad h^* = F^{-2} h_\alpha d\tilde{x}^\alpha, \quad \tilde{h} = \frac{1}{2}[h^\alpha]_{y^i y^j} dx^i \otimes dx^j \otimes \frac{\partial}{\partial \tilde{x}^\alpha}.$$

A minimal immersion f is called *stable* if the second variation of volume is always nonnegative with respect to all smooth variations that are compactly supported in M. Clearly, any totally geodesic submanifold in the Minkowski space must be stable.[16] Moreover, it can be proved that any minimal graph in Minkowski space $(\widetilde{V}^{n+1}, \widetilde{F})$ is stable. A Bernstein type theorem on complete stable minimal hypersurfaces in $(\widetilde{V}^{n+1}, \widetilde{F})$ was obtained in [16].

5. Minimal Immersions in (α, β)-Spaces

5.1. *(α, β)-Minkowski spaces*

Let $\alpha = \sqrt{a_{ij}(x) y^i y^j}$ and $\beta = b_i(x) y^i$ be a Riemannian metric and an 1-form on M. An (α, β)*-metric* is a Finsler metric F with the following form

$$F = \alpha\phi(s), \qquad s = \frac{\beta}{\alpha}, \tag{24}$$

where the function $\phi(s)$ satisfies

$$\phi(s) > 0, \quad (\phi(s) - s\phi'(s)) + (b^2 - s^2)\phi''(s) > 0.$$

Here s and $b = ||\beta||_\alpha$ (the norm of β with respect to α) satisfy

$$|s| \leq b < b_0 \qquad \text{for a constant } b_0.$$

In particular, if $\phi(s) = 1 + s$, then $F = \alpha + \beta$, which is called a *Randers metric*. This is a very important class of Finsler metrics in Finsler geometry.[6]

If $\{a_{ij}\}$ and $\{b_i\}$ are constants, i.e., α is Euclidean and β is constant, then $F = \alpha\phi(s)$ defined by (24) is a Minkowski metric, which is called an (α, β)-*Minkowski metric*. A vector space $\widetilde{V}$ endowed with an (α, β)-Minkowski metric is called an (α, β)-*Minkowski space*. Similarly, one can defines a *Randers-Minkowski space*. The following proposition is evident.[22]

Proposition 2. *Let F be an (α, β)-metric. Then the flag curvature of F vanishes identically if and only if it is an (α, β)-Minkowski space.*

In the following we consider the volume forms of an (α, β)-metric $F = \alpha\phi(s)$ defined by (24). A direct computation shows that the BH-volume form and the HT-volume form of $F = \alpha\phi(s)$ are respectively [22]

$$dV_F^{BH} = \sigma_{BH}(b)dV_\alpha \quad \text{and} \quad dV_F^{HT} = \sigma_{HT}(b)dV_\alpha, \tag{25}$$

where dV_α is the Riemannian volume form of α, functions $\sigma_{BH}(t)$ and $\sigma_{HT}(t)$ are defined respectively by

$$\begin{aligned} \sigma_{BH}(t) &:= \frac{\sqrt{\pi}\Gamma(\frac{n-1}{2})}{\Gamma(\frac{n}{2})}\Big[\int_0^\pi \frac{\sin^{n-2}\theta}{\phi^n(t\cos\theta)}d\theta\Big]^{-1}, \\ \sigma_{HT}(t) &:= \frac{\Gamma(\frac{n}{2})}{\sqrt{\pi}\Gamma(\frac{n-1}{2})}\int_0^\pi T(t\cos\theta)\sin^{n-2}\theta d\theta, \\ T(s) &:= \phi(\phi - s\phi')^{n-2}[\phi - s\phi' + (t^2 - s^2)\phi'']. \end{aligned} \tag{26}$$

Here $\Gamma(t) = \int_0^{+\infty} x^{t-1}e^{-x}dx$ is the Γ-function, $0 \leq t \leq b = ||\beta||_\alpha$.

5.2. *Some rigidity results*

Proposition 3. *Let $F = \alpha\phi(s)$ be an (α, β)-metric. If*

$$\phi(s) = (1 + h(s))^{-\frac{1}{n}} \tag{27}$$

for a smooth odd function $h(s)$, then $\sigma_{BH} = 1$, so that $dV_{BH} = dV_\alpha$.

Proof. As well known, for any odd function $h(s)$, we have

$$\int_0^\pi h(b\cos\theta)\sin^{n-2}\theta d\theta = 0$$

for any real number b. Substituting (27) into the first formula of (26) and noting that $\int_0^\pi \sin^{n-2}\theta d\theta = \sqrt{\pi}\Gamma(\frac{n-1}{2})/\Gamma(\frac{n}{2})$, we obtain directly $\sigma_{BH} = 1$. □

Consider the following system

$$\begin{cases} \phi(\phi - s\phi')^{n-1} = 1 + p(s) + s^2 q(s), \\ \phi(\phi - s\phi')^{n-2}\phi'' = q(s), \end{cases} \tag{28}$$

where $p(s)$ and $q(s)$ are arbitrary smooth functions. By (28), a similar calculus shows the following [7]

Proposition 4. *Let $F = \alpha\phi(s)$ be an (α, β)-metric. If ϕ satisfies (28), then $\sigma_{HT} = 1$, so that $dV_{HT} = dV_\alpha$.*

For the Randers metric $F = \alpha + \beta$, this proposition was proved in [14].

Let $(\widetilde{V}, \widetilde{F})$ be an (α, β)-Minkowski space with $\widetilde{F} = \tilde{\alpha}\phi(\frac{\tilde{\alpha}}{\tilde{\beta}})$, where $\tilde{\alpha} = \sqrt{\tilde{a}_{\alpha\beta}\tilde{y}^\alpha\tilde{y}^\beta}$ and $\tilde{\beta} = \tilde{b}_\alpha\tilde{y}^\alpha$. Consider an isometric immersion $f : (M, F) \to (\widetilde{V}, \widetilde{F})$. It is easy to see that $F = f^*\widetilde{F}$ is an (α, β)-metric, which has the following form in a local coordinate chart

$$F = \alpha\phi(s), \quad s = \frac{\alpha}{\beta}, \quad \alpha = \sqrt{a_{ij}y^iy^j}, \quad \beta = b_iy^i,$$

where

$$a_{ij}(x) = f_i^\alpha f_j^\beta \tilde{a}_{\alpha\beta}, \quad b_i(x) = f_i^\alpha \tilde{b}_\alpha, \quad f_i^\alpha := \frac{\partial f^\alpha}{\partial x^i}. \tag{29}$$

By Proposition 3 and 4, we have immediately

Theorem 4. ([7]) *Let $(\widetilde{V}, \widetilde{F})$ be an $(n+1)$-dimensional (α, β)-Minkowski space with $\widetilde{F} = \tilde{\alpha}\phi(\frac{\tilde{\alpha}}{\tilde{\beta}})$, where $\phi(s)$ satisfies (27) (resp. (28)). If $n \leq 7$, then any complete BH-(resp. HT-)minimal graph in $(\widetilde{V}, \widetilde{F})$ must be an n-dimensional linear subspace in $\widetilde{V}$.*

Theorem 5. ([7]) *Let $(\widetilde{V}, \widetilde{F})$ be an 3-dimensional (α, β)-Minkowski space with $\widetilde{F} = \tilde{\alpha}\phi(\frac{\tilde{\alpha}}{\tilde{\beta}})$, where $\phi(s)$ satisfies (27) (resp. (28)). Then any complete stable BH-(resp. HT-)minimal surface in $(\widetilde{V}, \widetilde{F})$ must be a plane.*

For the Randers metric $F = \alpha + \beta$ and the HT-volume form, these results were proved in [14].

5.3. *Non-trivial examples*

Let $(\widetilde{V}, \widetilde{F})$ be an (α, β)-Minkowski $(n+1)$-space with $\widetilde{F} = \tilde{\alpha}\phi(\frac{\tilde{\alpha}}{\tilde{\beta}})$. Given a n-plane $\mathcal{P}$ through the origin O of $\widetilde{V}$, we choose an orthonormal basis $\{e_1, \cdots, e_n, e_{n+1}\}$ with respect to $\tilde{\alpha}$ in $\widetilde{V}$ such that $\mathcal{P} = \text{span}\{e_1, \cdots, e_n\}$. Let $\{\tilde{x}^1, \cdots, \tilde{x}^{n+1}\}$ be the coordinate system in $\widetilde{V}$ with respect to this orthonormal basis. By using the Euclidean metric $\tilde{\alpha}$, we put $\theta_\alpha := \angle_\alpha(\tilde{e}, e_\alpha)$, where $\tilde{e}$ is the dual vector field to $\tilde{\beta}$. Thus, on putting $\tilde{y} = \tilde{y}^\alpha \frac{\partial}{\partial \tilde{x}^\alpha}$ and $\tilde{b} = ||\tilde{\beta}||_{\tilde{\alpha}}$, we have

$$\tilde{\alpha} = \sqrt{\sum_{\alpha=1}^{n+1} (\tilde{y}^\alpha)^2}, \quad \widetilde{F}(\tilde{y}) = \tilde{\alpha}\phi(s), \quad s = \frac{\tilde{b}\lambda_\alpha \tilde{y}^\alpha}{\tilde{\alpha}}, \quad \lambda_\alpha := \cos\theta_\alpha.$$

For an isometric immersion $f : (M, F) \to (\widetilde{V}, \widetilde{F})$ locally given by $\tilde{x}^\alpha = f^\alpha(x^1, \cdots, x^n)$, the induced (α, β)-metric $F = f^*\widetilde{F}$ is given by (29). In the following, denote by $\sigma_F(t)$ either $\sigma_{BH}(t)$ or $\sigma_{HT}(t)$ defined in (26). Define a function for $0 \le t \le \tilde{b}^2 = ||\tilde{\beta}||^2_{\tilde{\alpha}}$ by

$$\Phi(t) := 2(\sigma_F(\sqrt{t}))'(\tilde{b}^2 - t) + \sigma_F(\sqrt{t}). \tag{30}$$

Suppose that $f : \mathcal{D} \to (\widetilde{V}, \widetilde{F})$ is a graph over a connected domain $\mathcal{D}$ in the hyperplane $\mathcal{P}$ which is defined by

$$f(x^1, \cdots, x^n) = (x^1, \cdots, x^n, u(x^1, \cdots, x^n)). \tag{31}$$

It can be proved ([7]) that f is BH-(resp. HT-)minimal if and only if

$$\sum_{i,j} \Big\{\Phi(b^2)(\delta_{ij} - \frac{u_i u_j}{W^2}) + 2\tilde{b}^2\Phi'(b^2)(\lambda_i + q\frac{u_i}{W^2})(\lambda_j + q\frac{u_j}{W^2})\Big\} u_{ij} = 0, \tag{32}$$

where

$$W^2 := 1 + \sum_i u_i^2, \qquad b = ||\beta||_\alpha,$$

$$q := \lambda_{n+1} - \sum_k \lambda_k u_k, \qquad u_{ij} := \frac{\partial^2 u}{\partial x^i \partial x^j}.$$

If the hyperplane $\mathcal{P}$ is perpendicular to $\tilde{e}$ with respect to $\tilde{\alpha}$, then $\tilde{e}$ is in the direction of e_{n+1}, so that $\lambda_i = 0$ and $\lambda_{n+1} = 1$. In such a case, (32) is reduced as

$$\sum_{i,j} \Big\{\Phi(b^2)(\delta_{ij} - \frac{u_i u_j}{W^2}) + 2\tilde{b}^2\Phi'(b^2)\frac{u_i u_j}{W^4}\Big\} u_{ij} = 0, \tag{33}$$

where Φ is defined by (30).

An interesting problem is to study surfaces which are both BH-minimal and HT-minimal in a 3-dimensional (α,β)-Minkowski space $(\widetilde{V}^3,\widetilde{F})$. Obviously, the plane and the helicoid generated by lines screwing about x^3-axis are both BH-minimal and HT-minimal. Conversely, a surface in $\widetilde{V}^3$ can be locally viewed as a graph over a connected domain in a certain plane $\mathcal{P}$. Without loss of generality, it can be assumed that either $\mathcal{P}=\{x^3=0\}$ or $\mathcal{P}=\{x^1=0\}$. We suppose that $\Phi'(t)/\Phi(t)$ is different from σ_F which is either σ_{BH} or σ_{HT}.

If $\mathcal{P}=\{x^3=0\}$, it follows from (33) that

$$\begin{cases} u_{11}+u_{22}=0, \\ u_1^2u_{11}+u_2^2u_{22}+2u_1u_2u_{12}=0. \end{cases} \tag{34}$$

If $\mathcal{P}=\{x^1=0\}$, we have $\lambda_1=1$ and $\lambda_2=\lambda_3=0$, so that (33) is reduced as

$$\begin{cases} \sum_{i,j}(\delta_{ij}-\frac{u_iu_j}{W^2})u_{ij}=0, \\ u_{11}+\frac{u_1^2}{W^4}\sum_{i,j}u_iu_ju_{ij}-\frac{2}{W^2}\sum_j u_1u_ju_{1j}=0. \end{cases} \tag{35}$$

Since (34) and (35) have been explicitly solved in [27], we have immediately the following

Theorem 6. ([7]) *Let $(\widetilde{V}^3,\widetilde{F})$ be an (α,β)-Minkowski 3-space with $\tilde{e}$ dual to $\tilde{\beta}$ in the direction of x^3-axis. Then the helicoid generated by lines about x^3-axis is both BH-minimal and HT-minimal. Moreover, if $\Phi'(t)/\Phi(t)$ is different from σ_F, then any local surface which is both BH-minimal and HT-minimal must be a piece of either plane or helicoid generated by lines about x^3-axis.*

For the Randers metric $\widetilde{F}=\tilde{\alpha}+\tilde{\beta}$, the theorem was proved in [27].

Example 2. ([25]) Let $\left(\widetilde{V}^3,\widetilde{F}=\sqrt{\sum_{\alpha=1}^3(\tilde{y}^\alpha)^2}+\tilde{b}\tilde{y}^3\right)$ be a Randers–Minkowski 3-space, and $f:\mathbb{R}^2\to(\widetilde{V}^3,\widetilde{F})$ be a graph over the x^1x^2-plane given by $f(x^1,x^2)=(x^1,x^2,u(x^1,x^2))$. Substituting $\sigma_F(t)=(1-t^2)^{\frac{3}{2}}$ into (30), we get

$$\Phi(t)=(1-t)^{\frac{1}{2}}(1-3\tilde{b}^2+2t), \qquad \Phi'(t)=\frac{3[1-t+(\tilde{b}^2-t)]}{2(1-t)^{\frac{1}{2}}}, \quad \tilde{b}\in[0,1).$$

It is easy to see that $\Phi'(t) \geq 0$ if $t \in [0, \tilde{b}^2]$. If $\frac{1}{\sqrt{3}} < \tilde{b} < 1$, then $\Phi(t_o) = 0$ for $t_o = \frac{3\tilde{b}^2-1}{2}$. Hence, the following cone

$$u(x^1, x^2) = \sqrt{\frac{3\tilde{b}^2 - 1}{1 - \tilde{b}^2}} \sqrt{(x^1)^2 + (x^2)^2}$$

is a BH-minimal surface with a singular point at the origin. If $\tilde{b} = \frac{1}{\sqrt{2}}$, then the cone $u(x^1, x^2) = \sqrt{(x^1)^2 + (x^2)^2}$ is BH-minimal.

Example 3. ([16]) Let $(\widetilde{V}^3, \tilde{\alpha}\phi(\frac{\tilde{\beta}}{\tilde{\alpha}}))$ be an (α, β)-Minkowski space, where $\tilde{\alpha} = \sqrt{\sum_\alpha (\tilde{y}^\alpha)^2}$, $\tilde{\beta} = \tilde{b}\tilde{y}^3$ and $\tilde{b} \neq 0$ is a constant. Let M be a helicoid defined by

$$f(u, v) = \{u \cos v, u \sin v, cv\}, \quad (u, v) \in \mathbf{E}^2, \tag{36}$$

where $c \neq 0$ is a constant. Then

$$F = f^* \tilde{F} = \alpha\phi(\frac{\beta}{\alpha}),$$

where $\alpha = \sqrt{(y^1)^2 + (u^2 + c^2)(y^2)^2}$, $\beta = c\tilde{b}y^2$, and we have

$$h_{11} = 0, \quad h_{12} = \frac{-c}{\sqrt{u^2 + c^2}}, \quad h_{22} = 0.$$

It follows from that

$$\xi|_{\Sigma_x} = (\phi - \beta\phi')(\phi - \beta\phi' + (\tilde{b}^2 - \beta^2)\phi''),$$

which is a function only depending on y^2. Hence, the helicoid M is HT-minimal in (α, β)-Minkowski space $(\widetilde{V}^3, \tilde{\alpha}\phi(\frac{\tilde{\beta}}{\tilde{\alpha}}))$.

Example 4. ([7]) Let $f : \mathbb{R}^2 \supset \mathcal{D} \to \left(\widetilde{V}^3, \widetilde{F} = \frac{\tilde{\alpha}^2}{\tilde{\alpha} - \tilde{\beta}}\right)$ be a graph defined on $\mathcal{D}$, where $\tilde{\alpha} = \sqrt{\sum_{\alpha=1}^3 (\tilde{y}^\alpha)^2}$ and $\tilde{\beta} = \tilde{b}\tilde{y}^3$. By (30) we have

$$\Phi(t) = \frac{2(1 - \tilde{b}^2) + 3t}{(2 + t)^2}, \qquad \Phi'(t) = \frac{2 + \tilde{b}^2 + 3(\tilde{b}^2 - t)}{(2 + t)^3}.$$

If $\tilde{b} \in [0, \frac{1}{2})$ and $t \in [0, \tilde{b}^2]$, then $\frac{\Phi'(t)}{\Phi(t)} > 0$. Hence, $f(\mathcal{D})$ is a BH-minimal graph.

Example 5. ([7]) Let $f : \mathcal{D} \to (\widetilde{V}^{n+1}, \widetilde{F})$ be a graph defined on the hyperplane $\mathcal{P}$ in the (α, β)-Minkowski $(n+1)$-space, where $\mathcal{P}$ is perpendicular to $\tilde{e}$ with respect to $\tilde{\alpha}$, and f is defined by (31). Suppose that there is a $t_o \in (0, \tilde{b}^2)$ such that $\Phi(t_o) = 0$, where Φ is defined by (30). We then see that

$$u(x^1, \cdots, x^n) = \sqrt{\frac{t_o}{\tilde{b}^2 - t_o}} \sqrt{(x^1)^2 + \cdots + (x^n)^2}$$

is a both BH-minimal and HT-minimal graph with a singular point at the origin.

References

1. W. Barthel, *Über die Minimalflächen in gefaserten Finsler-Räumen*, Ann. Mat. Pura Appl., **36**(1954), 159-190.
2. ———, *Variationsprobleme der Oberflächenfunktion in der Finslerschen Geometrie*, Math. Z., **62**(1955), 23-36.
3. D. Bao, S. S. Chern and Z. Shen, *An Introduction to Riemann-Finsler Geometry*, GTM **200**, Springer-Verlag, 2000.
4. D. Burago and S. Ivanov, *Isometric embeddings of Finsler manifolds* (in Russian), Algebra i Analiz, **5**(1993), 179-193.
5. I. Comic, *Induced and intrinsic curvature tensors of a subspace in the Finsler space*, Publications de L'Institute Mathmétiane **23**(**37**) (1978), 67-74.
6. X. Y. Cheng and Z. Shen, *Finsler Geometry—An Approach via Randers Spaces*, Scicence Press, Beijing, 2012.
7. N. W. Cui and Y. B. Shen, *Bernstein type theorems for minimal surfaces in (α, β)-spaces*, Publ. Math. Debrecen, **74/3-4**(2009), 383-400.
8. ———, *Minimal rotational hypersurfaces in Minkowski-(α, β)-space*, Geom. Dedicatae, **151**(2011), 27-39.
9. E. T. Davies, *Subspaces of a Finsler space*, Proc. London Math. Soc., **49**(1947), 19-39.
10. M. Gromov, *Partial differential relations*, Springer, New York, 1986.
11. C. H. Gu, *Imbedding of a Finsler manifold in a Minkowski space*, Acta Math. Sinica, **6**(1956), 215-232. (in Chinese)
12. ———-, *Imbedding of Finsler manifolds in a Minkowski space*, Acta Math. Sinica, **8**(1958), 272-275. (in Chinese)
13. Q. He and Y. B. Shen, *On the mean curvature of Finsler submanifolds*, Chin. J. Contem. Math., **27**(2006), 431-442.
14. ———, *On Bernstein type theorems in Finsler spaces with the volume form induced from the projective sphere bundle*, Proc. AMS, **134**(2006), 871-880.
15. ———, *Some properties of harmonic maps for Finsler manifolds*, Houston J. Math., **33**(2007), 683-699.

16. Q. He and W. Zhao, *On the stability of minimal immersions into Finsler manifolds*, Diff. Geom. Appl., **30**(2012), 438-449.
17. R. S. Ingarden, *Über die Einbetting eines Finslerschen Rauumes in einan Minkowskischen Raume*, Bull. Acad. Polon. Sci., III **2**(1954), 305-308.
18. M. Matsumoto, *Foundations of Finsler geometry and special Finsler spaces*, Kaiseisha Press, Japan 1986.
19. J. Nash, *The imbedding problem for Riemannian manifolds*, Ann. of Math., **73**(1957), 20-37.
20. H. Rund, *The Differential Geometry of Finsler Spaces*, Springer, GMW 101, 1959.
21. L. Simon, *Equations of mean curvature type in 2 independent variables*, Pac. J. Math., **69**(1977), 245-268.
22. Y. B. Shen and Z. Shen, *Introduction to Modern Finsler Geometry*, Higher Eduction Press, Beijing, 2012. (in Chinese)
23. Z. Shen, *On Finsler geometry of submanifolds*, Math. Ann., **311**(1998), 549-576.
24. ———, *Differential Geometry of Spray and Finsler Spaces*, Kluwer Acad. Publ., 2001.
25. M. Souza and K. Tenenblat, *Minimal surfaces of rotation in a Finsler space with a Randers metric*, Math. Ann., **325**(2003), 625-642.
26. M. Souza, J. Spruck and K. Tenenblat, *A Bernstein type theorem on a Randers space*, Math. Ann., **329**(2004), 291-305.
27. B. Wu, *A local rigidity theorem for minimal surfaces in Minkowski 3-space of Randers type*, Ann. Glob. Anal. Geom., **31**(2007), 375-384.
28. M. Y. Zhang, *Mean curvature of a subspace in a Finsler space*, Ann. Mat. Pura Appl., **31**(1950), 97-302.

Remarks on Volume Growth for Minimal Graphs in Higher Codimension

Yuanlong Xin

Institute of Mathematics, Fudan University, Shanghai 200433, China
ylxin@fudan.edu.cn

A sufficient condion for minimal graphs in the Euclidean space of higher codimension having Euclidean volume growth has been proved.

Volume growth is important for minimal submanifolds. It is well known that for minimal submanifolds the minimum one is the Euclidean volume growth, which is just the case for minimal graphic hypersurfaces. Here the conditions, graph and codimension one, both are crucial. The present note is devoted to the volume growth for minimal graphs in higher codimension.

Let $\Omega \subset \mathbb{R}^n$ be an open domain and $f : \Omega \to \mathbb{R}^m$ be a smooth map with $p = \min(m, n)$. The induced metric tensor on M can be described as

$$g_{ij} = \delta_{ij} + \sum_\alpha \frac{\partial f^\alpha}{\partial\partial x^i}\frac{\partial f^\alpha}{\partial\partial x^j} = (\sec^2\theta_i)\delta_{ij},$$

where $\lambda_i^2 = \tan^2\theta_i$ are eigenvalues of $\sum_\alpha \frac{\partial f^\alpha}{\partial x^i}\frac{\partial f^\alpha}{\partial x^j}$. If f satisfies the minimal submanifold equations

$$\begin{cases} \sum_i \frac{\partial}{\partial x^i}\left(\sqrt{g}g^{ij}\right) = 0, \\ \sum_{i,j} \frac{\partial}{\partial x^i}\left(\sqrt{g}g^{ij}\frac{\partial f}{\partial x^j}\right) = 0, \end{cases}$$

where $(g^{ij}) = (g_{ij})^{-1}$ and $g = \det(g_{ij})$, the M is a minimal graph of dimension n and codimension m.

We have

$$|\nabla f|^2 = \sum_{\alpha,i,j} g^{ij}\frac{\partial f^\alpha}{\partial x^i}\frac{\partial f^\alpha}{\partial x^j} = \sum_i \cos^2\theta_i\tan^2\theta_i = \sum_i \sin^2\theta_i.$$

Denote v to be the volume element of the induced metric on M. Then,

$$v = (\det(g_{ij}))^{\frac{1}{2}} = \prod_i \sec\theta_i.$$

Lemma 1.

$$v \leq |\nabla f|^2 v + 1. \tag{1}$$

Proof. When the codimension $m = 1$, (1) is obviously. In general, we consider

$$A = \prod_i \cos\theta_i + \sum_i \sin^2\theta_i$$

on $0 \leq \theta_i \leq \frac{\pi}{2}$.

$$A_i = \frac{\partial A}{\partial \theta_i} = \sin\theta_i (2\cos\theta_i - \prod_{j\neq i}\cos\theta_j).$$

When $A_i = 0$, then $\sin\theta_i = 0$, or $2\cos\theta_i = \prod_{j\neq i}\cos\theta_j$.

If exist $\theta_1, \cdots, \theta_{i_k}$, such that $\sin\theta_{i_s} = 0$ and its complements $\theta_{j_1}, \cdots, \theta_{j_{p-k}}$ with

$$2\cos\theta_{j_l} = \prod_{j\neq j_l}\cos\theta_j,$$

then

$$A = \prod_{j_l}\cos\theta_{j_l} + \sum_{j_l}\sin^2\theta_{j_l}.$$

In the case $p - k \geq 2$,

$$2\cos\theta_{j_1} = \cos\theta_{j_2}\cdots\cos\theta_{j_{p-k}} = \frac{1}{2}\cos\theta_{j_1}\cos^2\theta_{j_3}\cdots\cos^2\theta_{j_{p-k}}$$

$$\Rightarrow \cos\theta_{j_1} = 0, \qquad \theta_{j_1} = \frac{\pi}{2},$$

similarly,

$$\theta_{j_2} = \cdots\theta_{j_{p-k}} = \frac{\pi}{2}.$$

Therefore,

$$A = \sum_{j_l}\sin^2\theta_{j_l} = p - k \geq 2.$$

If $p - k = 1$,

$$A = \cos\theta + \sin^2\theta = 1 - \cos^2\theta + \cos\theta \geq 1.$$

If $p-k=0,\ \theta_1=\cdots=\theta_p=0$,

$$A=\prod_i \cos_i = 1.$$

The above discussion implies that the extremal values of A is not less than 1. We thus proved

$$A \geq 1.$$

This means

$$v^{-1}+|\nabla f|^2 \geq 1, \tag{2}$$

which implies (1). □

It is well known that the coordinate functions of a minimal submanifold in Euclidean space is harmonic (see Ref. 1, for example). In particular,

$$\Delta f^\alpha = 0.$$

Let $B^{m+n}(R)$ be the Euclidean ball of radius R and centered at the origin. Denote the extrinsic ball by

$$D(R) = M \cap \mathbb{R}^{m+n}(R).$$

For $R>0$, define for each α

$$f_R^\alpha(x)=\begin{cases} R, & \text{if} \quad f^\alpha > R; \\ f^\alpha(x), & \text{if} \quad |f^\alpha(x)| \leq R; \\ -R, & \text{if} \quad f^\alpha < -R. \end{cases}$$

For any $\delta > 0$ define a non-negative cut-off function on $\mathbb{R}^n$ given by

$$\phi=\begin{cases} 1, & \text{on} \quad B^n(R); \\ \frac{(1+\delta)R-r(x)}{\delta R}, & \text{on} \quad B^n((1+\delta)R)\backslash B^n(R); \\ 0, & \text{on} \quad \mathbb{R}^n\backslash B^n((1+\delta)R), \end{cases}$$

where $r(x)$ is the Euclidean distance to the origin of $\mathbb{R}^n$. Note that $\phi f_R^\alpha = 0$ on $\partial(\Omega \bigcap B^n((1+\delta)R)$. Denote

$$M(R) = (x, f(x),\ x \in \Omega \cap B^n(R)).$$

Then for each α

$$\int_{M((1+\delta)R)} \phi f_R^\alpha \Delta f^\alpha * 1 = 0$$

and

$$\int_{M((1+\delta)R)} \nabla(\phi f_R^\alpha, \nabla f^\alpha) - \int_{M((1+\delta)R)} f_R^\alpha \langle \nabla\phi, \nabla f^\alpha \rangle$$
$$- \int_{M((1+\delta)R)} \phi \langle \nabla f_R^\alpha, \nabla f^\alpha \rangle = 0.$$

Summing up over α gives

$$\begin{aligned}\int_{\Omega\cap B^n((1+\delta)R)} \phi| \langle \nabla f_R, \nabla f \rangle |vdx &\le \int_{\Omega\cap B^n((1+\delta)R)} |f_R^\alpha|| \langle \nabla\phi, \nabla f^\alpha \rangle |vdx \\ &\le R \int_{\Omega\cap B^n((1+\delta)R)} |\nabla\phi||\nabla f|vdx,\end{aligned}$$

where vdx is the volume form of M. Since

$$|\nabla f| \le \sqrt{\sum_i \sin\theta_i} \le \sqrt{p},$$
$$|\nabla\phi| \le \sqrt{\sum_{i,j} g^{ij} \frac{\partial\phi}{\partial x^i}\frac{\partial\phi}{\partial x^j}} \le \max_i\{\cos\theta_i\}|\nabla\phi|_{\mathbb{R}^n} = \max_i\{\cos\theta_i\}\frac{1}{\delta R},$$

we then have

$$\begin{aligned}&\int_{\Omega\cap B^n((1+\delta)R)} \phi| \langle \nabla f_R, \nabla f \rangle |vdx \\ &\le \frac{\sqrt{p}}{\delta} \int_{\Omega\cap(B^n((1+\delta)R)\backslash B^n(R))} \max_i\{\cos\theta_i\} \prod_j \sec\theta_j dx. \qquad (3)\end{aligned}$$

If we assume that all the eigenvalues but the largest one of $\sum_\alpha \frac{\partial f^\alpha}{\partial x^i}\frac{\partial f^\alpha}{\partial x^j}$ are bounded by a constant C'. Since $D(R) \subset M \cap (B^n(R) \times [-R, R]^m)$ and using (1)and (3),

$$\begin{aligned}V(D(R) &\le \int_{\Omega\cap B^n(R)\cap\{|f^r|\le R\}} vdx \\ &\le \int_{\Omega\cap B^n(R)\cap\{|f^r|\le R\}} |\nabla f|^2 vdx + \int_{\Omega\cap B^n(R)\cap\{|f^r|\le R\}} dx \\ &\le \frac{\sqrt{p}C}{\delta} V(\Omega\cap B^n((1+\delta)R\backslash B^n(R))) + V(\Omega\cap B^n(R)),\end{aligned}$$

where C is dependent on C' and p.

Since,

$$V(\Omega\cap B^n(R)) \le \omega_n R^n,$$
$$V(\Omega\cap B^n(((1+\delta)R)\backslash B^n(R))) \le \omega_n((1+\delta)^n R^n - R^n).$$

We have

$$V(D(R)) \leq \frac{\sqrt{p}C}{\delta}(\omega_n((1+\delta)^n R^n - R^n) + \omega_n R^n$$

and

$$V(D(R)) \leq \lim_{\delta \to 0} \frac{\sqrt{p}C}{\delta}(\omega_n((1+\delta)^n R^n - R^n) + \omega_n R^n = \sqrt{p}C(n+1)\omega R^n.$$

In summary, we have

Proposition 1. *Let M be a minimal graph defined by a smooth map $f : \mathbb{R}^n \to \mathbb{R}^m$. If all the eigenvalues but the largest one of $\sum_\alpha \frac{\partial f^\alpha}{\partial x^i}\frac{\partial f^\alpha}{\partial x^j}$ are uniformly bounded, then M has Euclidean volume growth.*

Remark 1. If the codimension $m = 1$, there is only one eigenvalue of $\left(\frac{\partial f}{\partial x^i}\frac{\partial f}{\partial x^j}\right)$. This is just the minimal graphic hypersurfaces, which have Euclidean volume growth.

References

1. Yuanlong Xin, Minimal submanifolds and related topics. World Scientific Publ. (2003).

Separation of Variables for the Lax Pair of the Bogomolny Equation in 2+1 Dimensional Anti-de Sitter Space-time

Zi-Xiang Zhou*

School of Mathematical Sciences, Fudan University,
Shanghai 200433, China
zxzhou@fudan.edu.cn

A Lax system separating all the variables is reduced from the Lax pair of the Bogomolny equation for the Yang–Mills–Higgs field in 2+1 dimensional anti-de Sitter space-time. It also contains the Bogomolny equation in 2+1 dimensional Minkowski space-time (Ward equation) under further reduction or limit. The Darboux transformation can be constructed in the usual way to get explicit solutions of the Bogomolny equation.

1. Introduction

The Bogomolny equation for the Yang–Mills–Higgs field in the 2+1 dimensional anti-de Sitter space-time and Minkowski space-time are known to be integrable.[8,15] They have been studied and solved explicitly by several ways and various kinds of solutions have been obtained[5,8–10,14,15,17,18]

On the other hand, a lot of integrable systems in 2+1 dimensions without spectral parameters can be reduced by separating variables in the Lax pair. A typical way is the nonlinear constraint method,[1,2,11] in which a 2+1 dimensional integrable system can be reduced to several 1+1 dimensional systems.[7] Under this kind of constraint, we can obtain explicit solutions like the dromions[19] and the quasi-periodic solutions[1,20] of the 2+1 dimensional integrable system,

In this paper, we will separate the variables for the Lax pair (with a spectral parameter) of the Bogomolny equation in 2+1 dimensional anti-de

*This work was supported by the National Natural Science Foundation of China (11171073) and the Key Laboratory of Mathematics for Nonlinear Sciences of Ministry of Education of China.

Sitter space-time. The derived system is written in 1+1 dimensional form. The Darboux transformation can be applied to this derived system.

By various limit or reduction process, the results are also valid for the Bogomolny equation in 2+1 dimensional Minkowski space-time[3,5,8] (i.e. Ward equation, or the 2+1 dimensional principal chiral field with Wess–Zumino–Witten term[15]) and the standard 1+1 dimensional principal chiral field.[6,12,13,16]

2. Bogomolny Equation in 2+1 Dimensional Anti-de Sitter Space-time

The 2+1 dimensional anti-de Sitter space-time (AdS) of constant curvature $-1/\rho^2$ ($\rho > 0$) is the hyperboloid

$$U^2 + V^2 - X^2 - Y^2 = \rho^2 \tag{1}$$

in $\mathbf{R}^{2,2}$ with the metric

$$\mathrm{d}s^2 = -\mathrm{d}U^2 - \mathrm{d}V^2 + \mathrm{d}X^2 + \mathrm{d}Y^2. \tag{2}$$

By defining

$$r = \frac{\rho}{U+X} - \rho + 1, \quad y = \frac{Y}{U+X}, \quad t = -\frac{V}{U+X}, \tag{3}$$

a part of the AdS with $U+X > 0$ is represented by the Poincaré coordinates (r, y, t) with $r > -\rho + 1$ and the metric is

$$\begin{aligned} \mathrm{d}s^2 &= \frac{\rho^2}{(r+\rho-1)^2}(-\mathrm{d}t^2 + \mathrm{d}r^2 + \mathrm{d}y^2) \\ &= \frac{\rho^2}{(x_2+\rho-1)^2}(\mathrm{d}x_2^2 - 4\,\mathrm{d}x_1\,\mathrm{d}x_3) \end{aligned} \tag{4}$$

where $x_1 = (y+t)/2$, $x_2 = r$, $x_3 = (t-y)/2$.

For simplicity, denote

$$\gamma = \frac{x_2 + \rho - 1}{\rho}. \tag{5}$$

Here are some basic notions for the Yang–Mills–Higgs field.

Let G be an $N \times N$ matrix Lie group, $\mathcal{G}$ be its Lie algebra. The Yang–Mills–Higgs field in AdS is given by the Yang–Mills potentials $\{A_\mu\}$ and the Higgs field Φ with $A_\mu \in \mathcal{G}$, $\Phi \in \mathcal{G}$. Denote $\partial_\mu = \partial/\partial x_\mu$ where $\mu = 1, 2, 3$. The covariant derivative of an N-dimensional vector ψ on which $\mathcal{G}$ acts is $D_\mu\psi = \partial_\mu\psi + A_\mu\psi$. The strength of the Yang–Mills field is defined by

$F_{\mu\nu} = [D_\mu, D_\nu] = \partial_\mu A_\nu - \partial_\nu A_\mu + [A_\mu, A_\nu]$, and the covariant derivative of the Higgs field Φ is defined by $D_\mu \Phi = \partial_\mu \Phi + [A_\mu, \Phi]$.

For $g(x) \in G$, the transformation

$$A'_\mu = gA_\mu g^{-1} - (\partial_\mu g)g^{-1}, \quad \Phi' = g\Phi g^{-1} \tag{6}$$

is called a gauge transformation. Under the gauge transformation, the strength is changed to

$$F'_{\mu\nu} = gF_{\mu\nu}g^{-1}. \tag{7}$$

The Bogomolny equation with the metric (4) is[18]

$$D_1\Phi = \gamma F_{12}, \quad D_2\Phi = \frac{\gamma}{2}F_{13}, \quad D_3\Phi = \gamma F_{23}. \tag{8}$$

By (6) and (7), a gauge transformation transforms a solution of (8) to another solution of it.

Lemma 1. *For any Yang–Mills–Higgs field (A_μ, Φ) satisfying the Bogomolny equation (8) on AdS, there exists a gauge transformation such that $A'_1 = 0$, $A'_2 = \gamma^{-1}\Phi'$ where γ is defined by (5).*

Proof: After the gauge transformation given by $g \in G$,

$$\begin{aligned} &A'_1 = gA_1g^{-1} - (\partial_1 g)g^{-1}, \\ &A'_2 - \gamma^{-1}\Phi' = g(A_2 - \gamma^{-1}\Phi)g^{-1} - (\partial_2 g)g^{-1}. \end{aligned} \tag{9}$$

If we want $A'_1 = 0$, $A'_2 = \gamma^{-1}\Phi'$, then g should satisfy

$$\partial_1 g = gA_1, \quad \partial_2 g = g(A_2 - \gamma^{-1}\Phi). \tag{10}$$

The integrability condition of (10) is

$$\partial_2 A_1 - \partial_1(A_2 - \gamma^{-1}\Phi) - [A_1, A_2 - \gamma^{-1}\Phi] = 0, \tag{11}$$

which is the first equation of (8). Hence g always exists. Moreover, since $A_1 \in \mathcal{G}$ and $A_2 - \gamma^{-1}\Phi \in \mathcal{G}$, we can want $g \in G$. The lemma is proved.

Thus, we shall always choose the gauge so that $A_1 = 0$, $A_2 = \gamma^{-1}\Phi$ hold identically.

It was known that (8) has a Lax pair[18]

$$\begin{aligned} &((x_2 + \rho - 1)D_2 + \rho\Phi - (\rho\lambda - 2x_1)D_1)\eta = 0, \\ &\left(D_3 - \frac{\rho\lambda - 2x_1}{x_2 + \rho - 1}D_2 + \frac{\rho(\rho\lambda - 2x_1)}{(x_2 + \rho - 1)^2}\Phi\right)\eta = 0 \end{aligned} \tag{12}$$

which was presented by Ref. 15. This means that (8) is the integrability condition of (12).

Considering the gauge fixed in Lemma 1, (12) can be written as

$$\begin{aligned}\partial_2\eta &= \sigma(\lambda)\partial_1\eta - P\eta,\\ \partial_3\eta &= \sigma(\lambda)\partial_2\eta - Q\eta\end{aligned} \tag{13}$$

where

$$\sigma(\lambda) = \frac{\rho\lambda - 2x_1}{x_2+\rho-1}, \tag{14}$$

$$P = 2\gamma^{-1}\Phi = 2A_2, \quad Q = A_3. \tag{15}$$

There are two special cases. When $\rho \to +\infty$, the space-time tends to the Minkowski space-time, and $\sigma(\lambda) \to \lambda$. When $\rho = 1$, the space-time becomes the AdS with curvature -1, and $\sigma = \dfrac{\lambda - 2x_1}{x_2}$

3. Separation of Variables

Now we introduce a new Lax system

$$\partial_j\Psi = \sum_{\alpha=1}^{r} \frac{\sigma_\alpha^{j-1}B_\alpha}{\lambda-\lambda_\alpha}\Psi \quad (j=1,2,3). \tag{16}$$

Here $\lambda_1, \cdots, \lambda_r$ are distinct complex numbers, $\sigma_\alpha = \sigma(\lambda_\alpha)$, $B_1, \cdots, B_r$ are $N \times N$ matrix functions to be determined.

The integrability conditions of (16) are

$$\begin{aligned}&\partial_l(\sigma_\alpha^{k-1}B_\alpha) - \partial_k(\sigma_\alpha^{l-1}B_\alpha)\\ &+\sum_{\beta\neq\alpha} \frac{\sigma_\alpha^{k-1}\sigma_\beta^{l-1} - \sigma_\alpha^{l-1}\sigma_\beta^{k-1}}{\lambda_\alpha-\lambda_\beta}[B_\alpha, B_\beta] = 0\end{aligned} \tag{17}$$

$(1 \le k < l \le 3)$. In fact, (17) is equivalent to two equations within it with $k=1, l=2$ and $k=2, l=3$ respectively:

$$\begin{aligned}\partial_2 B_\alpha &= \partial_1(\sigma_\alpha B_\alpha) + \left[B_\alpha, \sum_{\beta=1}^{r}\gamma^{-1}B_\beta\right],\\ \partial_3(\sigma_\alpha B_\alpha) &= \partial_2(\sigma_\alpha^2 B_\alpha) + \left[\sigma_\alpha B_\alpha, \sum_{\beta=1}^{r}\gamma^{-1}\sigma_\beta B_\beta\right].\end{aligned} \tag{18}$$

B_α's can be constructed from the solutions of the original Lax pair (13) and its conjugate, as in the nonlinear constraint method for other equations.[2]

Suppose that F_α is an $N \times p$ solution $(1 \le p \le N)$ of (13) with $\lambda = \lambda_\alpha$, that is, F_α satisfies

$$\begin{aligned} \partial_2 F_\alpha &= \sigma_\alpha \partial_1 F_\alpha - P F_\alpha, \\ \sigma_\alpha \partial_3 F_\alpha &= \sigma_\alpha^2 \partial_2 F_\alpha - \sigma_\alpha Q F_\alpha. \end{aligned} \tag{19}$$

Suppose also that G_α is an $N \times p$ solution of

$$\begin{aligned} \partial_2 G_\alpha &= \partial_1(\sigma_\alpha G_\alpha) + P^T G_\alpha, \\ \sigma_\alpha \partial_3 G_\alpha &= \partial_2(\sigma_\alpha^2 G_\alpha) + \sigma_\alpha Q^T G_\alpha, \end{aligned} \tag{20}$$

which is conjugate to (19).

Let $B_\alpha = \kappa_\alpha F_\alpha G_\alpha^T$ where κ_α is a complex constant, then B_α satisfies

$$\begin{aligned} \partial_2 B_\alpha &= \partial_1(\sigma_\alpha B_\alpha) + [B_\alpha, P], \\ \partial_3(\sigma_\alpha B_\alpha) &= \partial_2(\sigma_\alpha^2 B_\alpha) + [\sigma_\alpha B_\alpha, Q]. \end{aligned} \tag{21}$$

Comparing (21) with (18), we know that when

$$P = \gamma^{-1} \sum_{\beta=1}^{r} B_\beta, \quad Q = \gamma^{-1} \sum_{\beta=1}^{r} \sigma_\beta B_\beta, \tag{22}$$

(18) is consistent with (21).

Neglecting the special expressions of B_α's, we have the following general theorem.

Theorem 1. *Suppose $\lambda_1, \cdots, \lambda_r$ are distinct complex constants, and $B_1, \cdots, B_r$ are $N \times N$ matrices satisfying the integrability conditions (18) of (16), then*

$$\begin{aligned} &A_1 = 0, \quad A_2 = \frac{1}{2}\gamma^{-1} \sum_{\beta=1}^{r} B_\beta, \\ &A_3 = \gamma^{-1} \sum_{\beta=1}^{r} \sigma_\beta B_\beta, \quad \Phi = \frac{1}{2} \sum_{\beta=1}^{r} B_\beta \end{aligned} \tag{23}$$

give a solution of the Bogomolny equation (8).

Proof: Since the gauge has already been fixed so that $A_1 = 0$, $A_2 = \gamma^{-1}\Phi$, we have

$$D_1 \Phi = \gamma \partial_1 A_2 = \gamma F_{12}, \tag{24}$$

which is the first equation of (8). Now we verify the other two equations of (8). Summing up (18) on α, we have

$$2D_2\Phi = 2\partial_2\Phi = \sum_{\alpha=1}^{r} \partial_2 B_\alpha = \gamma \partial_1 A_3 = \gamma F_{13}, \tag{25}$$

$$\begin{aligned} 2D_3\Phi &= 2\partial_3\Phi + 2[A_3,\Phi] = \partial_3\Big(\sum_{\alpha=1}^{r} B_\alpha\Big) - \Big[\sum_{\alpha=1}^{r} B_\alpha, \sum_{\beta=1}^{r}\gamma^{-1}\sigma_\beta B_\beta\Big] \\ &= \partial_2\Big(\sum_{\alpha=1}^{r}\sigma_\alpha B_\alpha\Big) + \sum_{\alpha=1}^{r}(\partial_2\sigma_\alpha)B_\alpha = \partial_2(\gamma A_3) - \rho^{-1}A_3 \\ &= \gamma\partial_2 A_3 = \gamma F_{23} + D_3\Phi. \end{aligned} \tag{26}$$

So $D_3\Phi = \gamma F_{23}$. The theorem is proved.

The solutions of (18) can be obtained by standard ways in 1+1 dimensions. In the next section, we will use the Darboux transformation method to solve them.

Remark 1. (16) can be generalized to $j = 1, \cdots, n$ with any positive integer n. For $n = 2$, $r = 2$, $\lambda_1 = 1$, $\lambda_2 = -1$, and $\rho \to +\infty$, the Lax system is

$$\begin{aligned} \partial_1\Psi &= \Big(\frac{B_1}{\lambda-1} + \frac{B_2}{\lambda+1}\Big)\Psi, \\ \partial_2\Psi &= \Big(\frac{B_1}{\lambda-1} - \frac{B_2}{\lambda+1}\Big)\Psi. \end{aligned} \tag{27}$$

This is the well-known Lax pair of the 1+1 dimensional principal chiral field (harmonic map from $\mathbf{R}^{1,1}$ to the Lie group G). A lot of explicit solutions of it have been known already.[6,12,13,16]

4. Darboux Transformation

We use the general procedure of constructing Darboux transformation for (16).[4]

Let $\mu_1, \cdots, \mu_N$ be non-zero complex numbers such that $\mu_1, \cdots, \mu_N$, $\lambda_1, \cdots, \lambda_r$ are distinct, h_j be a solution of (16) with $\lambda = \mu_j$ ($j = 1, \cdots, N$). Let

$$\begin{aligned} &\Lambda = \mathrm{diag}(\mu_1, \cdots, \mu_N), \quad H = (h_1, \cdots, h_N), \\ &S = H\Lambda H^{-1}. \end{aligned} \tag{28}$$

Then, for any solution Ψ of (16), $\widetilde{\Psi} = (\lambda I - S)\Psi$ satisfies

$$\partial_j\widetilde{\Psi} = \sum_{\alpha=1}^{r}\frac{\sigma_\alpha^{j-1}\widetilde{B}_\alpha}{\lambda-\lambda_\alpha}\widetilde{\Psi} \tag{29}$$

where

$$\widetilde{B}_\alpha = (\lambda_\alpha I - S)B_\alpha(\lambda_\alpha I - S)^{-1} \tag{30}$$

and S satisfies

$$\partial_j S = \sum_{\alpha=1}^{r} \sigma_\alpha^{j-1}(B_\alpha - \widetilde{B}_\alpha). \tag{31}$$

(31) can also be written as

$$\sum_{\alpha=1}^{r} \sigma_\alpha^{j-1}\widetilde{B}_\alpha = \sum_{\alpha=1}^{r} \sigma_\alpha^{j-1} B_\alpha - \partial_j S. \tag{32}$$

From (23), this gives a Darboux transformation as stated in the following theorem.

Theorem 2. *Suppose B_α's are solutions of (18), S is constructed as above, (A_1, A_2, A_3, Φ) are given by (23), then*

$$\begin{aligned} &\widetilde{A}_1 = 0, \quad \widetilde{A}_2 = A_2 - \frac{\rho}{2(x_2+\rho-1)}\partial_1 S, \\ &\widetilde{A}_3 = A_3 - \frac{\rho}{x_2+\rho-1}\partial_2 S, \quad \widetilde{\Phi} = \Phi - \frac{1}{2}\partial_1 S \end{aligned} \tag{33}$$

also give a solution of the Bogomolny equation (17).

This is the Darboux transformation of degree one. An infinite series of solutions can be obtained by successive actions of Darboux transformations of degree one.

Example. To use the Darboux transformation, we need a seed solution of (18). Suppose the seed solution is $B_\alpha = u_\alpha(x_1, x_2, x_3)C_\alpha$ $(\alpha = 1, \cdots, r)$ where u_α's are functions, C_α's are constant $N \times N$ matrices such that $[C_\alpha, C_\beta] = 0$ for $\alpha, \beta = 1, \cdots, r$. By (18), u_α's satisfy

$$\begin{aligned} &\partial_2 u_\alpha = \partial_1(\sigma_\alpha u_\alpha), \\ &\partial_3(\sigma_\alpha u_\alpha) = \partial_2(\sigma_\alpha^2 u_\alpha). \end{aligned} \tag{34}$$

The general solution is

$$u_\alpha = \Big(\frac{x_2+\rho-1}{\rho\lambda_\alpha - 2x_1}\Big)^2 f_\alpha(\theta_\alpha(x_1, x_2, x_3)) \tag{35}$$

where

$$\theta_\alpha(x_1, x_2, x_3) = \frac{(x_2+\rho-1)^2}{2(\rho\lambda_\alpha - 2x_1)} + x_3 - \frac{\rho-2}{2\lambda_\alpha}, \tag{36}$$

f_α's are arbitrary meromorphic functions.

Remark 2. In (36), the constant $-\dfrac{\rho-2}{2\lambda_\alpha}$ is added so that θ_α converges as $\rho \to +\infty$. Hence, in the Minkowski space-time, $u_a = \lambda_\alpha^{-2} f_\alpha(x_3 + \lambda_\alpha^{-1}x_2 + \lambda_\alpha^{-2}x_1)$.

(35) gives the seed solution of (18) for the Darboux transformation, and the corresponding

$$\begin{aligned} &A_1 = 0, \quad A_2 = \frac{\rho}{2(x_2+\rho-1)} \sum_{\alpha=1}^{r} u_\alpha C_\alpha, \\ &A_3 = \sum_{\alpha=1}^{r} \frac{\rho(\rho\lambda_\alpha - 2x_1)}{(x_2+\rho-1)^2} u_\alpha C_\alpha, \quad \Phi = \frac{1}{2} \sum_{\alpha=1}^{r} u_\alpha C_\alpha. \end{aligned} \tag{37}$$

The fundamental solution of (16) is

$$\Psi = \exp\left(\sum_{\alpha=1}^{r} \frac{F_\alpha(\theta_\alpha) C_\alpha}{\lambda - \lambda_\alpha} \right) \Psi_0 \tag{38}$$

where Ψ_0 is an $N \times N$ constant matrix, $F'_\alpha = f_\alpha$.

With the seed solution (37), we can get new explicit solutions according to Theorem 2.

Especially, take $G = SU(2)$, $C_\alpha = \mathrm{diag}(\mathrm{i}c_\alpha, -\mathrm{i}c_\alpha)$ where c_α's are constant real numbers. Then, as usual construction for $SU(2)$ solutions, we should take $\mu_2 = \bar{\mu}_1$ and $h_2^* h_1 = 0$. Hence

$$H = \begin{pmatrix} \alpha_1 g & -\bar{\alpha}_2 \bar{g}^{-1} \\ \alpha_2 g^{-1} & \bar{\alpha}_1 \bar{g} \end{pmatrix} \tag{39}$$

where

$$g = \exp\left(\mathrm{i} \sum_{\alpha=1}^{r} \frac{F_\alpha(\theta_\alpha) c_\alpha}{\mu_1 - \lambda_\alpha} \right), \tag{40}$$

α_1 and α_2 are complex constants.

$$\begin{aligned} S &= H \Lambda H^{-1} \\ &= \frac{\mu_1 + \bar{\mu}_1}{2} I + \frac{\mu_1 - \bar{\mu}_1}{2} \frac{1}{|\alpha_1|^2 |g|^2 + |\alpha_2|^2 |g|^{-2}} \cdot \\ &\cdot \begin{pmatrix} |\alpha_1|^2 |g|^2 - |\alpha_2|^2 |g|^{-2} & 2\alpha_1 \bar{\alpha}_2 g \bar{g}^{-1} \\ 2\bar{\alpha}_1 \alpha_2 \bar{g} g^{-1} & -|\alpha_1|^2 |g|^2 + |\alpha_2|^2 |g|^{-2} \end{pmatrix}, \end{aligned} \tag{41}$$

and

$$\begin{aligned} &\widetilde{A}_1 = 0, \\ &\widetilde{A}_2 = \frac{\rho}{2(x_2+\rho-1)} \sum_{\alpha=1}^{r} u_\alpha C_\alpha - \frac{\rho}{2(x_2+\rho-1)} \partial_1 S, \\ &\widetilde{A}_3 = \sum_{\alpha=1}^{r} \frac{\rho(\rho\lambda_\beta - 2x_1)}{(x_2+\rho-1)^2} u_\alpha C_\alpha - \frac{\rho}{x_2+\rho-1} \partial_2 S, \\ &\widetilde{\Phi} = \frac{1}{2} \sum_{\alpha=1}^{r} u_\alpha C_\alpha - \frac{1}{2} \partial_1 S. \end{aligned} \tag{42}$$

References

1. C.W. Cao, Y.T. Wu and X.G. Geng, Relation between the Kadomtsev-Petviashvili equation and the confocal involutive system, J. Math. Phys. **40** (1999) 3948-3970.
2. Y. Cheng and Y.S. Li, The constraint of the KP equation and its special solutions, Phys. Lett. **A157** (1991) 22-26.
3. B. Dai and C.L. Terng, Backlund transformations, Ward solitons, and unitons, J. Diff. Geom. **75** (2007) 57-108.
4. C.H. Gu, On the Darboux form of Bäcklund transformations, in: X.C. Song (Eds.), *Integrable System*, World Scientific (1989) 162-168.
5. C.H. Gu, Darboux transformation and solitons of Yang–Mills–Higgs equations in $R^{2,1}$, Science in China **A45** (2002) 706-715.
6. C.H. Gu and H.S. Hu, Construction of unitons via purely algebraic algorith, Chin. Ann. Math. **20B** (1999) 1-6.
7. C.H. Gu and Z.X. Zhou, On Darboux transformations for soliton equations in high dimensional space-time, Lett. Math. Phys. **32** (1994) 1-10.
8. N.J. Hitchin, G.B. Segal and R.S. Ward: *Integrable systems, Twistors, loop groups and Riemann surfaces*, (Clarendon Press, Oxford), 1999.
9. T. Ioannidou, Bogomolny Yang–Mills–Higgs solutions in (2+1) anti-de Sitter space, Nonlinearity **15** (2002) 1489-1497.
10. T. Ioannidou, Nontrivial soliton scattering in planar integrable systems, Intl. J. Mod. Phys. **A18** (2003) 4975-4998.
11. B. Konopelchenko, J. Sidorenko and W. Strampp, (1+1)-dimensional integrable systems as symmetry constraints of (2+1) dimensional systems, Phys. Lett. **A157** (1991) 17-21.
12. A.T. Ogielski, K. Prasad, A. Sinha and L.L. Chau, Bäcklund transformations and local conservation laws for principal chiral fields, Phys. Lett. **B**91 (1980) 387-391.
13. K. Uhlenbeck, Harmonic maps into Lie groups: classical solutions of the chiral model, J. Diff. Geom. **38** (1980) 1-50.
14. R.S. Ward, Soliton solutions in an integrable chiral model in 2+1 dimensions, J. Math. Phys. **29** (1988) 386-389.
15. R.S. Ward, Two integrable systems related to hyperbolic monopoles, Asian J. Math. **3** (1999) 325-332.
16. V.E. Zakharov and A.V. Mikailov, Relativistically invariant two-dimensional models of field theory which are integrable by means of the inverse problem scattering method, Sov. Phys. JETP **47** (1978) 1017-1027.
17. Z.X. Zhou, Solutions of the Yang-Mills-Higgs equations in 2+1-dimensional anti-de Sitter space-time, J. Math. Phys. **42** (2001) 1085-1099.
18. Z.X. Zhou, Relation between the solitons of Yang-Mills-Higgs fields in 2+1 dimensional Minkowski space-time and anti-de Sitter space-time, J. Math. Phys. **42** (2001) 4938-4946.
19. Z.X. Zhou, Localized solitons of hyperbolic su(N) AKNS system, Inverse Problems **14** (1998) 1371-1383.

20. Z.X. Zhou, W.X. Ma and R.G. Zhou, Finite-dimensional integrable systems associated with Davey-Stewartson I equation, Nonlinearity **14** (2001) 701-717.